Lecture Notes in Electrical Engineering   871

Series Editors

The book series *Lecture Notes in Electrical Engineering* (LNEE) publishes the latest developments in Electrical Engineering—quickly, informally and in high quality. While original research reported in proceedings and monographs has traditionally formed the core of LNEE, we also encourage authors to submit books devoted to supporting student education and professional training in the various fields and applications areas of electrical engineering. The series cover classical and emerging topics concerning:

- Communication Engineering, Information Theory and Networks
- Electronics Engineering and Microelectronics
- Signal, Image and Speech Processing
- Wireless and Mobile Communication
- Circuits and Systems
- Energy Systems, Power Electronics and Electrical Machines
- Electro-optical Engineering
- Instrumentation Engineering
- Avionics Engineering
- Control Systems
- Internet-of-Things and Cybersecurity
- Biomedical Devices, MEMS and NEMS

For general information about this book series, comments or suggestions, please contact leontina.dicecco@springer.com.

To submit a proposal or request further information, please contact the Publishing Editor in your country:

China

Jasmine Dou, Editor (jasmine.dou@springer.com)

India, Japan, Rest of Asia

Swati Meherishi, Editorial Director (Swati.Meherishi@springer.com)

Southeast Asia, Australia, New Zealand

Ramesh Nath Premnath, Editor (ramesh.premnath@springernature.com)

USA, Canada

Michael Luby, Senior Editor (michael.luby@springer.com)

All other Countries

Leontina Di Cecco, Senior Editor (leontina.dicecco@springer.com)

** This series is indexed by EI Compendex and Scopus databases. **

Qilian Liang · Wei Wang · Jiasong Mu · Xin Liu ·
Zhenyu Na
Editors

Artificial Intelligence in China

Proceedings of the 4th International Conference
on Artificial Intelligence in China

 Springer

Editors
Qilian Liang
Department of Electrical Engineering
University of Texas at Arlington
Arlington, TX, USA

Jiasong Mu
Tianjin Normal University
Tianjin, China

Zhenyu Na
School of Information Science
and Technology
Dalian Maritime University
Dalian, China

Wei Wang
Tianjin Normal University
Tianjin, China

Xin Liu
Dalian University of Technology
Dalian, China

ISSN 1876-1100   ISSN 1876-1119 (electronic)
Lecture Notes in Electrical Engineering
ISBN 978-981-99-1258-2   ISBN 978-981-99-1256-8 (eBook)
https://doi.org/10.1007/978-981-99-1256-8

This Springer imprint is published by the registered company Springer Nature Singapore Pte Ltd.
The registered company address is: 152 Beach Road, #21-01/04 Gateway East, Singapore 189721, Singapore

Contents

Traffic Flow Prediction Based on ACO-BI-LSTM

Guo Jincheng and Pan Weimin[⊠]

College of Computer Science and Technology, Xinjiang Normal University, Urumchi 830054, China
379483304@qq.com

Abstract. For traffic flow forecasting task exists for the future traffic flow prediction accuracy is low, lack of generalization and the deep learning model and such problems as incomplete, is put forward based on the two-way LSTM traffic flow prediction model of ant colony algorithm, ant colony algorithm for Bi - LSTM network layers, the number of neurons, batch size, number of training to optimize the parameters. In this paper, experiments are carried out on two public data sets of daily traffic flow in Bao'an District published by California Highway and Shenzhen government open platform. RMSE and MAE are taken as evaluation indexes. The results show that ACO-BI-LSTM model has strong optimization ability and better prediction performance.

Keywords: Traffic flow prediction · Ant colony algorithm · Bidirectional short and long time memory network

1 Introduction

For the traffic flow of a city or region, the prediction of people flow can prevent the occurrence of some sudden accidents. With the great development of artificial intelligence, many machines learning and deep learning models have emerged, but traffic flow data has the characteristics of large scale, high dimensionality and dynamic change over time, and it is a great challenge to achieve accurate traffic flow prediction [1]. Early traffic flow prediction was mainly based on mathematical methods such as mathematical statistics and calculus, such as historical averaging model, time series model, Kalman filter model, parametric regression model [2] Zhong et al. proposed a genetically optimized artificial neural network (GA-ANN) model developed and tested, the former predicting traffic flow every 5 min within 30 min, and the latter being used to predict the traffic every 5 min on the four lanes of Beijing's urban arterial roads [3]. The Bi-LSTM highway traffic flow prediction model based on multiple factors proposed by Zhang Wei et al. has stronger adaptability and higher accuracy in the short-term prediction of expressway [4].

The above methods improve the accuracy of traffic flow prediction, but the parameters mainly rely on expert knowledge, and subjective factors have a great impact on the model effect. Traffic flow data has time series, and the time-dependent extraction effect of the current model on the data is not good, resulting in the general prediction effect of time series data. Aiming at these two problems, this paper proposes a traffic flow prediction model based on ACO optimized Bi-LSTM. Therefore, it has its unique advantages in traffic flow prediction problems, and combines the global optimization ability of ant colony optimization algorithm to optimize the Bi-LSTM model. By comparing HA, SVR, BP, LSTM, GRU and other methods, the results show that the ACO-Bi-LSTM model has better prediction results and stronger generalization ability.

2 Ant Colony Optimization Algorithms

The ant colony optimization algorithm is a kind of simulation optimization algorithm that simulates the foraging behavior of ants, and the basic idea of applying the ant colony algorithm to solving optimization problems is: the walking path of the ant is used to represent the feasible solution of the problem to be optimized, and all the paths of the entire ant colony constitute the solution space of the problem to be optimized [5]. Ants with shorter paths released more pheromones, and with the passage of time, the accumulated pheromone concentration on the shorter path gradually increased, and the number of ants choosing this path increased. In the end, the entire ant will concentrate on the optimal path under the action of positive and negative feedback, and the optimal solution to the problem to be optimized will be corresponded [6].

2.1 State Transition Rules

The ant's choice of the next walking path is determined by the probability formula:

$$P_{ij}^k(t) = \int_{0, j \notin J_k(i)} \frac{[\tau_{ij}(t)]^\alpha [n_{ij}(t)]^\beta}{\sum [\tau_{ij}(t)]^\alpha [n_{ij}(t)]^\beta} \, j \epsilon J_k(i) \tag{2.1}$$

$P_{ij}^k(t)$: The probability that the kth ant in the t generation chooses from i to j;
α: Pheromone factors;
β: The relative importance of the heuristic;
n_{ij}: Heuristics;
$J_k(i)$: Ant K currently has a choice of cities;

α and β represent the relative influence of pheromone and heuristic information on the construction of candidate solutions.The higher the α value is, the more likely the ant is to choose the previous path. The higher the β value is, the more likely the state transition probability is to be greedy [7].

2.2 Pheromone Update Rules

When all ants complete an iteration, the pheromone concentration on the path changes, so the information needs to be updated.

$$\tau_{ij}(t+1) = (1 - \rho)\tau_{ij}(t) + \Delta\tau_{ij} \tag{2.2}$$

where $\tau_{ij}(t)$, t represents the pheromone concentration on the path at time t, ρ is a pheromone volatile factor, $1 - \rho$ Indicates a residual factor, ρ The size determines whether the path is continued to be searched. $\Delta\tau_{ij} = \sum_{i=1}^{m} \Delta\tau_{ij}^{k}$, Represents the sum of the pheromones left by m ants on the path from i to j.

3 Bi-LSTM Model

The bidirectional long-short-time memory network is a combination of forward LSTM and backward LSTM.Models in LSTM Only the information "above" is used, and the information of "below" is not taken into account, in a real scenario, the prediction task needs to take into account the information of the entire input sequence. Bi-LSTM combines the information of the input sequence in both forward and backward directions based on LSTM, for the output of the t-moment, the forward layer has the information in the input sequence at the t-moment and before the t-moment, and the backward layer has the information at the t-moment and after the t-moment (Fig. 1).

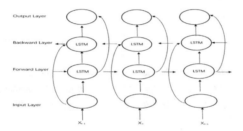

Fig. 1. Bi-LSTM structure diagram

The Bi-LSTM core unit is the LSTM structure, which consists of three gates and a memory unit,As shown in Fig. 2, the three gates are the input gate, the forgotten gate, and the output gate,Ct is a cellular state.The forgetting gate determines what information is discarded from the cellular state, reading the output of the previous cell and the input of the current cell,Output a value between 0 and 1 to the neural unit of Ct-1,where '1' means full retention and '0' means complete abandonment [8].

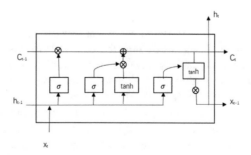

Fig. 2. The internal structure of the LSTM

At present, the Bi-LSTM model has been successfully applied to named entity recognition tasks [9], Trajectory prediction tasks [10], Problem classification [11], Sentiment analysis [12] fields. The bi-LSTM model also applies the two-way memory ability of time series to the traffic flow prediction task, and deeply extracts the temporal characteristics of traffic flow data to improve the prediction effect.

4   ACO-Bi-LSTM Forecasting Model

In response to the above question 1: The traffic flow data with time series attributes is not predicted well; Problem 2: The parameter setting of deep learning neural network relies on expert knowledge, and subjective factors have a great impact on the model effect and lack of generalization. Construct an ACO-Bi-LSTM model in which the Bi-LSTM submodule extracts the time dependence of time series data, and the ACO module performs a tune on the Bi-LSTM module. Improve the accuracy of traffic flow data prediction, and different data sets optimize parameters through ACO Algorithm to Improve generalization.

4.1   Model Framework

The main process steps of the model are: 1) Construct ant solutions and initialize ant populations; 2) Update the pheromones of each ant's path, the local pheromones; 3) The mean squared error of the Bi-LSTM neural network is used as a function of fitness; 4) Pheromones are updated globally, and if the fitness value is optimal, proceed to the next step, otherwise return to step 1; 5) Enter the optimal hyperparameters into the Bi-LSTM for training.

4.2   Ant Colony Algorithm Optimizes Model Parameters

The hyperparameters associated with the Bi-LSTM model are the number of neural units, the number of hidden layers, the activation function, the optimization algorithm, the batch size, the number of iterations, and so on. The ant colony optimization algorithm is used to find the number of optimized hyperparameter solutions to the space [Bi-LSTM neurons (B_nums), Bi-LSTM hides the number of layers(B_deeps), Batch size(batch_size), Number of trainings(epochs)].

4.3   Model Training

Divide the processed dataset into training and test sets on an 8:2 basis and enter the Bi-LSTM network.The first stage of training the Bi-LSTM network is to find the optimal solution space as an adaptability function, aiming for the minimum mean squared error (MSE).The optimal solution space obtained in the first stage, that is, the network hyperparameters, is fed into the Bi-LSTM network for the second stage of training, and the model is trained using the training set, the performance of the model is verified on the test set and the mean squared error is output for model evaluation. Set dropout to 0.2 and use L2 regularization to prevent overfitting during training, the training process is as follows:

1. Build training and testing sets in an 8:2 ratio
2. Initialize the initial solution of the ant colony system
3. Bi-LSTM is trained as an adaptability function, iterating with the MSE minimum as the goal
4. Outputs the optimal hyperparametric solution space for the Bi-LSTM network
5. The optimal hyperparameter Bi-LSTM network is trained, and loss convergence stops training
6. Test set testing, output model evaluation criteria

5 Experimental Analysis

6 Data Preprocessing

(1) Dataset processing

The California Highway dataset is collected every five minutes by sensors set up every five minutes around the highway, and the experiments in this paper set the time step according to the time interval of this dataset. The traffic data of Bao'an District has been collected intermittently since 2017 and is currently updated to November 4, 2021, with a total of 1268 pieces. The data items include data time, traffic flow, average speed, head distance, average lane occupancy, long traffic, medium traffic, etc. This paper extracts the traffic flow data for experiments. For the sake of experimental uniformity, the daily data of the dataset is processed into a five-minute interval by averaging.

(2) Data normalization

The normalization process of data is to map the data uniformly to the [0, 1] interval, in order to remove the unit limit of the data, it is converted into a dimensionless pure value, which is convenient for different units or magnitudes of indicators to operate [14]. In this paper, the min-max normalization method is used to process the data, that is, for the original traffic flow datax1, x2, x3,..., xn:

$$y_i = \frac{x_i - \min_{1 \le j \le n} \{x_j\}}{\max_{1 \le j \le n} \{x_j\} - \min_{1 \le j \le n} \{x_j\}} \tag{5.1}$$

y_i is the result of normalization, $y_i \in [0, 1]$. The convergence speed and accuracy of the normalized model are significantly improved.

7 Model Evaluation

This experiment uses two widely used machine learning evaluation metrics to evaluate the prediction model: (1) Root Mean Square Error (RMSE): RMSE $= \sqrt{\frac{1}{m} \sum_{i=1}^{m} (y_i - y^*_i)^2}$; (2) Average Absolute Error (MAE): MAE $= \frac{1}{m} \sum_{i=1}^{m} |y_i - y^*_i|$; Where y_i and y^*_i represents the true and predicted values, respectively, m is the number of time steps for the forecast [5].

8   Analysis of Results

The Bi-LSTM network input time step is 12, which is a one-step prediction using data from the past hour. Figures 3(a) and (b) show the loss curve obtained by the ACO-Bi-LSTM model with MSE as a loss function on the training and test sets. Figures 4(a) and (b) compare the model prediction results with the real data, and intuitively show the fitting effect of the model, which has a good fit effect on the conventional data in addition to the extremely high and low values in the data set, which proves that the model has a strong fit ability for time series data.

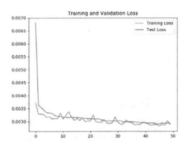

(a)   Loss curve of the California dataset

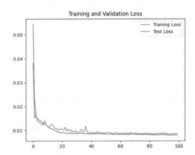

(b) Traffic flow data loss curve in Bao'an District

Fig. 3. Loss curve

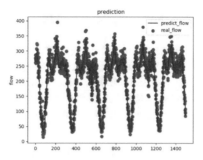

(a) California prediction and true data fit curve

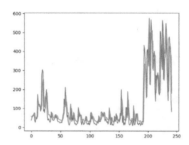

(b) Prediction and true data fit curves for Bao'an District

Fig. 4. Data fitting curve

The model in this paper is experimentally compared with the following baseline model:

HA [13], SVR [14], BP [5], LSTM [15], GRU [16].

The population size of the ant colony optimization algorithm is generally set to 1.5 times the parameters to be optimized, and the parameter settings are as shown in Table 1:

Table 1. Ant colony algorithm parameter setting table

Parameter	value
Populations are small and large	6
Number of iterations	100
Pheromone factors	[1, 4]
Pheromone volatile factor	[0.2, 0.8]
Heuristics	[2, 5]
The importance of the heuristic factor	2

Table 2 is the comparison results of RMSE and MAE on two datasets of the five models, and both RMSE and MAE maintain the best performance on the data set. On the California highway dataset, ACO-Bi-LSTM decreased by 2.78% and 2.30% compared to the optimal LSTM models RMSE and MAE, respectively,On the traffic flow dataset in Bao'an District, RMSE decreased by 6.00% compared with the best BP and MAE was reduced by 12.71% compared with the best GRUs. This paper shows that the ant colony algorithm can improve the performance of the Bi-LSTM model.

Table 2. Comparison of effects of different models

California Transportation Dataset			Traffic data set of Bao'an District	
model	RMSE	MAE	RMSE	MAE
HA	71.17	54.74	64.86	36.81
SVM	22.48	17.45	41.40	30.02
BP	22.26	17.24	37.79	29.15
LSTM	21.54	16.47	40.79	25.92
GRU	21.73	16.69	41.06	25.32
This article model	**21.65**	**16.33**	**36.27**	**24.79**

Figure 5 is a fitting curve of six models and real data, which can be seen that the ACO-Bi-LSTM model has stronger fitting ability and can better capture the change trend of real data.

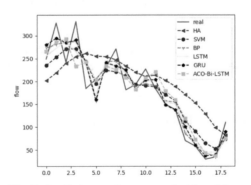

Fig. 5. Prediction results compared with real data

9 Conclusion

In this paper, an ACO-Bi-LSTM model based on ant colony algorithm optimization of bidirectional long-short-time memory network is proposed, which optimizes the hyper-parameters of bidirectional long-short-time memory networks by ant colony algorithm to obtain an optimal set of hyperparameter solution spaces. By comparing the baseline models such as HA, SVM, BP, LSTM and GRU on the California Highway Dataset and the Bao'an District Traffic Dataset, it is found that the model has the best effect under the two evaluation indicators of RMSE and MAE, and the prediction accuracy is higher. The ACO-Bi-LSTM model has broader applicability and greater accuracy in data prediction. The ACO-Bi-LSTM model only makes one-step predictions, and does not take into account the influence of multivariate factors on traffic flow, and future work will consider multivariate factors input to neural networks for single- and multi-step predictions to further improve accuracy. In addition, traffic flow data has spatio-temporal properties, and how to extract temporal and spatial correlations is also the focus of future work.

Acknowledgments. 国家自然科学基金项目 (项目批准号: 62162061), 项目名称: 道路交叉口交通异常事件检测与信号灯配时方法研究。

References

1. 刘世泽,秦艳君,王晨星,苏琳,柯其学,罗海勇,孙艺,王宝会.: 基于深度残差长短记忆网络交通流量预测算法. 计算机应用, 41(06), 1566–1572 (2021)
2. 张晓利.: 基于非参数回归的短时交通流量预测方法研究. 天津大学 (2007)
3. Chan, K.Y., Dillon, T., Chang, E., et al.: Prediction of short-term traffic variables using intelligent swarm-based neural networks. IEEE Trans. Control Syst. Technol. **21**(1), 263–274 (2013)
4. Raza, A., Zhong, M.: Hybrid artificial neural network and locally weighted regression models for lane-based short-term urban traffic flow forecasting. Transp. Plan. Technol. **41**(8), 901–917 (2018)
5. 张维,袁绍欣,陶建军,周晨蓉,阿合提⊃杰恩斯.: 基于多元因素的Bi-LSTM高速公路交通流预测. 计算机系统应用 **30**(06), 184–190 (2021)
6. 潘昕, 吴旭升, 侯新国,等.: 一种基于遗传蚁群算法的无人机全局路径规划方法: CN107229287A (2017)
7. 郁磊,史峰,王辉,胡斐编著. MATLAB智能算法30个案例分析 (第2版) : 北京航空航天大学出版社 09 (2015)
8. 张松灿,孙力帆,司彦娜,普杰信.: 单种群自适应异构蚁群算法的机器人路径规划[J/OL]. 计算机科学与探索: 1–15(2021–10–21)
9. 林原, 李家平, 许侃,等.: 基于多头注意力的双向LSTM情感分析模型研究. 山西大学学报: 自然科学版 **43**(1), 7 (2020)
10. Le, T.A., Arkhipov, M.Y., Burtsev, M.S.: Application of a hybrid Bi-LSTM-CRF model to the task of Russian named entity recognition. In: Filchenkov, A., Pivovarova, L., Žižka, J. (eds.) AINL 2017. CCIS, vol. 789, pp. 91–103. Springer, Cham (2018). https://doi.org/10.1007/978-3-319-71746-3_8
11. Liu, C., Li, Y., Jiang, R., Lu, Q., Guo, Z.: Trajectory-based data delivery algorithm in maritime vessel networks based on Bi-LSTM. In: Yu, D., Dressler, F., Yu, J. (eds) Wireless Algorithms, Systems, and Applications. WASA 2020. Lecture Notes in Computer Science, vol 12384. Springer, Cham (2020). https://doi.org/10.1007/978-3-030-59019-2_25
12. Zhang, Q., Mu, L., Zhang, K., Zan, H., Li, Y.: Research on question classification based on Bi-LSTM. In: Hong, J.-F., Su, Q., Wu, J.-S. (eds.) CLSW 2018. LNCS (LNAI), vol. 11173, pp. 519–531. Springer, Cham (2018). https://doi.org/10.1007/978-3-030-04015-4_44

13. 罗浩然,杨青.: 基于情感词典和堆积残差的双向长短期记忆网络的情感分析. 计算机应用:, 1–11 (2021)
14. 张维,袁绍欣,陶建军,周晨蓉,阿合提⊃杰恩斯.: 基于多元因素的Bi-LSTM高速公路交通流预测. 计算机系统应用 **30**(06), 184–190(2021)
15. Zhou, F., Yang, Q., Zhong, T., Chen, D., Zhang, N.: Variational graph neural networks for road traffic prediction in intelligent transportation systems. IEEE Trans. Ind. Inf. **17**(4), 2802–2812 (2021). https://doi.org/10.1109/TII.(2020).3009280
16. Smith, B.L., Demetsky, M.J.: Traffic flow forecasting: comparison of model-ing approaches. J. Transp. Eng. **123**(4), 261–266 (1997)

Research on Natural Language Processing Technology in Construction Project Management

Xiangshu Peng[1], Zhiming Ma[1(✉)], Ping Wang[2], Yaoxian Huang[3], and Lixia Zhang[1]

[1] School of Computer Science and Technology, Xinjiang Normal University, Urumchi 830054, China
2298882664@qq.com
[2] School of Electricity and Computer, Jilin JianZhu University, Changchun 130000, China
[3] School of Computing, Harbin Finance University, Harbin 150000, China

Abstract. This paper analyses the research and application of Natural Language Processing (NLP) technology in the management of construction projects, aiming at the problems such as the expansion of the scale of construction projects and the increasing number of document information, the difficulty and complexity of their management, the inefficiency and complexity of traditional manual collation and induction, the inability of technicians to fully, timely and efficiently consult, modify and improve document information, resulting in the degradation of project performance. Convert unstructured documents in the field of architecture into structured information to solve common problems in architecture.

Keywords: Natural language processing · Construction project management · Contract risk · Knowledge graph

1 Introduction

With the rapid development of artificial intelligence technology, natural language technology is becoming more and more mature, computers understand human language through machine learning. From basic semantic similarity, dependent syntax analysis to applied human-machine interaction and report analysis, NLP has shown great application prospects in various fields. How to better apply natural language processing technology to construction document management is a meaningful research that can solve many practical problems.

Document management of construction projects is the process of collecting and managing various types of project documents throughout the construction life cycle. Project document is a very important data resource and data wealth in the life cycle of a building. By building a construction project management system, the differences between project document management based on natural language processing technology and manual project document processing in the key indicators of recall rate (1),

Q. Liang et al. (Eds.): AIC 2022, LNEE 871, pp. 11–17, 2023.
https://doi.org/10.1007/978-981-99-1256-8_2

accuracy (2), timeliness, F1 value (3), etc., are verified, which provides strong support for subsequent research.

$$Recall(\text{R}) = \frac{\text{count(correct)}}{\text{count(correct)}+\text{count(lose)}} * 100\% \tag{1}$$

$$Precision(\text{P}) = \frac{\text{count(correct)}}{\text{count(correct)}+\text{count(false)}} * 100\% \tag{2}$$

$$F1 = \frac{2*\text{P}*\text{R}}{\text{P}+\text{R}} \tag{3}$$

2 Introduction to Natural Language Processing

Natural language processing refers to the computer processing of natural language in all aspects, including the input, output, recognition, analysis, understanding, generation of text, and so on.

(1) Named Entity Recognition: Named Entity Recognition (NER) refers to the recognition of entities with specific meaning in the text. It is a basic task in natural language processing. In short, it identifies named designations from the text and paves the way for subsequent tasks.
(2) Part-of-speech labeling: also known as part-of-speech labeling, refers to labeling the part-of-speech of a word resulting from a word division according to its original meaning; It can be determined that each word is a part-of-speech process such as a noun or verb or an adjective.
(3) Synonym Analysis: Common methods of synonym analysis include dictionary-based synonym analysis and dictionary construction using a variety of open source dictionary expectations.
(4) Dependent grammar analysis: The analysis of the dependency relationships among the components of a language unit to explain their syntax structure.
(5) Word Position Analysis: By modeling the position of words in an article, different weights are given to different positions, which can better vectorize the article
(6) Semantic normalization technology: refers to the identification of words or phrases with the same meaning from the article, and then the digestion of the reference, making a referential transformation from different descriptions with the same meaning in the document to the same meaning.
(7) Text Error Correction Technology: To correct all kinds of errors in text, first find the errors (such as misspellings, disorder and incomplete information), and then make different corrections for the errors.
(8) Tag extraction: Use phrases or words as a summary of the main content of a document.
(9) Text similarity technology: Find similar parts of text or similar content in text, and use semantic normalization technology to process.
(10) Document classification technology: Under the conditions of a specific classification system, the process of automatically determining the type of text based on the content of the text to understand the meaning of the full text and the theme of the text.

The main methods used in NLP are rule-based, statistics-based and in-depth learning. Rule-based methods require technicians to manually write knowledge representation and invoke rules to develop algorithms. The statistical method is to try to build a large number of corpus and lexicon so that the machine can learn the text features according to the probability model and train the language model. Depth-based learning is a feature-based learning method, which obtains the distributed eigenvalue representation of data through the non-linear processing of multiple hidden layers. Table 1 shows the common NLP methods.

Table 1. Common NLP methods

Common method	Rule-based	Statistics-based	Depth Learning-based
Algorithm model	If-then expert system	KNN	Word2vec
	Regular expression matching	SVM	CNN
		NBC	LSTM
		DT	Transformer
		HMM	BERT
		CRF	

3   Application of NLP Technology

3.1   Application of Risk in Construction Contract

From project initiation to construction, operation and maintenance of a project generally go through many stages, such as project initiation, feasibility study, conceptual design, further design, construction, operation and maintenance. Among them, the construction stage is often the easiest to lead to project contract disputes. Because of the complexity of the project itself, as well as reasons such as inadequate communication, market fluctuations, construction project contract disputes often occur during the long implementation process. Through the legal clauses and causes of action cited in the litigation cases of contract disputes, this paper explores the weak links in risk management of construction contracts, identifies the frequent risk points of legal risks in construction contracts, and gives corresponding preventive and control measures against the identified risk factors. In order to provide reference and reference for risk management and signing of construction contracts in practice.

First, based on the analysis of keyword statistics, the keyword weights are calculated by TF-IDF algorithm. TF-IDF index can be considered as the product of TF value and IDF value, where TF is the frequency of words (the frequency of words appearing in articles) and IDF is the frequency of reverse files (the fewer articles containing a word, the larger the IDF, the better the category discrimination of this word). Considering the particularity of project text, the importance of text with different characteristics varies according to the location, length and part of speech of the vocabulary in the text. Therefore, the TF-IDF algorithm modified by Python is used to extract the text keywords of engineering projects. Formula reference (4), formula (5) formula (6) are calculated. In the TF-IDF model, the importance of a word is related to two factors: the first is the number of times it appears in a document, and the second is the number of times it appears in the corpus, which together determine the importance of a word.

$$TF_w = \frac{\text{The number of times the word w appears in a document}}{\text{Number of all words in this document}} \tag{4}$$

$$IDF_w = \log\left(\frac{\text{Number of documents in the corpus}}{\text{Number of documents containing the word w+1}}\right) \tag{5}$$

$$TF - IDF_w = TF_w * IDF_w \tag{6}$$

Then, based on the analysis of the citation frequency of legal provisions, through the statistics of the legal provisions cited in the judgment documents, the frequency of the 20 provisions with the highest frequency accounts for more than 80% of the total frequency. Through the analysis of the 20 provisions, it is found that the project payment disputes, qualification problems and contract invalidity are the main reasons for the construction contract disputes of construction projects.

Finally, the statistics of qualification cases and the analysis of relevant cases show that qualification problems often involve the identification of contract effectiveness. At the same time, it is necessary to reasonably allocate the number of fault behaviors of both parties in the invalid contract, so as to allocate fault liability. According to the statistics of project payment disputes, it is found that project payment disputes are one of the main risk sources leading to contract litigation disputes; In the case of project payment dispute, because the facts of the case are generally clear, the claim for the payment of project payment is easy to be supported by the court with sufficient evidence. In view of the above risks, by improving the qualification system, strengthening supervision and review, improving the deposit system, preventing project payment disputes, adjusting measures to "things", optimizing the design of contract terms, establishing and improving risk prevention and control measures, we can help the healthy and sustainable development of construction contracts.

3.2 Visual Analysis of Building Safety Knowledge Atlas

The characteristics of the construction industry, such as high risk, disorder and complexity, environmental dynamics and large amount of information, make it one of the industries with the highest incidence of safety accidents. At present, the research on construction safety is mostly carried out in the form of text combined with model, lacking the intuitive expression of the current situation analysis and development trend of this field, and the research of econometric analysis and visual analysis. The knowledge map can form a graph combining the relationship between knowledge development and structure, mine the implicit information of data, and integrate a large number of literature content to obtain visual information. As shown in Fig. 1, the type distribution diagram of engineering safety accidents is shown. CiteSpace III software is the most popular

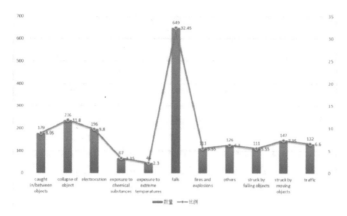

Fig. 1. Distribution of types of engineering safety accidents

research tool, which can draw a dynamic knowledge map in multiple and time-sharing, In order to deeply understand the research status in the field of building safety in China.

Using quantitative methods, CiteSpace III software is used for literature visual analysis to identify the trend and dynamics of literature research. For the visual analysis of building safety knowledge map, based on the bibliometric analysis of literature data, the annual average number of documents issued in the field of building safety is classified and counted. It is found that the number of documents issued is increasing year by year, It shows that scholars pay more attention to this field and become mature and perfect in the application of theories and technical methods. Based on the statistics of the research paper authors, the domestic research authors in the field of building safety are counted, the cooperative relationship between the authors is clarified by using the knowledge map, the research direction of important teams is tracked, and the research hotspots in the field of building safety are found. Through analysis, it is found that the representatives of domestic building safety research are concentrated in Colleges and universities, and college personnel are the main force in the field of building safety research. Based on the analysis of major research institutions, CiteSpace III's "institution" is used to analyze the distribution of research institutions in the sample data. It is found that the frequency of research institutions has a significant relationship with the institutions to which important authors belong. The number of main authors has a positive effect on the publication frequency of institutions, and important authors in institutions are an important factor affecting their number of papers, Domestic research institutions in the field of building safety are mainly composed of major universities. Based on the distribution analysis of published journals, the published journals are classified, and it is concluded that the research papers in the field of building safety are mainly concentrated in safety science and building science, followed by university journals. The research on building safety has gradually become a trend.

3.3 Research on BIM Visualization Based on Knowledge Map

BIM (building information modeling) is a landmark intelligent innovation platform in the construction industry. The core of BIM is to provide a complete and consistent construction engineering information base for the model by establishing a virtual three-dimensional model of construction engineering and using digital technology. Therefore, based on the theory of literature statistics, CiteSpace III software is used to sort out and analyze the relevant research in BIM field in SCI and core journals at home and abroad in recent 8 years, So as to timely grasp the latest direction and practical application of BIM research, Fig. 2 shows the map of hot knowledge in BIM field, So as to strengthen the application research of standards, specifications and building information management in BIM field, Continuously improve the theoretical framework related to BIM field in China; It provides guidance and reference for the theoretical research and practical application in the field of BIM.

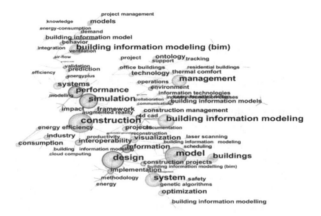

Fig. 2. Map of hot research knowledge in BIM field

The development and flow of information are abstract. CiteSpace III software is used to analyze the current situation and development trend of domestic construction safety field, analyze the research literature in the field of construction safety through the knowledge map theory, and draw the knowledge map in a visual way. By drawing the visual knowledge map of literature from the aspects of data measurement of literature samples, author cooperation, organization distribution and journal distribution, we find the influential authors, institutions, journals and their cooperative relations in this research field; Using keyword atlas analysis to find the hot issues in this field; Using Timeline view and time zone view, combined with the analysis of emerging words, to clarify the research frontier and development trend in this field, which will provide reference and reference for the future research trend of domestic construction safety management.

4 Conclusion

NLP technology has made some progress in the direction of construction engineering management, both at the technical and application levels. It transforms unstructured documents in the construction field into structured information, which is convenient for efficient induction and management of construction project document information and realizes the automatic management of construction documents; Accelerate the progress of domestic application research and implementation, and the construction project management based on NLP technology will bring greater benefits to the construction industry. This paper provides reference for the follow-up application research of NLP technology in the field of engineering management.

References

1. Wang, Y., Hui, D., Xiaoyao, L., Yichuan, D.: A review on the application of natural language processing technology in construction engineering. J. Graphics **41**(04), 501–511 (2020)
2. Wang, X.: Research on the application of natural language processing technology in project document management. Beijing University of Posts and Telecommunications (2019)
3. Huang, Y.: Research on risk of construction engineering safety accident report based on natural language processing. Huazhong University of Science and Technology (2019)
4. Kun, L., Sun Jun, Y., Dayan, Y.: Research on construction contract risk of construction project based on semantic retrieval. Constr. Econ. **40**(02), 106–110 (2019)
5. Wang, X.: Research on text classification and Knowledge Q & a system of construction quality complaints. Huazhong University of Science and Technology (2018)
6. Wu, S.: Research on BIM construction process safety risk inspection integrating ontology and natural language processing. South China University of Technology (2018)
7. He, Q., Qian L., Duan, Y., et al.: Current situation and barriers of BIM implementation. J. Eng. Manag. **26**(1), 12–16 (in Chinese)
8. Jiwei, Z., Yali, J., Zhu, Z., Qin, Z.: Research comparison of BIM field at home and abroad based on knowledge atlas. J. Civ. Eng. **51**(02), 113–120 (2018)
9. Dan, W., Jingjing, G.: Visualization research in domestic construction safety field based on knowledge atlas. J. Eng. Manag. **30**(06), 43–48 (2016)
10. Hu, H.: Research on knowledge modeling and extraction of construction engineering quality acceptance specification. Huazhong University of Science and Technology (2014)
11. Yijing, S.: Research on document management in engineering project management. Shandong Indust. Technol. **14**, 244 (2018)
12. Jiao, Y.: Application Research of Project Management Office of Y Company. Jilin University (2016)
13. Bu, Y., Liu, F., Chen, M., et al.: Visual analysis of foreign natural language processing research topics. Chinese Med. Guide **20**(05), 302–307 (2018)
14. Han, Z., Zhichang, Z.: Application analysis of natural language processing in information retrieval. CompuT. Fan **02**, 199 (2018)
15. Sheng, Y.: Research on contract risk assessment based on natural language processing. Harbin Engineering University (2017)

Exploring the Application of SketchUp and Twinmotion in Building Design Planning

Xiangshu Peng[1], Zhiming Ma[1(✉)], Ping Wang[2], Yaoxian Huang[3], and Menglin Qi[1]

[1] School of Computer Science and Technology, Xinjiang Normal University, Urumchi 830054, China
2298882664@qq.com
[2] School of Electricity and Computer, Jilin JianZhu University, Changchun 130000, China
[3] School of Computing, Harbin Finance University, Harbin 150000, China

Abstract. With the rapid development of industry 4.0 technology, the demand for real-time 3D rendering technology in all walks of life is unprecedented, especially in architectural design and planning, In view of the problems existing in traditional design planning, such as creativity is difficult to express immediately and intuitively, 2D drawings are complex and difficult to understand, workflow is chaotic, and it is difficult for all parties to coordinate and review, This paper analyzes the real-time seamless collaborative creation of SketchUp and twinmotion as a reference to solve the above problems, and uses three cases to analyze the media output from preliminary conceptual design to film and television photo level, in order to build a concise architectural visualization workflow in the spirit of open innovation.

Keywords: Real time 3D visualization · SketchUp · Twinmotion

1 Introduction

The construction industry has entered the era of rapid development of informatization, and real-time 3D rendering technology is booming. More and more people in the architecture and construction (AEC) industry use this technology to assist in architectural design and planning. Sketchup (Sketch Master) is an excellent tool for the creation of 3D architectural design schemes selected by millions of designers around the world. It is directly oriented to the design scheme creation process, and visualizes ideas and creativity with ultra-intelligent 3D modeling and convenient operation. Twinmotion is the easiest, fastest, most intuitive and most innovative 3D visualization real-time rendering 3D interactive software today, dedicated to the fields of architecture, construction, urban planning and landscape design. Using the world's most open and advanced real-time 3D creation platform - Unreal Engine (Unreal Engine) as the main core engine, it can easily and quickly create high-quality images, panoramic images, specifications or 360-degree VR videos in seconds. In this paper, sketchup and Twinmotion are combined to create, develop and display projects. Through the analysis of three project cases, the relevant technical characteristics and innovations are analyzed to help the future development of architectural visualization.

Q. Liang et al. (Eds.): AIC 2022, LNEE 871, pp. 18–24, 2023.
https://doi.org/10.1007/978-981-99-1256-8_3

2 Modeling Tools - SketchUp

SketchUp was launched by Trimble and then acquired by Google. It is a conception and expression software that seems to be extremely simple, but actually contains a streamlined and robust toolset and an intelligent presumption guidance system. In the creative process, usually from a vague concept, with further design deliberation, gradually refined, SketchUp can instantly and intuitively express the creator's ideas and ideas, and can also be deepened as the project progresses to dynamically and creatively. Explore models, materials, components, etc. SketchUp is the leader in the exploration of creative expression and design process in the field of AEC software. Its core features are:

(1) With a unique and concise visual operation interface, there is no need to learn a wide variety of instruction sets with complex functions, reducing investment in education and training exercises, and simplifying the 3D modeling process.
(2) It has a wide range of applications, and the software is free and easy to use. It is widely used in architecture, planning, gardening, landscape, interior and industrial design and other fields.
(3) Robust data conversion interface, strong compatibility, it is an open and extensible platform and supports rich plug-ins, which can quickly realize parametric modeling of various functions, and seamlessly interconnect with various platforms in real time.
(4) With a rich resource library, you can access the world's largest 3D model library. SketchUp has established a huge 3D model library, which gathers model resources from various countries around the world to form a sharing platform. There are tens of thousands of models available for free from 3D models. Warehouse download, you can directly call, insert, and copy existing models in the design, without starting from scratch.
(5) The requirements for computer hardware configuration are not high, and the smooth running speed enables people to observe and experience the effect of the work from any angle as they wish.

3 Real-Time Rendering Tool - Twinmotion

3.1 Introduction to Twinmotion

Twinmotion belongs to Epic Games, the most mainstream engine developer today, and is different from traditional real-time rendering software: Unity 3D requires C# programming foundation, VRay for 3ds Max is complicated to operate and requires a certain foundation, Renderman, Mental Ray and other plug-ins are mainly used in animation and film and television. Special effects and other fields, while Twinmotion only focuses on the fields of design, visualization and architectural interaction. It does not require complex operations or codes, and its intuitive icon-style Chinese interface only requires dragging modules and setting simple parameters to create real-time immersive 3D architectural visualization.

3.2   Research on the Specific Function of Twinmotion

Supports the study of environmental systems such as lighting, seasons and weather time in Twinmotion, and can control weather effects such as clouds, wind, and rain in real time; quickly add various plants through the vegetation painting tool, quickly create forest and grass greening, and adjust one or the whole piece. Forest leaf color, tree age, growth status, etc. Add a 2D static or dynamic one or a group of characters or animals according to your needs; build a complex traffic system, place various vehicles (bicycles, cars, boats, planes, etc.), choose colors and directions, choose a car or add traffic animation, draw the vehicle path and turn on the travel mode, adjust the traffic flow and its speed; sculpt and texture the terrain, automatically generate the simple building complex of the project by entering the date and precise geolocation, including the local environmental conditions, with (walking, vehicle speed), flight) mobile mode to observe items from various perspectives. Just need one step, all media are easy to share, create Twinmotion scene images, animations, panoramas, VR videos and customer demo mode in seconds, support VR virtual reality operation, and more unique presenter cloud function, no need to download any large file or prepare a workstation with powerful performance, share the demo file to all parties through a link, easy demonstration and review; there are also a large number of exquisite and realistic materials to choose from, a unique material extractor, which can absorb any model material and adjust the appropriate parameters (UV texture, size, color, etc.), which is convenient and fast, brings unlimited creative freedom of materials, and many options and functions.

3.3   Real-Time Interconnection Between SketchUp and Twinmotion Data

The development of building digitization is accelerating day by day. Thanks to computer graphics technology to assist building design and planning to achieve rapid iterative workflow, most of the rendering tools currently used are closed frameworks, and there is a lack of communication and interaction between software and links. Aiming at the bottleneck problem of using different software in different links between data transmission pipelines, Twinmotion uses the DIRECT LINK plug-in to synchronize with the mainstream architectural 3D modeling software Revit, SketchUp, 3ds Max, C4d, Rhino 3D, etc. Original surfaces and 3D objects are automatically replaced with objects that can interact with the environment. Thanks to the seamless real-time interconnection of SketchUp and Twinmotion data pipelines, the above problems are perfectly solved; the next stage of digital transformation in the construction industry is to develop seamless and open real-time pipelines.

3.4   Unreal Engine Detailed Design

Started in Twinmotion and finished in Unreal Engine. This is the unique advantage of the solution. Develop basic creations in Twinmotion, express initial conceptual ideas, and when the limit is reached, pass the project to the visualization through the extended pipeline plug-in Bridge to Unreal Engine, as shown in Fig. 1. Further development in expert Unreal Engine. Then add accurate lighting and material schemes, use higher-quality animations and top-of-the-line rendering capabilities, and make visualizations better than any other real-time rendering software can do.

Fig. 1. Bridge to Unreal Engine plugin

4   The Application of SketchUp and Twinmotion in the Project

4.1   Local Regional Planning of Huadian City, Jilin Province

In the early stage of planning and design, information collection and analysis are carried out by means of satellite, aerial photography, on-site inspection photography, etc., and CAD drawings for local planning of Huadian City are drawn, as shown in Fig. 2, including various facilities in the city (buildings, landscape greening, road traffic system, etc.). Optimize the drawings, clean up redundant layers and line segments, import them into SketchUp software, set equal-scale units, improve modeling accuracy, and cleverly use the blank library and various plug-ins (more than 160 plug-ins) through simple plane push-pull tools. More than 400 functions can be realized). It is convenient and quick to establish the required model, optimize the model, and use the map method for secondary buildings to increase the fidelity of the model, reduce the file size, reduce the nesting of components, and classify the components by category to improve the efficiency of later detailed design. Figure 3 is the SketchUp model. After the modeling stage is completed, synchronize it to Twinmotion through the DIRECT LINK plug-in to establish real-time interconnection, adjust the import settings (retain or merge materials, reference axis and placement, etc.), the next stage is landscape layout, Real-time rendering visualization creation of building materials, traffic, climate system, etc., as shown in Figs. 4 and 5 for the output renderings. The effect has been fully affirmed and praised by relevant departments of Huadian City.

4.2   Architectural Design and Planning of a Commercial Center in Changchun

In the early stage of architectural design, plan and design the floor plan according to the type of use of the building and the needs of Party A. Figure 6 is the floor plan of a commercial center in Changchun. The corresponding 3D model is established through SketchUp. The main buildings need to be modeled in detail. The surrounding buildings are represented by translucent squares, which can highlight the main buildings and reduce the size of the model file. As shown in Fig. 7 the fine adjustment of the building glass curtain wall material is carried out in Twinmotion. The unique material absorber is used to absorb the satisfactory glass material and apply it to the curtain wall system, and add appropriate greening, traffic and character embellishment. According to Requirements

Fig. 2. CAD drawings

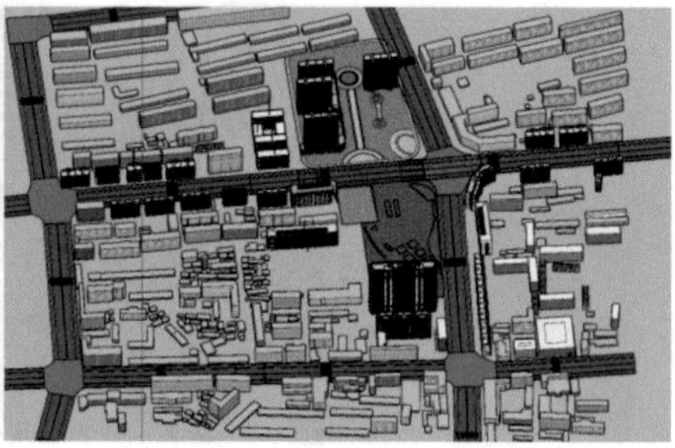

Fig. 3. SketchUp model

Fig. 4. Design sketch

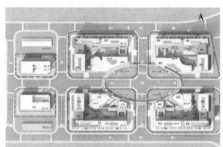

Fig. 5. Design sketch

and content output different media files. Figures 8 and 9 shows the architectural rendering of a commercial center in Changchun. From the general structural framework of the building to the local details, real-time modification and real-time rendering always run through the entire project process, and the intuitive and convenient form of expression provides a reference for the finalization of the scheme.

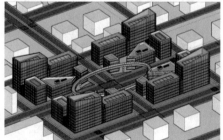

Fig. 6. Building floor plan

Fig. 7. SketchUp model

4.3 Architectural Design and Planning of a Commercial Center in Changchun

In order to improve the effect and rationality of street landscape design and improve urban supporting facilities, the landscape design is mainly the spatial combination of

Fig. 8. Design sketch   **Fig. 9.** Design sketch

trees, public facilities and roads, so that each space can be opened and closed to a certain degree, interspersed with each other, and imported CAD drawings into SketchUp for modeling. Use the road generation plug-in to create the corresponding road, download the required park facility models (such as landscape corridors, pavilions, sculptures, etc.) from the 3D model library, as shown in Fig. 10 is the SketchUp model, and perform space in Twinmotion The deduction and generation of the effect, and the rationality of the layout design and facility construction are explored. Figure 11 is the output effect diagram.

Fig. 10. SketchUp model   **Fig. 11.** SketchUp model

5   Conclusion

The 21st century is the era of digitalization, and the application of computer graphics and real-time rendering technology to assist in diversified design and performance is the future development trend. In the field of architectural visualization, the real-time interconnection and collaborative creation of Sketchup and Twinmotion, the perfect combination of lightweight, simple and fast modeling technology with photo-level real-time rendering and dynamic visualization media technology, fully explain the creative concept and design, allowing customers to visualize in real-time 3D Ways to participate in the design and appreciate the shine of the project. Through the analysis of technical theory and project practice, it will help the sustainable development of architectural visualization.

References

1. Huang, J., Yang, Y., Hu, S., Mai, X.: Application of sketch up and Lumion in 3D visualization design of Hainan shared farm. Trop. Agric. Sci. **40**(08), 79–84 (2020)
2. Ma, S., Li, D., Li, X.: A brief analysis of the application of 3D software in the generation and performance of landscape schemes - taking Sketchup and Lumion software as examples. Art and Des. (Theory) **2**(12), 83–85 (2020)
3. Jiang, J.: The application of Sketch up combined with Lumion in architectural design courses. Technol. Econ. Mark. **04**, 183–184 (2018)
4. Feng, Y., Wei, J., Hu, K.: Comparison of SketchUp and Lumion in landscape planning and design. J. Wuhan Eng. Univ. **37**(06), 46–50 (2015)
5. Di, Y.: SketchUp+VRay Interior Design Renderings. People's Posts and Telecommunications Press (2015)
6. Peng, S.: Research and application of computer-aided design software in architectural design. Shanghai Jiaotong University (2009)
7. Ma, C., Zhu, S., Wang, M.: Research on the application of machine learning technology in architectural design. Southern Archit. (02), 121–131 (2021)
8. Kong, Y., Xin, S., Zhang, N.: Translation and reconstruction: the application of traditional construction wisdom in architectural design. J. Archit. (02), 23–29 (2020). https://doi.org/10.19819/j.cnki.ISSN0529-1399.202002004
9. Wang, J., Chen, Y., Wei, J., Zhu, L., Dong, R.: Application of BIM technology in cost control of construction projects. Value Eng. **38**(35), 230–232 (2019). https://doi.org/10.14018/j.cnki.cn13-1085/n.2019.35.092
10. Jiang, S.: Research on the application of Sketchup software in architectural animation design. Art Appreciat. (09), 181–182 (2018)
11. Peng, S.: Application of SketchUp in architectural design. Shanxi Archit. (06), 365–367 (2008)

Lifecycle-Based Software Defect Prediction Technology

Xiangshu Peng[1], Zhiming Ma[1(✉)], Ning Zhang[2], Yaoxian Huang[3], and Menglin Qi[1]

[1] School of Computer Science and Technology, Xinjiang Normal University, Urumchi 830054, China
2298882664@qq.com

[2] School of Electricity and Computer, Jilin JianZhu University, Changchun 130000, China

[3] School of Computing, Harbin Finance University, Harbin 150000, China

Abstract. In order to improve the efficiency and quality of software testing, aiming at various factors affecting software reliability, how to find defective modules and optimize them in the early stage of software development has become an urgent problem to be solved, This paper introduces the software defect prediction technology based on life cycle. According to the measurement elements affecting software reliability, relevant internal indicators and design defects, find the defect module, lock it in advance, adopt machine learning technology and reasonably allocate limited resources, which is conducive to evaluate the software design scheme, optimize the design strategy, reduce design changes and improve the software operation process, It plays a role in cost evaluation, resource management, scheme determination and quality prediction in software management. It is hoped to provide some theoretical support and practical reference for the development of software defect prediction.

Keywords: Software defect predict · Software metric · Software lifecycle · Introduction to the study

1 Introduction

With the popularization of information technology in all walks of life, the number and complexity of all kinds of computer systems and software are growing rapidly, its development technology is constantly updated and developed, the operating environment is becoming more and more complex and severe, and if the software is defective in the running process, it may lead to serious consequences, and the reliability of software operation has become an important guarantee for the sustainable development of various industries. Software reliability mainly refers to the prediction of software defects. In each stage of the whole software life cycle (mainly including market customer demand analysis stage, content design stage, coding implementation stage and operation and maintenance stage), defects are not evenly or randomly distributed in the software; Aiming at the prediction technology of defect distribution, this paper uses machine learning technology to construct software defect prediction model, in order to improve software testing efficiency and ensure software reliability.

Q. Liang et al. (Eds.): AIC 2022, LNEE 871, pp. 25–31, 2023.
https://doi.org/10.1007/978-981-99-1256-8_4

2   Analysis of Software Prediction Technology

2.1   Definition of Software Prediction Technology

Software prediction is not complex in the actual development process. It mainly needs to run the analysis of a certain program under the specified conditions, find out whether there are certain design errors in the program in the actual development process, test the software quality accordingly, and evaluate whether the program meets the design requirements in the actual operation process, Therefore, software prediction technology will become more important in the actual development process. The update, improvement and innovation of software prediction technology play a very important role in improving the quality of software. The purpose of software prediction is: (1) to analyze and observe defects to improve the quality of software. (2) Verify whether certain requirements are met in the actual development process. (3) In the actual development process, the forecaster should have a certain confidence in the software quality itself.

2.2   Principles of Software Prediction Technology

(1) the test shows that there are defects; When testing, the scope of testing should be comprehensive. (2) Early detection and intervention (low cost and high efficiency). (3) Defect clustering: 80% defects are concentrated in 10% modules. (4) In the actual development process, when a detection method fails to find relevant defects in time, the test cases should be replaced in time. (5) There must be a certain relationship between testing activities and testing background. (6) No shortcomings are fallacies (useful systems) in the actual development process. When testing software, we should analyze its advantages and disadvantages, find shortcomings and problems, and make relevant improvement and innovation in time.

2.3   Main Workflow of Software Prediction Technology

Carry out reasonable planning and control of the software in the actual test process, reasonably plan the test scope, classify and innovate the test plan, and analyze and count the relevant data of the project to be tested, such as the input and output documents, product description, main functions of the software in the actual development process, It includes reasonable analysis and control of the output documents, integration and statistical analysis of all resources (human, material, financial, etc.) invested in the whole test plan, and certain evaluation and prediction of relevant risks. In addition, the development environment of the project should be considered when selecting the training set to further promote software quality and prediction efficiency.

3   Software Reliability Measurement

To predict software defects, the first is to select the measurement element of software reliability, which should focus on simplicity and practicality, face each stage of software development, run through the whole development process (demand analysis, design, implementation and testing), and focus on reliability. Software reliability is mainly composed of three activities: software error proofing, finding and troubleshooting faults, and measuring software to achieve the best reliability; The measurement of software reliability is mainly analyzed from the technical measurement, As shown in Table 1 is Technical metrics for each stage of the software lifecycle. And then subdivided from the four stages of software development process. (1) Requirements reliability measurement: customer requirements include software technology, quality and various functions. The key lies in the preparation of requirements description. Finding and correcting requirements errors will consume huge human and material resources and reduce the quality of software. Therefore, requirements description requires structure, stability, high quality, easy understanding and easy application. (2) Design and implement reliability measurement: Based on the complexity and scale of the module, analyze the code structure and function system, and determine the appropriate development language and format, so as to find and predict the defects and errors of that module. The evaluation of the combination of scale and complexity is more effective. (3) Test reliability measurement: in order to ensure the integrity of software development function, the evaluation of test plan is adopted to reduce the lack and error of expected function.

Table 1. Technical metrics for each stage of the software lifecycle

Software lifecycle phase	Technical metrics
The requirements analysis phase	1: The team's level of development skills and experience 2: The correctness and rationality of demand decomposition
Content design phase	1: The average number of fan-ins of the module 2: The average number of fan-outs of the module 3: The average ring complexity of the module 4: The average coupling of the module
The coding implementation phase	1: The length of the module code 2: Develop language types 3: Coding rules

4 Lifecycle-Based Software Defect Prediction Technology

4.1 Existing Research

Life cycle software defect prediction technology is a very important technology, which plays a very important role in ensuring and improving software quality. Software defect prediction technology based on life cycle belongs to an important type of this technology. Learn from and learn other types of software defect prediction technology, analyze and improve it, so as to better promote the development of technology. Document [6] proposed a software defect prediction method based on software development cycle and PCA-BP fuzzy neural network. The results show that compared with the prediction method based on BP neural network, the PCA-BP Neural Network prediction method combined with principal component analysis method has faster convergence speed and higher prediction accuracy. Document [1] proposed a software defect prediction model construction method for cost sensitive classification. The decision tree classifier for cost sensitive classification is constructed by bagging multiple random sampling training samples, and then the software module defect prediction is carried out after voting integration. The results show that this method significantly reduces the false positive rate on the premise of ensuring the defect prediction rate, AUC and F values are better than the existing methods. Document [2] proposed a cross project software defect prediction method based on multi-source data, obtained multi-source similar project data, and realized the prediction model based on Naive Bayesian algorithm. The results show that the performance of this method is better than the traditional WP method.

4.2 Experimental Analysis

The first is the collection of test data, collecting measurement data of 16 flight control software, totaling 613 modules. The sixteen software serves as training samples for the neural network, and the last eight software serves as the software to be predicted. Using PCA-BP network model and BP network model, defect prediction is made for several stages of software life cycle. We compare and analyze the convergence speed of the neural network and the accuracy of software defect prediction. The convergence speed of the neural network is reflected by the mean of the sum of the errors between the expected output and the real output of the network. The formula is as follows (1).

$$e = \frac{1}{m} \sum_{j=1}^{n} \left(Y_j - Y_{j(\text{prediction})}\right)^2 \qquad (1)$$

Formula: m denotes the number of sample software, Y_j represents the actual value of the jth software defect density, $Y_{j(\text{prediction})}$ represents the predicted value of the jth software defect density, and the e represents the average of the sum of squared errors between the predicted value and the actual value. The training error line graph is shown in Fig. 1 and Fig. 2. The horizontal coordinate represents the number of training sessions and the vertical coordinate represents the average of the sum of squares of training errors. The results show that the convergence speed of PCA-BP network is not different from that of BP network in the stage of requirement analysis, but PCA-BP network converges faster than BP network during the training of coding phase.

4.3 Index Analysis

Two indicators are designed: number of developers (NOD) and coding quality lock (locq). Bayesian network is used to analyze the relationship between each index and error trend. The results show that RFC (class response), LOC (number of lines of code) and locq are the most effective metrics, and CBO (coupling between objects), WMC (weighting method per class) and LCOM (lack of method cohesion) have little effect on defect tendency. Two initial feature subsets are obtained by lg based filtering method

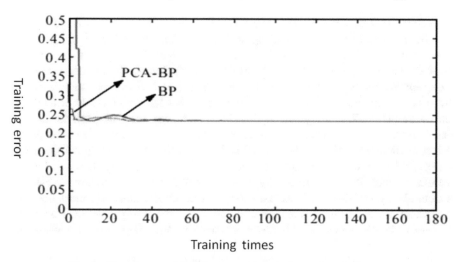

Fig. 1. Neural network training error curve in demand analysis stage

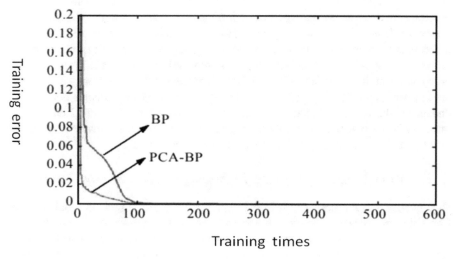

Fig. 2. Training error curve of neural network in coding stage

(MIM) and correlation coefficient based filtering method (SBS); Then, these subsets are combined into global optimization. (GA) defs (differential evolution case feature selection) or PSO (particle warming optimization) are applied to the combination subset respectively, and an effective feature subset is obtained by taking variance as the stop criterion.

4.4 Method Analysis

The filtering method is to calculate the chi square test (CS) or information gain (Ig) or correlation coefficient score or Pearson correlation value of each measurement element, and select the feature according to these values. The selection process is independent of subsequent learners. The packaging method takes the learner's performance as the evaluation criterion for selecting the best feature subset. The encapsulation method takes the learner's performance as the evaluation function, so the performance of the encapsulation method is better than the filtering method. However, the time cost of encapsulation method is large, while the filtering method has nothing to do with learners and the cost is small, so the generalization ability is stronger than encapsulation method. Fesch (feature selection using clustering of hybrid data) is used to solve cross-project defect prediction problems. The method is divided into two phases. In the first stage, the original feature set is clustered by the Peak Density Clustering (DPC) method. In the second stage, local feature density (LDF), feature distribution similarity (SFD), and feature class correlation (rel) are used to deal with cluster defects. Defect life cycle is a whole process from the creation of a defect to the end of the defect. During this process, defects may undergo different processing scenarios depending on development and product strategies: continue to repair it).

Scenario 1: Confirm Error Resolution Test Submission Defect [new], Develop Confirm Defect [open], Develop Resolve Defect [fixed], Test Regression Defect, Close Defect [closed]; Scenario 2: Validation fails, defects still exist, Test submission defects, development confirmation defects [open], development resolution defects [fixed], testing regression defects, assigning development to resolve [reopen]; Scenario 3: The closing defect reappears and the tester reopens the closing defect; Scenario 4: Develop delayed processing test submission defects [new], develop validation defects [open], delayed processing (publishing); Scenario 5: Refuse to process test submission defects [new], develop validation defects [open], reject (reject). To understand software defects, first understand the concept of software defects, second understand the detailed characteristics of software defects, and finally understand the properties of software defects. The next higher level is learning to use tools to manage software defects.

5   Software Defect Prediction Technology Defect Report Analysis

High quality defect reports can help developers quickly locate problems and fix them. Convenient for testers to count, analyze, track and manage defects; It is an important communication tool between testers and developers. Therefore, if our testers find a defect during the test, they need to record the defect and submit a defect report. It mainly includes recording software defects, then classifying and summarizing the defects, then tracking software defects, and finally analyzing and summarizing the defects. Important components of a defect report are number (defect ID), Title (summary), date of detection, module to which the defect belongs (subject), release of detection, and assignment of tasks. Defect states include; (1) Describe the status of the defect at this time: when the tester finds a defect and submits it to the development manager. (2) Open: The state after the development manager acknowledges and accepts the defect (if the development manager finds that it is not a defect, it will reject it and the defect state will be rejected). (3) Fix: The developer receives the bug and the status of the fix. (4) Off: Testers perform tests to repair defects and verify that the status is tested (if the tester verifies that a defect has not been repaired, that is, if the test fails, the defect status will be reopened and the developer will continue to repair it). (5) Defect severity.

6 Conclusion

This paper mainly introduces the concept of software defect prediction technology based on life cycle, the selection of measurement elements and the early prediction method of software reliability using PCA-BP fuzzy neural network, which plays a great role in improving and ensuring software quality and sustainable development, but there are still some problems in the research of software defect prediction technology, We also need to constantly analyze and summarize experience in the event, so as to promote the continuous development and progress of the industry. I hope this paper can provide certain theoretical support and practical reference for relevant staff.

References

1. Yong, L., Zhiqiu, H., Bingwu, F., Yong, W.: Software defect prediction method for cost sensitive classification. Comput. Sci. Explor. **8**(12), 1442–1451 (2014)
2. Li, Y., Liu, Z., Zhang, H.: Overview of cross project software defect prediction methods. Comput. Technol. Dev. **30**(03), 98–103 + 121 (2020)
3. Yong, L., Zhiqiu, H., Yong, W., Bingwu, F.: Cross project software defect prediction based on multi-source data. J. Jilin Univ. (Engineering Edition) **46**(06), 2034–2041 (2016)
4. Li, Y.: Software defect prediction combined with under sampling and integration. Comput. Appl. **34**(08), 2291–2294 + 2310 (2014)
5. Li, Y., Liu, Z., Zhang, H.: Overview of integrated classification algorithms for unbalanced data. Comput. Appl. Res. **31**(05), 1287–1291 (2014)
6. Wu Chao, X., Jianping, C.L.: Software defect prediction technology based on life cycle. Comput. Eng. Des. **30**(12), 2956–2959 (2009)
7. Tao, M.: Research on feature selection method for software defect prediction. Jilin University (2020)
8. Lina, G., Shujuan, J., Li, J.: Research progress of software defect prediction technology. J. Softw. **30**(10), 3090–3114 (2019)
9. Cai, L., Fan, Y., Meng, Y., Xia, X.: Research progress of real-time software defect prediction. J. Softw. **30**(05), 1288–1307 (2019)
10. Shen, P.: Research on software defect prediction method based on machine learning. Southwest University (2019)
11. Wang, T.: Research on software defect prediction based on measurement. Wuhan University (2018)
12. Li, L.: Research on cross version software defect prediction technology. Nanjing University of Aeronautics and Astronautics (2018)
13. Zou, J.: Research and application of feature selection method for software defect data. China University of Petroleum (East China) (2017)
14. Lu, G.: Research on software defect prediction technology based on deep learning. Nanjing University of Aeronautics and Astronautics (2017)
15. Cheng, M.: Research on some key technologies of software defect prediction. Wuhan University (2016)

Unsupervised Anomaly Detection Method Based on DNS Log Data

Wang Jiarong[1], Liang Zhongtian[1,2], Qi Fazhi[1], Yan Tian[1], Liu Jiahao[3], and Zhou Caiqiu[1(✉)]

[1] Institute of High Energy Physics, Chinese Academy of Sciences, Beijing 100049, China
zhoucq@ihep.ac.cn
[2] University of Chinese Academy of Sciences, Beijing 100049, China
[3] School of Cyber Science and Engineering, Zhengzhou University, Zhengzhou 450001, China

Abstract. In order to solve the problem of network attack by malicious code using Domain Name System (DNS), on the basis of analyzing the characteristics of malicious code lines and abnormal operation behaviors, this paper proposes an unsupervised abnormal IP detection method based on DNS log data. Through the construction of DNS fingerprint characteristics, it is used to demonstrate the DNS behavior characteristics of IP to the greatest extent. The detection model is constructed by using isolated forest and local outlier factor algorithm, and the anomaly score of IP is obtained. The experimental results show that the detection method designed in the paper can well detect the attack exceptions and operation exceptions in the network environment. With the help of whitelist, the accuracy of the method can reach more than 90% after selecting the appropriate anomaly score threshold.

Keywords: Domain Name System · malicious code · DNS fingerprint · anomaly detection · anomaly score

1 Introduction

Domain Name System (DNS) is one of the core components of the Internet. The operation of various network services is inseparable from the DNS system. In addition, the security defects of DNS itself have attracted many network attackers to carry out network attacks directly or indirectly by using DNS. Most of the traditional protection methods are to filter the queried domain name and the returned IP address based on the black-and-white list, and detect them by using the way of signature matching. With the progress of attacker technology, attackers try their best to bypass the blockade, which greatly increases the protection cost of traditional means, and it is difficult to achieve complete protection. In recent years, on the basis of making full use of DNS log data, scholars have carried out in-depth research from the aspects of DNS query behavior, malicious domain name detection, anomaly IP detection and so on. Lin Chenghu et al. [1] analyzed and counted DNS server queries every 10 s and extracted the data, manually set the weights of each feature, and used W-kmeans algorithm to identify DNS traffic anomalies in the form

Q. Liang et al. (Eds.): AIC 2022, LNEE 871, pp. 32–43, 2023.
https://doi.org/10.1007/978-981-99-1256-8_5

of binary classification; By analyzing the DNS log, Wang Qi et al. [2] analyzed and extracted the entropy, length, sub domain name and other relevant information of the secondary domain name as the characteristics, took the DNS cache hit rate as the key characteristic, and used the random forest algorithm to establish a model to detect the DNS tunnel; Wang Jingyun et al. [3] believe that the anomaly degree of the source IP can be measured by calculating the relative density of the DNS behavior characteristics of the source IP, distinguish between normal IP and anomaly IP by outlier detection, and extract 9 features based on the relationship between DNS log and DNS behavior to describe the DNS behavior of the source IP.

However, most of these researches are mining and analyzing DNS traffic or log data to detect a specific form of attack, which use supervised and semi supervised methods [4]. The detection ability depends on the integrity and reliability of the data source used in training, and it has weak adaptability to changing multi-terminal and rapidly generating and disappearing malicious code.

Based on this situation, the paper proposes an unsupervised anomaly detection method base on DNS log data. On the basis of analyzing the basic behavior of DNS, combining the analysis of malicious code behavior and anomaly running performance of programs to build features, extracting DNS fingerprints from DNS log data over a period of time, and inputting the anomaly detection model constructed by the isolated forest (Iforest) and local outlier factor (LOF) algorithm to give the IP anomaly score, and finally anomaly IP in the network environment is found.

2   Anomaly Detection Based on DNS Log Data

The traditional network attacks based on DNS log data are mostly aimed at a specific type of attack detection, while the use of supervision or unsupervised machine learning, neural network and other methods [5–7]. The detection ability depends on high-quality labeled data sets, leading to excess detection ability depends on the quality of tag data sets, and the detection is single and unable to detect multiple types of attacks.

In order to make up for the defects of traditional detection, an unsupervised anomaly detection method based on DNS log data is proposed in the paper. Firstly, the DNS server log data is obtained from the data platform. Secondly, feature engineering is performed on the obtained DNS log data, and DNS fingerprint is extracted with IP as the core. Then, the DNS fingerprint data is input into the isolated forest algorithm and LOF algorithm respectively for anomaly detection. The values used by the two algorithms to measure the degree of IP anomaly are mapped to the same interval and added together to obtain the final anomaly score of IP. The higher the anomaly score is, the higher the anomaly degree of IP is. Finally, anomaly IP addresses are found according to the anomaly score ranking. The algorithm flow is shown in Fig. 1.

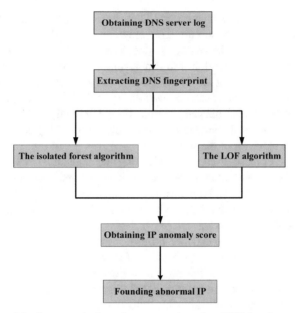

Fig. 1. anomaly detection process based on DNS log data

2.1 Feature Extraction

In order to detect abnormal DNS behavior, this paper extracts 10 behavior features, as shown in Table 1, which can reflect whether a specific IP is infected with malware.

Table 1. characteristic items of DNS fingerprint

Parameter name	Features
total	Query times
nxdomain	Number of query failures
nxrate	Failure success ratio
anomaly	Number of query anomaly domain names
PTR	PTR query times
PTRtarget	PTR target number
SITEcount	Query the number of domain names
SITEentropy	Query domain name entropy
SITEsingle	Maximum number of single domain name queries

1) Number of queries: The number of DNS queries generated during normal network activities is not too high. When malicious codes are infected, the number of DNS queries increases significantly. Figure 2 shows the statistics of the number of DNS server queries in the real environment. The number of DNS queries of most IPS is less than 10000. In the figure, there is an IP with nearly 80000 queries. This IP is obviously abnormal. After investigation, the host corresponding to the IP address is infected with a virus.

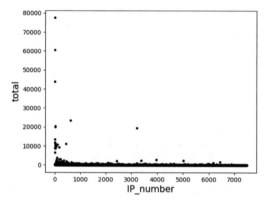

Fig. 2. Number of queries

2) Number of query failures: Normal DNS queries have a low probability of query failure, while anomaly IP addresses usually generate a large number of failed queries, especially those infected with malicious codes based on the DGA algorithm, which is a powerful indicator of network intrusion monitoring. Figure 3 shows the statistics of DNS server query failures in the real environment. The majority of IP query failures are close to 0. Among them, the IP address with a large number of DNS query failures is the anomaly host. After comparison, it is found that this IP address is the same as the host infected with virus in Fig. 2.

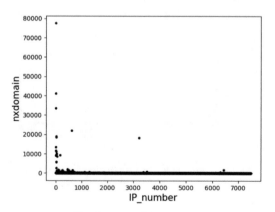

Fig. 3. Query failure times

3) Number of query anomaly domain names: Normal domain names have a certain format. Users may query a small number of anomaly domain names due to misoperations, but a large number of anomaly IP addresses may be accessed. Figure 4 shows the number of anomaly domain names queried by different IP addresses. It is obvious that a large number of queries were made from one anomaly IP address.

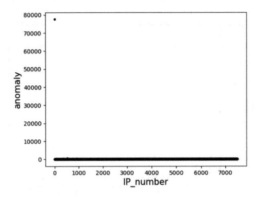

Fig. 4. Querying the number of anomaly domain names

5) Maximum number of queries for a single domain name: Malicious code based on fast-flux will change the domain name and IP mapping relationship in a relatively ordinary way, and frequent queries are needed to ensure the connection; some DNS tunnels will encapsulate the data in the DNS protocol and transmit the data through the controlled DNS server. Figure 5 shows the maximum number of queries for a single domain name by different IPs.

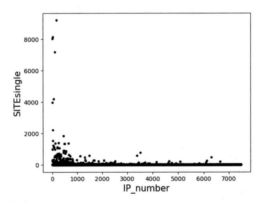

Fig. 5. Maximum number of single domain name queries

2.2 Anomaly DNS Behavior Detection

In order to get rid of the dependence on high-quality data sets, the paper selects unsupervised detection algorithm as detection means. The paper combines isolated forest algorithm and local outlier factor algorithm to detect abnormal DNS behavior, and considers global and local anomalies to improve the accuracy of anomaly detection.

2.2.1 Isolated Forest Anomaly Detection

Isolated forest algorithm is an unsupervised anomaly detection algorithm. It regards outliers as points with sparse distribution and far away from high-density groups [8–11]. The algorithm will randomly segment the data set, and outliers are usually isolated in the early stage. The actual operation is realized by constructing multiple binary trees (isolated trees). The specific process is as follows:

Firstly, m sample points are randomly selected as data subsets to establish the root node of binary tree.

Secondly, select a feature p randomly, and select a value q randomly within the value range of the feature.

Thirdly, put the sample points on feature p, which is less than q into the left child node, and those greater than q into the right child node of the tree.

Finally, repeat steps 2 and 3 until there is only one data in the child node or the isolated tree reaches the limited height.

2.2.2 LOF Anomaly Detection

LOF algorithm is a density-based anomaly detection algorithm, which is an unsupervised detection algorithm [12–14]. LOF algorithm calculates the outlier factor of each sample point, which reflects the similarity between each sample point z and the density of adjacent points, so as to judge whether sample point z is an outlier. If the outlier factor LOF is close to 1, it means that the local density distribution of the sample point is similar to the surrounding neighboring points, which is a normal point; on the contrary, it is an anomaly node. The specific calculation is as follows:

1) Find the k-th distance $d_k(z)$ from point z, which represents the k-th smallest Euclidean distance between z and other points.
2) Find the set of points $N_k(z)$ whose distance from point z is less than or equal to $d_k(z)$.
3) Calculate the local reachable density of point z, such as formula (1):

$$lrd_k(z) = \frac{|N_k(z)|}{\sum_{oN_k(z)} reach - d(z)o} \tag{1}$$

Among, $|N_k(z)|$ represents the number of points in $N_k(z)$, reach-d(z, o) represents the reachable distance from adjacent point o to point z, and the calculation equation is:

$$reach - d(z, o) = \max\{d_k(o), d(z, o)\} \tag{2}$$

4) calculate LOF, the equation is:

$$LOF_k(z) = \frac{\sum_{o \in N_k(z)} \frac{lrd_k(o)}{lrd_k(z)}}{|N_k(z)|} \tag{3}$$

It can be seen from the formula, the LOF of sample point z is actually the average ratio of the local reachable density to local reachable density of adjacent points. The closer the value is to 1, the more likely point z belongs to the same cluster as adjacent points.

2.2.3 Anomaly Score

According to the input isolated forest and LOF algorithm, the outlier scores are obtained respectively [15–18]. In order to combine local exceptions and global exceptions, the anomaly score is accumulated to obtain the final anomaly score. The larger the anomaly score of IP, the more likely it is to be an exception, the equation is (4):

$$y' = y_{min} + \frac{y_{max} - y_{min}}{x_{max} - x_{min}} * (x - x_{min}) \tag{4}$$

where y' is the scaled value, y_{min} and y_{max} are the maximum and minimum values of the scaled interval. In the paper, the two values are 0 and 5. We control the anomaly score of the final output in the range of 0 to 10. x_{min} and x_{max} is the maximum and minimum value of the current data value range, and x is any value of the current data.

3   Experiment and Result Analysis

The experiment is carried out in Anaconda 4.9.2 integrated Python environment. After extracting the DNS fingerprint from the DNS server log file by using python, the isolated forest algorithm and LOF algorithm are realized by using the python machine learning library scikit-learn 0.23.2, and the detection model, which is described in Sect. 2.2, is constructed.

3.1   Performance Validation of the Detection Model

The DNS log server data used in the experiment is from the local DNS server of the Institute of High Energy Physics, Chinese Academy of Sciences. In order to ensure the amount of information of data and avoid abnormal behavior being covered up by long-time DNS behavior, the paper selects the working time period to obtain data. We select the DNS server log of 9:00 on December 17, 2020 for detection and analysis. During this period, users have more active DNS behaviors. Among them, 7494 IP address are involved, which are replaced by serial numbers. A total of 980000 DNS queries are conducted. Finally, arrange the anomaly scores from large to small, and the data distribution is shown in Fig. 6.

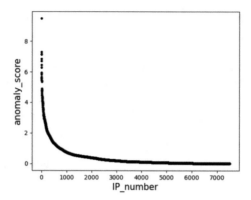

Fig. 6. Abnormal score data distribution (from

It can be seen from Fig. 6, most IP's anomaly scores are less than 2 and close to 0, while there are 9 IP's with anomaly scores greater than 6, which is far from the anomaly scores of other IPS. The specific scores and some characteristics are shown in Table 2.

Table 2. Nine IP addresses with the highest anomaly scores

IP order number	total	nxdomain	anomaly	PTR	PTR tardet	SITE count	SITE entropy	SITE single	NX rate	FIVE max	Anomaly score
12	77553	77553	77553	0	0	0	0	0	77553	7741	9.69
6	60565	33500	0	0	0	60545	15.89	2	1.24	6248	7.50
24	11183	11183	0	11183	2	2	0.85	8128	11183	958	7.27
13	11769	11698	0	11698	2	7	0.95	8023	162.47	1044	7.04
8	13534	13534	0	13534	18	18	3.19	3967	13534	1201	6.99
1	44026	41153	0	0	0	43996	15.42	2	14.32	4839	6.96
509	836	836	0	836	836	836	9.71	1	836	185	6.63
167	9243	9240	0	0	0	8	0.06	9198	2310	837	6.53
9	9158	9158	0	9158	48		5.34	999	9158	823	6.10

It can be seen from the Table 2 that among the ten features of the above IP, there are always one or two features with unusually high values, especially the total number of queries and the number of query failures. IP-12 with the highest anomaly score has the highest total query time. In addition, the queried domain names are abnormal that do not comply with the naming rules. The highest point in Fig. 5 is IP-12; IP-24, IP-13, IP-8 and IP-9 all have a very high number of query failures and PTR queries, but the number of queried domain names is very low, and the maximum number of requests for a single domain name is very high, indicating that these IPs are frequently performing IP backchecks on certain IPs. This behavior is obviously abnormal, which may be caused by anomaly application configuration; The number of IP-509 queries is not high, but its PTR target number is very high. The value is equal to the number of PTR queries, indicating that each DNS query is a reverse query of different IPs, which is also an obvious abnormal behavior; The maximum number of single domain name requests of IP-167 is 9198, which is very close to the total number of 9243 queries. Therefore, it is speculated that there is FASTFLUX behavior.

IP-6 and IP-1 have a very high number of query failures. Their values are 33500 and 41153 respectively, and the maximum number of single domain name requests is 2, indicating that the queried domain names of the two are almost different. Their values are 33500 and 41153 respectively. Therefore, it is speculated that they may be infected with malicious code using DGA algorithm. After counting the requested domain names of the two, it is found that the domain names accessed by IP-1 are similar, and the top-level domain names of the domain names are very random, as shown in Fig. 7, which proves that they do have DGA behavior.

pizzaman-dortmund.biz.dj
pizzaman-ffm.pp.tn
pizzaman-wuppertal.com.gf
pizzamancomics.in.ae
pizzamancorp.ac.fo
pizzamandala.edu.ls
pizzamandala.ind.my
pizzamandolino.co.ai

Fig. 7. IP-1 Some queried domain names

The top-level domain name and secondary domain name of IP-6 queried domain name are relatively fixed, but the randomness of subdomain name is very strong, as shown in Fig. 8. Therefore, it is speculated that there is FASTFLUX or DNS tunnel behavior.

029135.grab.com
029153.grab.com
030020.grab.com
032124.flare-on.com
032135.flare-on.com
033149.flare-on.com
033252.livestream.com

Fig. 8. IP-6 Some queried domain names

In general, after integrating the basic DNS behavior and malicious code behavior, the constructed features can better distinguish the IP abnormal behavior. After calculating the anomaly score through the anomaly detection model, the anomaly IP can be found effectively.

3.2   Performance Verification of Blacklist and Whitelist

In order to further evaluate the effectiveness of the detection method, the paper compares the detection results with the blacklist and whitelist, and calculates the accuracy and false positive rate of the detection method.

There are two data sources for the blacklist: the first one comes from two sets of commercial security detection platforms currently being used by the Computing Center of the Institute of High Energy Physics, Chinese Academy of Sciences. By searching on the two platforms, it is judged whether the detected IP has malicious behavior records. If an IP is detected malicious behavior, it is regarded as hitting the blacklist, and this blacklist is recorded as Blacklist 1; the other blacklist comes from the IP accumulated in the process of tracing the source of abnormal hosts during the operation of the Computing Center SOC, and this blacklist is recorded as Blacklist 2. These two types of blacklists correspond to attack exceptions and operation exceptions, respectively.

The data of the whitelist comes from the asset information of the Computing Center, which records the IPs of the functional servers in the network, and filters the detection results through the whitelist to avoid false positives.

According to the Table 2, when the anomaly score is more than 4, the difference between the anomaly scores of different IPs changes greatly. Therefore, when the thresholds of the anomaly scores (referred to as) are set to 4, 5, and 6, respectively, the output detection results are compared with the blacklist and whitelist, and the comparison results are shown in Fig. 9 below:

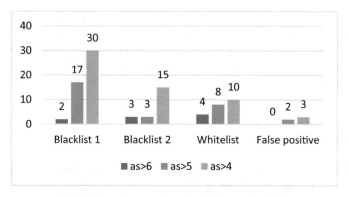

Fig. 9. Detection results under different thresholds

Misjudgment in the Fig. 9 refers to the number of IP records in the test results that do not exist in the three lists. After traceability analysis, it is found that the three misjudged IPs: one corresponding host is a new website server in the network; The other two are caused by a large number of queries on CDN domain names, which are misjudged as FASTFLUX by the detection algorithm.

By observing the changes in the number of hits of different detection results under different thresholds, it is found that with the decrease of anomaly score threshold, the number of IP in the detection results increases rapidly, the false positive rate increases slowly, while the hit rate of Blacklist 2 increases significantly, and the hit rate of Blacklist 1 slows down. It can be predicted that as the threshold decreases again, the number of IPs in the detection results increases significantly. At the same time, the number of whitelist detection reaches the peak, the hit rate of blacklist decreases gradually, and the number of misjudgments increases gradually, which greatly increases the time for security operators to check and correct errors.

Therefore, it is reasonable to set the anomaly score threshold to 4, which can ensure that the detection method can maintain high accuracy rate and low misjudgment rate with the help of whitelist. The specific performance is shown in Table 3 below:

Table 3. Detection results of as more than

as > 4	Detection number	Detection rate
Blacklist 1	30/58	51.72%
Blacklist 2	15/58	25.86%
Whitelist	10/58	17.24%
False positive	3/58	5.17%
Total detection (whitelist not filtered)	45/58	77.59%
Total detection (whitelist filtered)	45/48	93.75%

By observing the data in the table, it can be found that after reasonably setting the anomaly score threshold, with the help of the whitelist, the detection rate of the detection method designed in this paper can reach 93.75%, which improves the accuracy of nearly 16% compared with the detection results that are not suitable for whitelist filtering. Experiments show that this detection method can well find the attack anomalies in the network.

4 Conclusion

The paper proposes an unsupervised anomaly detection method based on DNS log data. In order to characterize the abnormal DNS behavior of malware, first extract the DNS fingerprint from the DNS log data, and then input the anomaly detection model to calculate the anomaly score. The anomaly detection model combines the isolated forest and local anomaly factor algorithm, and considers the global and local anomalies at the same time, which improves the detection accuracy. The experimental results on the real DNS server log data of the Institute of High Energy Physics, Chinese Academy of Sciences show that the anomaly detection method proposed in this paper can effectively find anomalies, detect malicious code behavior and system and application anomalies.

Funding Statement. This work was supported by the National Natural Science Foundation of China (NSFC) under Grant No. 61901447.

References

1. Lin, C.H., Li, X.D., JJ, et al.: DNS traffic anomaly detection based on w-kmeans algorithm. Comput. Eng. Des. **34**(006):2104–2108 (2013)
2. Wang, Q., Xie, K., Ma, Y., et al.: DNS tunnel detection based on log statistics. J. Zhejiang Univ. (Engineering Science edition) **2020**(9) (2020)
3. Wang, J.Y., Shi, J.T., Zhang, Z.X., et al.: Source IP anomaly detection algorithm for DNS request data flow based on relative density. High Technol. Commun. **26**(Z2), 849–856 (2016)
4. Mockapetris, P,, Dunlap, K.J.: Development of the domain name system. ACM SIGCOMM Comput. Commun. Rev. **25**(1) (2001)
5. Li, B.S., Chang, A.Q., Zhang, J.X.: IoT botnets threatened network infrastructure security seriously—analysis of Dyn attacked by Botnet. J. Inf. Secur. Res. **2**(11), 1042–1048 (2016)
6. Wang, J.Y., Shi, J.T., Zhang, Z.X., et al.: An algorithm for detection of source IP anomalies in DNS query based on relative density. Chinese High Technol. Lett. **26**(Z2), 849–856 (2016)
7. Ji, X., Huang, T., Hua, E.X., Sun, L.: A DNS query anomaly detection algorithm based on log information. J. Bei Jing Univ. Posts Telecommun. **41**(6), 83–89 (2018)
8. Singh, M., Singh, M., Kaur, S.: Detecting bot-infected machines using DNS fingerprinting. Digit. Investig. **28** (2018)
9. Lee, J., Lee, H.: GMAD: Graph-based Malware Activity Detection by DNS traffic analysis. Comput. Commun. **49**(Aug.1), 33–47 (2014)
10. Gu, Y.H., Guo, Z.Y.: Fast-flux botnet domain detection method based on network Traffic. J. Inf. Secur. Res. **006**(005), 388–395 (2020)
11. Woodbridge, J., Anderson, H.S., Ahuja, A., et al.: Predicting domain generation algorithms with long short-term memory networks. arXiv: 1611.00791 [cs], (2016)
12. Breunig, M.M., Kriegel, H.P., Ng. R.T., et al.: LOF:Identifying density-based local outliers. In: Proceedings of ACM SIGMOD Conference, pp. 427-438. ACM, New York (2000)
13. Liu, F.T., Ting, K.M., Zhou, Z.H.: Isolation forest. In: 2008 Eighth IEEE International Conference on Data Mining (2009)
14. Chandola, V., Banerjee, A., Kumar, V.: Anomaly Detection: A Survey. ACM Comput. Surv. **41**(3) (2009)
15. Antonakakis, M., Perdisci, R., Nadji, Y., et al.: From throw-away traffic to bots: detecting the rise of DGA-based malware. In: Usenix Conference on Security Symposium (2012)
16. Wang, Q., Xie, K., Ma, Y., et al.: Detection of DNS tunnels based on log statistics feature. J. Zhejiang Univ. (Engineering Science), **54**(9) (2020)
17. Buczak, A.L., Hanke, P.A., Cancro, G.J., et al.: Detection of tunnels in PCAP data by random Forests cyber In: Cyber and Information Security Research (CISR) Conference (2016)
18. Zuo, X.J., Dong, L.M., Qu, W.: Fast-flux botnet detection method based on domain name system traffic. Comput. Eng. **43**(09), 185–193 (2017)

Binary Hyperdimensional Computing for Image Encoding

Jinghan Li[1], Jin Chen[1(✉)], Jiahui Liang[1], Sen Li[1], Baozhu Han[1], and Hanlin Wu[2]

[1] Tianjin Key Laboratory of Wireless Mobile Communications and Power Transmission, Tianjin Normal University, Tianjin 300387, China
cjwoods@163.com

[2] Beijing Institute of Astronautical Systems Engineering, Beijing 100076, China

Abstract. Hyperdimensional computing uses hypervectors as basic patterns to construct cognitive codes to represent atomic entities through encodes different types of data into the same data structure based on hyperspace. In this paper, we exploit the reversibility of binary hyperdimensional computing to encode images to hypervectors and decode them back. We introduce turnover rate to properly separate the distance between adjacent values while maintaining the distance between them, so as to avoid the poor effect of segmentation or the direct generation that leads to the distance between adjacent hypervectors being too close to distinguish. We compared the performance of reversibility with the original hyperdimensional computing. The proposed approach has better performance.

Keywords: Binary Hyperdimensional Computing · Image Encoding · Hyperdimensional image information preserve

1 Introduction

Hyperdimensional computing uses hypervector to simulate neurons to construct cognitive code, uses combined elements to represent complex concepts, and generates complex concepts of the same structure while maintaining connections between elements. The use of hyperdimensional computing to simulate human neural activity has made breakthroughs in some areas, Including analog-based reasoning, language recognition [1, 2], predictive multimodal sensor fusion, speech recognition [3, 4], brain-computer interface, emotion recognition, epilepsy detection [5], DNA sequencing, human activity recognition, latent semantic analysis, text classification [6], etc. Some researchers have also made many improvements in improving the accuracy and energy efficiency of hyperdimensional computing, and proposed methods such as cooperative processing [7], dynamic hyperdimensional calculation [8], adaptive hyperdimensional calculation [9], semi-hypervised learning [10] and some in-memory calculation methods [11]. Every image contains rich information. Scientific research and statistics show that about 75% of the information obtained by human beings from the outside world comes from the visual system, which is obtained from images. The information in an image mainly includes the position of each pixel and the intensity value of the position of this pixel.

The relation between different pixels not only has the spatial position relation, but also has the digital size relation between the intensity value of different pixels. In the field of hyperdimensional computing, the existing coding methods for continuous values are to quantify a large number of continuous values to a smaller range and then encode, which leads to inaccurate coding and loss or weakening of part of the basic feature information. Another limitation of this method is that the reversibility of hyperdimensional computing is damaged. Once some reversible operations are performed on the encoded image hypervector, the original accurate value can only be restored to a range value. In this paper, we propose an image coding method suitable for hyperdimensional calculation to overcome the above limitations.

2 Hyperdimensional Computing

There are a large number of neurons and synapses in the nervous system, and hyperdimensional computing uses hypervectors to simulate neurons and synapses to represent some characteristics. When dimensions are measured in thousands we call them hyperdimensional, and hyperdimensional computing is based on such hyperdimensional vectors. High-dimensional modeling of neural circuits can be traced back decades to artificial neural networks, parallel distributed processing, and connection mechanisms, and these models are supported by features of high-dimensional space. [12] Take the binary vector space, for example. A 10,000 dimensional binary vector space contains $2^{10,000}$ independent vectors. This space forms a hypercube of higher dimensions, and the distribution of other points is always the same from any point. [13] Hyperdimensional computing is based on three basic operations. Addition, multiplication and permutation operations, all operations are based on d-dimensional hypervectors, operation results are also d-dimensional hypervectors. These three operations can be applied to both binary and non-binary hypervectors.

2.1 Addition (Bundling)

Hyperdimensional addition refers to the operation of adding two vectors bitwise, also known as bundling, commonly uses to represent a set. Suppose you have two hypervectors A and B, add A + B point by point to get an intermediate vector of A and B, and the resulting hypervector contains information from both vectors. When a series of hypervectors are added, the resulting hypervector is the representation of the set of these hypervectors. Compared with the randomly generated hypervector, the distance between this hypervector and either of the two hypervectors is the smallest, and the resulting hypervector is the most similar to each of the hypervectors contained in the addition. For the addition of two or more binary hypervectors, we use the majority principle to obtain the binary vector by thresholding the hypervector. However, for the addition of even number of vectors, the number of 0 and 1 May be equal. In this case, we randomly introduce a hyper vector to add it into an odd number of hyper vectors, and then perform addition operation according to the majority principle [14].

2.2 Multiplication (Binding)

The multiplication of hypervectors refers to the bitwise multiplication of two hypervectors, also known as binding, in order to form an association between a pair of related hypervectors. The multiplication results in a hypervector that is not similar to either of the two hypervectors involved in the multiplication, that is, a hypervector that is nearly orthogonal to both of the two hypervectors involved in the multiplication. This is also verified by 3000 point-by-point multiplications of two randomly generated 10,000-bit binary hypervectors in [15]. $X = A * B$ is used to represent the bit-multiplication of two hypervectors A and B to obtain X. For binary hypervectors, XOR operation (XOR) can be used to achieve this process, so it can also be written as $X = A \oplus B$. According to the commutative law of the XOR operation, it can be known that there are two hyper vectors $A \oplus B = B \oplus A$. According to the commutative law, it can be inferred that the XOR operation has reflexivity, that is, $A \oplus A = O$ (here O represents the vector of all zero). In addition, the XOR operation has the property that a vector XOR with any all-zero vector yields itself, namely $A \oplus O = A$. Due to generate vector and participate in the operation of two ultra vector is similar, also means in hyperspace, the hypervector is mapped to a place far away from the original hypervector, the multiplication operation is the purpose of binding together of pairs of vectors respectively mapped to different subspace of hyperspace, reduce the interference between the pairs of hypervector. Since the multiplication is reversible, no matter where the vector is mapped, it can be restored to the original hypervector by inverse operation [14]. The multiplicative mapping preserves position information, and the binding of $X * A$ can also be understood as $x = a$, thus linking each position to intensity value.

2.3 Permutation

Permutation operation is to reorder the components of the hypervector, commonly uses to present a sequence. Permutation is the only unitary operation among the three operations of hyperdimensional computing. It generates a hypervector that is nearly orthogonal to the one before the permutation by permutation of the hypervector itself. A mathematical permutation is achieved by multiplying a vector with a permutation matrix in which each row and column has one and only one 1. But because permutation itself is represented by matrix, not by vector addition and multiplication, the hypervector uses a list of $1, 2\ldots$ D integers to represent the permutation order. A random permutation of n! Choose one of the possible permutations at random. Permutation is distributed over vector addition, similar to multiplication, and it is distributed over any coordinate operation, including XOR multiplication – it preserves distance, and is invertible. But unlike multiplication, it is not its own inverse, and permutation of a permutation of a hypervector is not the original hypervector, which makes permutation suitable for encoding sequence data. In a geometric sense, permutation rotates a hypervector in space, which works well for sequences where only the order is different. In this paper, there is no permutation operation involved. We use two operations of multiplication and addition in hyperdimensionl to realize the improvement of image coding based on hyperdimensionl calculation, and use hamming distance as a measurement method to measure the distance between hypervectors after binary.

3 Proposed Method

When coding the intensity value, we generate two random hypervectors for the maximum value 255 and the minimum value 0 in the intensity value respectively, as the maximum hypervector and the minimum hypervector. According to the properties of hyperspace, the maximum hypervector and the minimum hypervector are nearly orthogonal to each other in the hyperspace. Generate the intermediate hypervector according to the maximum and minimum hypervector. In this step, first generate the intermediate hypervector representing the intensity value i, where i is any number from 1–254. Then randomly select $i \times d/255$ dimensional positions from the intermediate hypervector and perform XOR operation with 1. As a middle position vector in the dimension of number, query in the maximum hypervector position dimensions of number is not selected, and the numerical exclusive or operation with 1, as a middle position vector in the dimension of number, get the first i intensity values in the middle of the vector, and generate on behalf of all the hypervector between 1–254. On this basis we introduce a turnover rate p, the hyperparameter p is determined from 0–1 to be used as the ratio of flipping. After flipping all intermediate hypervectors, maximum hypervector and minimum hypervector, the intensity hypervector representing each intensity value is obtained. XOR operation is performed on the selected dimension and 1 to obtain the flipped hypervector. At this point, the final intensity value hypervector table is obtained. All the above hypervectors are binary hypervectors, and the number of dimension is d. The median hypervector that is close to the minimum value has more bits in the minimum value, and the remaining values are selected from the maximum value, and so on, all the hypervectors representing pixels are generated. As for positions, since there is no continuous relationship within them, we generate random hypervectors to build position hypervector table according to the size of images.

The pipline for image coding are as follows: based on the generation of the intensity value hypervector table and position hypervector table, read the image to be encoded. Starting from the first pixel, multiply the corresponding intensity hypervector and position hypervector as pixel hypervectors. Add up all the pixel hypervectors to obtain the integer image hypervector. Finally, binarization processing for it to get final binary image hypervectror. Taking 28*28 Gy image as an example, the process of hyperdimensional image coding is shown in Fig. 1.

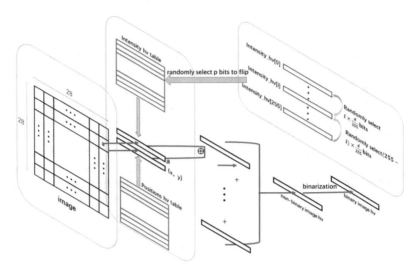

Fig. 1. The pipeline of image encoding

4 Experiment

4.1 Setup

We use software to verify the performance of the hyperdimensional computing image encoding method. We encode and restore the MNIST dataset image using Python on Intel Core I5 8265 CPU. Compare with existing methods. We compare the performance of our proposed method with the previous piecewise encoding method, and verify the advantages of our proposed encoding method through distance comparison and restoration of effect pictures.

4.2 Experimental Results

Table 1 shows the distance comparison between the current method and the adjacent hypervectors with different flipping probabilities proposed by us when encoding all 256 intensity values of the image. The existing methods usually quantize the image intensity value of 256 to ten levels, and the normalized Hamming distance of the level hypervectors is about 0.17. When the level of the original method is set to 256, it can be understood as the encoding of the intensity value whose turnover rate is $p = 0$. Table1 lists the normalized Hamming distances for different inversion rates. As shown in Table1, when p is set to 0.1, the effect of normalized Hamming distance between 256 levels is almost the same as that of 10 levels originally. However, if the original method is directly used, the space between the hypervectors will be too small, almost no discrimination.

We compared the image restoration effect when p is 0.05 and l is 10 in the original coding method. It can be seen from the Fig. 2 that the image information preservation effect is better after the introduction of turnover rate. Obviously the former performs better.

Table 1. The normalized Hamming distance of different inversion rates

$p = 0$	$p = 0.05$	$p = 0.1$	$p = 0.2$
0.0078	0.097	0.18	0.27

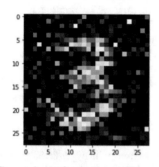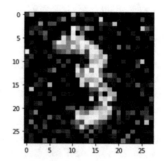

Fig. 2. Comparison of reduction performance between original method and proposed method

5 Conclusion

In this paper, we have proposed a flexible image encoding method suitable for hyperdimensional computing, which can preserve the information of each pixel and flexibly adjust the hypervector spacing with continuous value relationship. In addition, the image information encoded by our method has good reversibility. Experimental results show that our method is better than the current hyperdimensional encoding method in restoring information on image datasets.

Acknowledgment. The work was supported by the National Key Research and Development Plan Project: Robotic Systems for Agriculture (RS-Agri) NO. 2019YFE0125200.

References

1. Joshi, A., Halseth, J.T., Kanerva, P.: Language geometry using random indexing. In: de Barros, J., Coecke, B., Pothos, E. (eds.) Quantum Interaction. QI 2016. LNCS, vol. 10106, pp. 265–274. Springer, Cham (2016). https://doi.org/10.1007/978-3-319-52289-0_21
2. Rahimi, A., Kanerva, P., Rabaey, J.M.: A robust and energy-efficient classifier using braininspired hyperdimensional computing. In: Proceedings of the 2016 International Symposium on Low Power Electronics and Design, pp. 64–69 (2016)
3. Imani, M., Kong, D., Rahimi, A., et al.: Voicehd: Hyperdimensional computing for efficient speech recognition. In: 2017 IEEE International Conference on Rebooting Computing (ICRC), pp. 1–8 IEEE (2017)
4. Imani, M., Huang, C., Kong, D., et al.: Hierarchical hyperdimensional computing for energy efficient classification. In: 2018 55th ACM/ESDA/IEEE Design Automation Conference (DAC), pp. 1–6 IEEE (2018)

5. Rahimi, A., Kanerva, P., Benini, L., et al.: Efficient biosignal processing using hyperdimensional computing: Network templates for combined learning and classification of ExG signals. Proc. IEEE **107**(1), 123–143 (2018)

6. Najafabadi, F.R., Rahimi, A., Kanerva, P., Rabaey, J.M., et al.: Hyperdimensional computing for text classification 1

7. Nazemi, M., Esmaili, A., Fayyazi, A., et al.: SynergicLearning: neural network-based feature extraction for highly-accurate hyperdimensional learning. arXiv preprint arXiv:2007.15222 (2020)

8. Chuang, Y.C., Chang, C.Y., Wu, A.Y.A.: Dynamic hyperdimensional computing for improving accuracy-energy efficiency trade-offs. In: 2020 IEEE Workshop on Signal Processing Systems (SiPS), pp. 1–5. IEEE (2020)

9. Imani, M., Morris, J., Bosch, S., et al.: AdaptHD: Adaptive efficient training for brain-inspired hyperdimensional computing. In: 2019 IEEE Biomedical Circuits and Systems Conference (BioCAS), pp. 1–4 IEEE (2019)

10. Imani, M., Bosch, S., Javaheripi, M., et al.: SemiHD: semi-hypervised learning using hyperdimensional computing. In: ICCAD, pp. 1–8 (2018)

11. Gupta, S., Morris, J., Imani, M., et al.: THRIFTY: training with hyperdimensional computing across flash hierarchy. In: Proceedings of the IEEE/ACM 2020 International Conference on Computer-Aided Design (ICCAD) (2020)

12. Kanerva, P.: Hyperdimensional computing: an introduction to computing in distributed representation with high-dimensional random vectors. Cogn. Comput. **1**(2), 139–159 (2009)

13. Mitrokhin, A., Sutor, P., Summers-Stay, D., et al.: Symbolic representation and learning with hyperdimensional computing. Front. Robot. AI **7 (2020)**

14. Kanerva, P.: Computing with 10,000-Bit words. In: IEEE (2014). https://doi.org/10.1109/ALLERTON.2014.7028470

15. Ge, L., Parhi, K.K.: Classification using hyperdimensional computing: a review. IEEE Circuits Syst. Mag. **20**(2), 30–47 (2020)

Application of Artificial Intelligence in Water Supply Dispatching of Smart City

Pei Zhang[1], Hai Wang[2(✉)], Bo Zhang[3], Kai Ma[1,4], and Peng Xu[5]

[1] Production Technology Department, Tianjin Water Group Co., Ltd., Tianjin 300042, China
[2] Teaching Center for Experimental Electronic Information, College of Electronic Information and Optical Engineering, Nankai University, Tianjin 300071, China
wanghai@nankai.edu.cn
[3] Tianjin Water Group Co., Ltd., Dispatching Center, Tianjin 300042, China
[4] Tianjin Tap Water Group Co., Ltd., Post-Doctoral Research Center, Tianjin 300040, China
[5] Tianjin Public Utility Design and Research Institute, Tianjin 300100, China

Abstract. Smart water construction is not only an important part of the development process of smart city, but also a key link in the implementation of urban people's livelihood guarantee. Urban water supply dispatching realizes artificial intelligence, which can carry out real-time dynamic control and ensure the stability of the overall water supply pattern. At the same time, it is also the primary platform for the most direct business connection between production technology management and grass-roots units. It also establishes channels for various sections in the field of urban water supply, such as engineering management, measurement management, water quality management, water plant technical measures management and so on. With the continuous innovation and development of urban water supply dispatching, it has experienced the traditional manual experience stage, the primary information stage of production process monitoring and data transmission. At present, we are in the artificial intelligence construction stage including SCADA system, pipe network geographic GIS system and hydraulic model system as comprehensive dispatching auxiliary means.

Keywords: Artificial intelligence · Process control · Urban water supply · Plan library

1 Introduction

The application of artificial intelligence in the field of water supply dispatching is to model the experience of urban water supply operation and management, automatically collect, track and analyze various data, and give analysis reports and solutions [1–4]. Managers can modify the scheme and confirm the final scheme according to the actual working conditions [5–8]. On this basis, on the one hand, continuously improve the plan database and increase the accuracy and operability of the plan formulated through artificial intelligence, on the other hand, further improve the efficiency and scientificity of urban water supply dispatching [9–12]. For excellent urban water supply dispatching, the most important link is to predict and plan in advance, such as how to predict the change of pipe network pressure after one hour, the change range of water level after one day, the increase and decrease trend of water volume after one week, and so on [13–15]. The realization of artificial intelligence will intervene in advance according to the change of water supply working condition trend, analyze and judge from the perspectives of balancing raw water supply, water production of water plant and stability of pipe network, so as to ensure that the implementation effect of dispatching order is connected with the change of working condition, so as to continuously maintain the integrity and stability of the whole water supply ecological chain, and control the hidden dangers of water supply safety and various risks in the embryonic stage [16–18].

2 Conventional Intelligent Control and Early Warning Function of Urban Water Supply

Intelligent control mainly carries out overall control from three aspects: raw water configuration, water plant pressure and flow adjustment and maintaining the overall operation stability of the pipe network, and organically combines the three to carry out more refined water volume prediction and analysis, and predict and plan in advance. Continuously monitor the changes of various parameters (water pressure, water volume, water level, water quality, etc.) 24 h a day, implement key early warning for sensitive data, carry out intelligent work according to the changes of working conditions in different periods and the requirements of operation safety, adhere to the primary goal of ensuring the water supply safety of water plants in the city, never issue instructions at the cost of increasing hidden dangers of water supply safety, and take the safety of urban water supply as the top priority. Shown in Fig. 1, which means how is the free water pressure surface of urban water supply distributed.

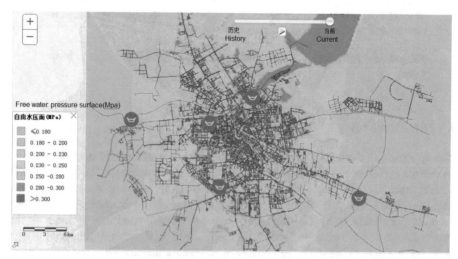

Fig. 1. Pressure distribution diagram of free water pressure surface of urban water supply

2.1 Overall Planning and Intelligent Dispatching of Urban Water Supply Facilities

Artificial intelligence can timely change the overall water supply dispatching mode of the city according to the layout of urban water plants, production and water supply capacity and the improvement and transformation process, so as to ensure that the pipe network pressure in the central city is not less than 0.20mpa. The matching of pump units shall be based on the principle of high efficiency first, near first and far later, balance the water supply of each region, and strive to reduce energy consumption and realize relatively economic operation on the premise of ensuring water quality and pressure. Shown in Fig. 2, which means how to optimize the pump set of water plant. At the same time, by integrating the change trend and the setting of upper and lower limits, we can realize artificial intelligence control, further refine the upper and lower limits of the factory pressure and flow of each water plant in the city, carry out more accurate process control and adjustment by time and water distribution volume, eliminate the potential safety hazards in the process of urban water supply in time, and improve the safety guarantee rate of water plant production and water supply and pipe network operation. In addition, through the continuous monitoring of the water purification process of the water plant, we strive to further tap the potential of the water plant in terms of water production capacity, regulation and storage of clean water reservoir and allocation of pump units, establish an operation mode of continuous optimization and adjustment, and maximize the water supply capacity of the whole city in the peak period of water supply.

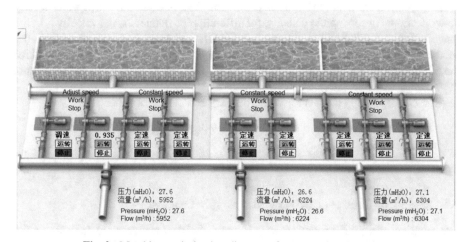

Fig. 2. Matching optimization diagram of pump set in water plant

2.2 Accurate and Intelligent Dispatching of Urban Water Supply Network

According to urban development and social needs, AI can focus on the targeted analysis of a small number of low-pressure areas and new water supply areas in the water supply network, and study new water matching schemes according to the water demand along the line. After checking the water volume of the system, the adjustment of the pipe network shall be checked.

$$\text{Specific flow qs} = (Q - \sum qt)/\sum I$$

Q: total water consumption of pipe network (L/s)
zat: total centralized water consumption of large users (L/s)
$\sum$: total length of main pipe.
Calculate the flow along the line ql = qs*1 1: the length of the pipeline
Calculate node flow qj = gt + $0.5\sum ql$

When calculating the node water pressure elevation of the pipe network, the head loss of each pipe section is usually calculated. Starting from the constant pressure node (generally from the most unfavorable point of the pipe network), the upstream node water pressure elevation is calculated by using Bernoulli equation. However, for the actual uniform water supply, the flow between the two nodes changes along the way. The flow velocity V1 of upstream node 1 is large and the flow velocity V2 of downstream node 2 is small. Therefore, the head loss HW between the two nodes is greater than the actual water pressure elevation difference,

$$hw = (Z1 + p1/y + v12/2g) - (Z2 + p2/y + v22/2g) > (Z1 + p1/y) - (Z2 + p2/y)$$

In this way, the water pressure elevation of the upstream node of the constant pressure node calculated by the above method is greater than the actual water pressure elevation of the pipe network node, resulting in error. However, because the urban water supply network adopts the economic velocity, for different pipe diameters, the velocity is in the range of 0.6–1.4 m/s, and the velocity head V22/2G changes in the range of 0.02–0.1 M, the difference between velocity and head is slightly flat, which is usually ignored for the actual pipe network. Therefore, it is feasible to use the node flow model to calculate the water pressure elevation. Again according to

The corresponding calculation result of Zeng Hai can be obtained

$$V = 0.44 * C * (Re/C)^{0.075} * (g * D * I)^{0.5}$$
$$Re = V * D/v$$

2.3 Intelligent Control of Urban Water Supply Leakage

The addition of artificial intelligence will make statistics and detailed analysis on the flowmeter data of water transmission network and water distribution network, so as to improve the accuracy of hydraulic model calculation and the reliability of working condition simulation application. Figure 3 shows accurately prediction the water volume and continuously check it According to the distribution of main pipelines and the setting of water distribution pipelines, combined with the design capacity, reasonably match the water demand of upper and end users of water supply pipelines, booster pump stations and upstream and downstream users. For those pipelines that have been operating at full capacity during peak hours, we can no longer increase users, so as to avoid new low-pressure areas of water supply, resulting in the forced booster operation of the water plant, and the leakage rate may also rise.

In addition to the direct water transfer of the water plant, there are also many secondary booster pump stations in the urban pipe network. Their water inlet and water supply are highly subjective and generally lack overall management and coordination, resulting in unreasonable water distribution in the local pipe network, large and small flow in the surrounding pipes, good and bad water inspection by users, which is not conducive to the smooth operation of the overall water supply pipe network. Artificial intelligence can effectively link the secondary dispatching platform of urban water supply, cooperate with the primary dispatching platform to monitor the water plant and pipe network, and carry out regional joint control over the water supply facilities such as the secondary booster pump station under its jurisdiction, so as to coordinate and use the water supply operation of each pump station in a planned way.

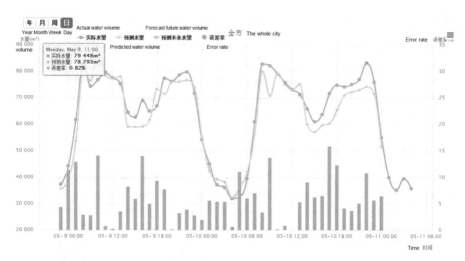

Fig. 3. Comparison between actual water volume and predicted water volume

3   Intelligent Application in Urban Water Supply Emergency

Urban water supply needs to be safe and stable all the time, but it is inevitable that there will be occasional equipment failures and improper operation in the operation process, resulting in chain reactions. When a water supply emergency occurs, artificial intelligence enables the urban water supply emergency to have the ability of rapid response and efficient disposal. It establishes a rich pre plan database in terms of raw water shortage, water plant power failure, pipeline explosion, etc., quickly retrieves the most matching corresponding disposal scheme from the pre plan database, and orderly eliminates the fault according to the type of accident, restores normal operation, or switches the water source, or adjusts the process and power supply equipment. After the accident is handled, summarize the data and operation methods, measure the actual use effect of the plan, add and correct the unqualified parts, and improve the guiding significance of timely disposal.

3.1 Emergency Intelligent Control of Urban Water Supply Facilities

When the water demand of the water supply pipe network is large, the urban water production scale is smaller than the pipe network demand, and the water production equipment is in the overload operation state, the situation that the supply is less than the demand in the city appears; Or the water production process of urban water supply fails, resulting in the decline of water production and the limited regulation and storage capacity of clean water reservoir, which may further lead to water quality problems and water supply hidden dangers of insufficient water. In order to solve this problem, the artificial intelligence system will generate early warning through different parameter changes and determine it as the corresponding emergency event. Once the system finds a plan with high matching, it will carry out step-by-step implementation according to the plan at the first time. For example, the relevant urban water plants are instructed to reduce the factory pressure and the amount of factory water to prevent the water level of the clean water reservoir from falling further. At the same time, the pressure balance and water supplement of the pipe network are carried out by simultaneously increasing the water supply of other water plants in the city, so as to buy time for emergency disposal and continue to ensure the safety and stability of urban water supply.

3.2 Emergency Intelligent Control of Urban Water Supply Network

In case of pipe explosion and water leakage in the urban water supply pipeline, the artificial intelligence will make a preliminary judgment in combination with the pressure monitoring points and flow meters around the leakage point, lock the approximate location and pipeline distribution that may cause water leakage, and immediately give the gate closing scheme of rapid water stop. Display detailed information such as the number of gates closed, the number of pipelines closed, the affected area and the number of users, and provide ideas for water supply adjustment after water stop. On the basis of relatively perfect response and emergency plans, the artificial intelligence system can even carry out the dispatching and control of urban water supply under the adverse conditions of lack of some monitoring data.

4 Conclusion

The development of artificial intelligence in the field of urban water supply dispatching is to integrate and apply multi-disciplinary and multi field knowledge, give innovative thinking, give full play to the technical capacity and plan library role of intelligent control in multiple links of urban water supply from the aspects of energy conservation and consumption reduction, quality and efficiency improvement and solving practical production problems, and provide solid technical support for the high-quality and sustainable development of water supply sector in smart city, Combined with the construction of smart operation, smart water plant and smart pipe network, we will realize a new situation of scientific urban water supply from digitization and informatization to intellectualization and intellectualization.

Acknowledgement. This work was supported by the 1st batch of Industry University Cooperative Education Projects of Ministry of Education in 2021 (202101186002 and 202101186014); the 2nd batch of Industry University Cooperative Education Projects of Ministry of Education in 2021 (202102296002); the 2021 Self-made Experimental Teaching Instrument and Equipment Project Fund of Nankai University (21NKZZYQ01); the 2022 Undergraduate Education Reform Project Fund of Nankai University and the 2022 Undergraduate Experimental Teaching Reform Project Fund of Nankai University.

References

1. Lu, M., Zhang, T., Zhao, H. modeling method for optimal operation of large-scale water supply system. Water Supply Drainage **27**(6), 81–86 (2001)
2. Yimei, T., Jiangtao, L., et al.: application of genetic algorithm in optimal operation of water supply system. China Water Supply Drainage **17**(12), 63–65 (2001)
3. Fang, Y., Hongwei, Z.: optimal operation of urban water supply network under water-saving conditions. China Water Supply Drainage **18**(3), 82–84 (2002)
4. Zheng, D., Wang, N., Yang, J.: Optimal operation of multi-source large-scale water supply network. J. Water Conserv. **2**, 1–13 (2013)
5. Zhiguang, N., Hongwei, Z.: Genetic algorithm for optimal operation of urban water supply system. China Water Supply Drainage **19**(4), 42–44 (2003)
6. Shan, J., Dai, X.: Using BP network to establish a prediction model of urban water consumption. China Water Supply Drainage **17**(8) (2001)
7. Liu, S.: Optimal distribution of water supply in pumping stations of multi-source water supply system. China Water Supply Drainage **5**(4) (1988)
8. Zhong, W., Xu, N.: Hierarchical optimal operation of urban water supply system with booster pump station. Syst. Eng. Theory Pract. **9**(5), 60–68 (1989)
9. Minnan, L.: hydraulic calculation and optimal operation of water supply network. J. Water Conserv. **9**, 32–39 (1995)
10. Mackle, G., Savie, D.A., et al.: Application of genetic algorithms to pumpscheduling for water supply. In: 1995 GALESIA, First International Conference on Genetic Algorithms In Engineering Systems: Innovations and Applications, pp. 400–405
11. Pezeshk, S., Helweg, O.J.: Adaptive search optimization in reducing pump operating costs. J. Water Resour. Plann. Manag. ASCE **122**(1), 57–63 (1996)
12. Niranjan Reddy, P.V., Sridharan, K., Rao, P.V.: WLS method for parameter estimation in water distribution networks. J. Water Resour. Plann. Manag. **122**(3). 157–164 (1996)
13. Scholkopf, B., Simard, P., Smola, A.J:. Prior knowledge in support vector kernels. In: Jordan, M., Kearns, M., Solla, S. (eds.) Advance inNeural Information Processing Systems, pp. 640–646. MIT Press, Denver (1998)
14. Sakarya, B.A., Mays, L.W.: Optimal operation of water distribution pumps considering water quality. J. Water Resour. Plann. Manag. ASCE **126**(4), 210–220 (2000)
15. Jowitt, P.W., Germanopoulos, G.: Optimal pump scheduling in water-supply networks. J. Water Resour. Plann. Manag. **118**(4). 406-422 (1992)
16. Lansey, K.E., Awumah, K.: Optimal pump operation considering pump switches. J. Water Resour. Plann. Manag. **120**(1), 17–35 (1994)

17. Ormsbee, L.E., Lansey, K.E.: Optimal control of water supply pumping systems. J. Water Resour. Plann. Manag. ASCE **120**(2), 237–252 (1994)
18. Jain, D.A., Joshi, U.C.: Short-term water demand forecasting using artificial neural networks: IIT Kanpur experience. In: Proceedings of 15th International Conference on Pattern Recognition, vol. 2, pp. 459–462 (1994)

Remote Sensing Image Object Detection Based on Improved SSD Algorithm

Xu Pan[1], Bingcai Chen[1,2(✉)], Menglin Qi[1], and Zeqiang Sun[1]

[1] School of Computer Science and Technology, Xinjiang Normal University, Urumchi 830054, China
px85@qq.com

[2] School of Electricity and Computer, Jilin JianZhu University, Changchun 130000, China

Abstract. Aiming at the problems of complex background, serious illumination and small target in optical remote sensing image, a remote sensing image object detection algorithm based on decoupling head and attention mechanism (DHA-SSD) is proposed. The algorithm is based on the combination of resnet-50 and SSD algorithm. Firstly, the decoupling head structure is added to decouple the classification task and positioning task of object detection, so as to alleviate the conflict between the two tasks; Secondly, in the multi-scale detection stage, SimAM attention module is introduced to improve the multi-scale detection performance of the model without adding parameters; Finally, the experimental results on the public RSOD optical remote sensing image data set show that the mAP value is improved by 4.4% compared with the benchmark algorithm.

Keywords: Optical Remote Sensing Image · Decoupling Head · Attention Module · SSD

1 Introduction

The object detection task on optical remote sensing image is often used in UAV reconnaissance, traffic monitoring, land survey, wildlife tracking, disaster observation and other application scenarios. This is the basic work of many follow-up studies [1].

In recent years, there are many object detection algorithms based on convolutional neural network (CNN [2]), such as SSD [3], Faster R-CNN [4], YOLOX [5], etc. These algorithms are usually used to solve these tasks. When moving to remote sensing object detection task, some new problems will appear. For example, the complex background and small targets in optical remote sensing images lead to the low accuracy of conventional object detection algorithms. Based on this, a large number of scholars have improved the traditional object detection algorithm, and achieved good results in optical remote sensing images.

Using the multi-scale features of different depths of the network, FPN generates a feature pyramid by transforming the feature mapping of the image, which effectively improves the detection accuracy [6]. Wang H T [7] by changing the preset anchor point of SSD algorithm, the adaptability of the algorithm to remote sensing targets is improved.

© The Author(s), under exclusive license to Springer Nature Singapore Pte Ltd. 2023
Q. Liang et al. (Eds.): AIC 2022, LNEE 871, pp. 59–66, 2023.
https://doi.org/10.1007/978-981-99-1256-8_8

Wang D L [8] use the improved dense connection network to replace the darknet53 network in YOLOv3, and add the feature enhancement module to improve the detection effect of remote sensing images. Ju M [9] add a high-resolution feature fusion line on the basis of YOLOv3 algorithm to enhance the effect of small object detection.

In this paper, SSD algorithm is used to realize object detection in optical remote sensing image. In order to make SSD algorithm more suitable for remote sensing image object detection task, the following improvements are made to the algorithm:

1) The decoupled detection head is used to decouple the classification and regression tasks to alleviate the degradation of detection performance caused by the conflict between the two tasks.
2) In the multi-scale detection stage, SimAM attention mechanism [10] is introduced to improve the learning and expression ability of the network.
3) Experiments are carried out on the public RSOD remote sensing image dataset [11] to verify the feasibility and effectiveness of the improved strategies.

2   Improved SSD Algorithm

2.1   Overall Network Structure Design

The background of optical remote sensing image is complex, and the targets are small and dense, which brings challenges to target classification and positioning. According to the characteristics of convolutional neural network, the local features learned in the shallow layer of the network are used to detect small targets, while the global features learned in the deep layer are used to detect large targets. In SSD algorithm, the output of the conv4_3 layer of backbone network and five additional convolution blocks are used to realize multi-scale target detection. Taking an image with input size of 300 * 300 as an example, the output feature size of conv4_3 layer in SSD algorithm is 38 * 38, while the output feature size of the additional five convolution blocks are 19 * 19, 10 * 10, 5 * 5, 3 * 3 and 1 * 1 respectively. Six default boxes with different scales are constructed at each point on feature maps, Then match these default boxes to the ground truth boxes, and predict both the shape offsets and the confidences for all object categories. Finally, the default boxes are combined, and use non maximum suppression (NMS) to generate the final detection result.

Due to the lack of learning and expression ability of SSD algorithm in remote sensing image object detection task, it is can not to satisfy people's needs. Therefore, this paper takes the improved version of SSD algorithm, which uses resnet-50 as the backbone, as the benchmark algorithm. Then the SimAM attention mechanism and decoupling detection head are introduced to obtain better detection effect. The network structure of the improved SSD algorithm is shown in Fig. 1.

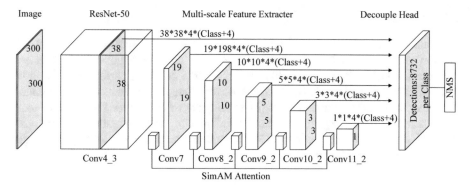

Fig. 1. The Structure of The Improved SSD Algorithm

2.2 SimAm Attention Mechanism

Facts have proved that attention mechanism is a potential means to enhance deep CNN. Attention mechanism can help the model give different weights to different parts of neurons and extract more interesting information. In order to enhance the multi-scale detection ability of SSD algorithm, attention mechanism is added to the multi-scale detection network.

SimAM attention module [10] consists of spatial attention and channel attention. The module first calculates the linear separability between neurons, then uses the energy function to calculate the energy of different neurons, and weights each neuron according to the energy. Because neurons add weight to each position in the feature map, the network can focus on the target of interest and ignore or weaken irrelevant information. Because the attention mechanism of SimAM is nonparametric, it will hardly bring additional calculation and storage to the model (Fig. 2).

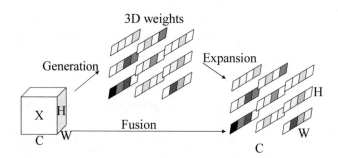

Fig. 2. SimAM attention module

2.3 Decoupling Head

The detection head in SSD algorithm uses the extracted multi-scale features for classi-fication and location tasks at the same time. Because the coupling of classification task and location task will lead to conflict, and affect the detection performance. In order to satisfy the decoupling of two tasks and the real-time requirements of the algorithm, this paper Construction the decoupling head using convolution layer. For the multi-scale features of the input, the prediction head corresponding to the classification and posi-tioning tasks is 1×1 convolution and 3×3 combination of convolutions. The structure of the decoupling head is shown in Fig. 3.

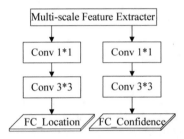

Fig. 3. Decoupling head module

3 Materials and Methods

3.1 RSOD Dataset

Wuhan University released the RSOD remote sensing image dataset in 2016 [11], which has 800 images, including four types of targets: aircraft, playground, overpass and oil-tank. This study takes RSOD dataset as the research object. In order to increase the training samples, the pictures are randomly rotated and flipped during training.

3.2 Evaluating Indicator

The experiment mainly uses map (mean average precision), AP (average precision) and FPS (frames per second) as evaluation indexes.

Precision (P) is used to measure the proportion of all positive samples detected that are also labeled as positive samples, also known as precision. Recall (R) measures the ratio of the number of positive samples correctly detected to the number of all positive samples in the dataset, also known as recall.

$$Precision = \frac{TP}{TP + FP}$$
$$Recall = \frac{TP}{TP + FN}$$

(1)

In formula 1, TP is true positives, representing the number of samples correctly detected as positive; FP is false positives, representing the number of negative samples misjudged as positive samples; FN is false negatives, which represents the number of positive samples misjudged as negative samples.

P and R conflict with each other, and a single index cannot comprehensively evaluate the advantages and disadvantages of the algorithm, so AP is used as the evaluation index. AP is the offline area of (PRC, precision–recall curve). The higher the AP value, the better. Map is the average value of AP of different categories. As shown in formula 2.

$$AP = \int_0^1 P(r)dr$$
$$mAP = \frac{\sum_1^n AP_i}{n}$$

(2)

where n represents the number of target categories.

FPS is used to evaluate the speed of the model, indicating the number of pictures detected in the unit second of the model.

4 Experiment and Result Analysis

This study discusses the influence of learning rate on the convergence speed and accuracy of the model, and explores the influence of attention mechanism and decoupling head on the detection performance of the model through experiments. In order to verify the effectiveness of the improved algorithm, the improved algorithm is compared with the classical object detection algorithm. The comparison algorithm includes two-level object detection algorithm Faster R-CNN and single-level object detection algorithm SSD and YOLOv3.

4.1 Experimental Environment

The software and hardware configuration of the experiment is as follows: the CPU is R7–5800, the graphics card is NVIDIA RTX3060, the operating system is windows 10, the deep learning platform is pytorch, and the development language is python.

In order to verify the effectiveness of the improved model, comparative experiments are carried out on the same dataset, and the parameter settings of the comparative algorithm refer to the original text. The parameter settings of SSD algorithm are as follows: select SGDM optimizer, set the optimization model and momentum to 0.9, and set the learning rate to 0.001. In order to achieve better results, crossentropy loss is selected as the confidence loss and smoothl1loss is selected as the positioning loss.

4.2   Comparative Test of Different Algorithm

In order to compare the performance of each comparison algorithm, the evaluation index mentioned in Sect. 2.2 is used to verify the performance of the model. The results are shown in Tables 1 and 2.

Table 1. AP(%) and mAP(%) Comparison of SSD and DHA-SSD on RSOD dataset

Model	AP				mAP
	playground	Overpass	Oiltank	Aircraft	
SSD	97.2%	80.5%	88.2%	79.8%	86.4%
DHA-SSD	99.5%	89.6%	93.3%	80.7%	90.8%

As shown in Table 1. Compared with the benchmark algorithm, DHA-SSD algorithm improves greatly in the categories of overpass and oiltank, which are 9.1% and 5.1% respectively. Because the background in the overpass image is complex, while the targets in the oiltank image are dense, which verifies the effectiveness of the improved strategy in complex background and dense target scenes. The improvement in the aircraft category is small, about 0.9%, mainly because the background of this category of images is complex, there are many and dense small targets, which is a great challenge to the object detection algorithm.

Table 2. mAP(%) and FPS comparison of algorithms

Model	mAP	FPS
YOLOv3-Tiny	63.9%	40
YOLOv3	78.6%	38.6
Faster R-CNN	93.2%	3.4
SSD	86.4%	23
DHA-SSD	90.8%	21

By analyzing the experimental results in Table 2, it can be seen that the DHA-SSD algorithm proposed in this paper is superior to the single-stage object detection algorithms YOLOv3, YOLOv3-Tiny, and SSD algorithm in detection performance, and is 26.9%, 12.2% and 4.4% higher in average accuracy (map) respectively. The map of the improved algorithm is slightly worse than the Faster R-CNN algorithm, with a gap of 2.4%, but the detection speed is about 6 times that of the Faster R-CNN algorithm. The test results are shown in Fig. 4.

Thanks to the advanced nature of the baseline algorithm and the effectiveness of the improved strategy for remote sensing object detection, the improved model can meet the real-time and accuracy requirements of remote sensing image detection.

Fig. 4. Prediction Results of DHA-SSD in RSOD Dataset

5 Conclusion

Aiming at the problems of complex image background, small and dense targets in remote sensing image detection, an improved method based on decoupling head and attention mechanism is proposed, which provides a way to solve the problems in remote sensing image detection. This method selectively focuses on the region of interest in the feature map through the attention module, and uses the decoupling head to decouple and sum the classification task and regression task. Several groups of comparative experiments show that the proposed method has good performance in remote sensing image detection task. This has certain significance for solving the problem of object detection in remote sensing images in the future. This paper does not consider more remote sensing image datasets. In the future, relevant experiments of multi-source data fusion will be carried out to further improve the performance of the model.

References

1. Oksuz, K., Cam, B.C., Akbas, E., et al.: Generating positive bounding boxes for balanced training of object detectors. In: Proceedings of WACV, pp. 883–892 (2020)
2. Krizhevsk, Y.A., Sutskever, I., Hinton, G.: ImageNet classification with deep convolution neural network. Adv. Neural. Inf. Process. Syst. **25**, 1106–1114 (2012)
3. Liu, W., Anguelov, D., Erhan, D., Szegedy, C., Reed, S., Fu, C.-Y., Berg, A.C.: SSD: single shot MultiBox detector. In: Leibe, B., Matas, J., Sebe, N., Welling, M. (eds.) ECCV 2016. LNCS, vol. 9905, pp. 21–37. Springer, Cham (2016). https://doi.org/10.1007/978-3-319-464 48-0_2

4. Ren, S., He, K., Girshick, R., et al.: Faster R-CNN: towards real-time object detection with region proposal networks. IEEE Trans. Pattern Anal. Mach. Intell. **39**(6), 1137–1149 (2017)

5. Ge, Z., Liu, S., Wang, F., et al.: Yolox: Exceeding yolo series in 2021. arXiv preprint arXiv: 2107.08430 (2021)

6. Lin, T.Y., Dollar, P., Girshick, R., et al.: Feature pyramid networks for object detection. In; 2017 IEEE Conference on Computer Vision and Pattern Recognition (CVPR). IEEE Computer Society (2017)

7. Wang, H., Guo, Z.: Target detection of SSD aircraft remote sensing images based on anchor frame strategy matching. J. Front. Comput. Sci. Technol. 1–15 (2021)

8. Wang, D., Du, W., et al.: Remote sensing images detection based on dense connection and feature enhancement. Comput. Eng. 1–9 (2021). https://doi.org/10.19678/j.issn.1000-3428. 0061482

9. Ju, M., Luo, H., Wang, Z., et al.: The application of improved YOLOV3 in multi-scale target detection. Appl. Sci. **9**(18), 3775 (2019)

10. Yang, L., Zhang, R.Y., Li, L., et al.: SimAM: a simple, parameter-free attention module for convolutional neural networks. In: International Conference on Machine Learning. PMLR 2021, pp. 11863–11874 (2021)

11. Yang, L., et al.: Accurate object localization in remote sensing images based on convolutional neural networks. IEEE Trans. Geosci. Remote Sens. **55**(5), 2486–2498 (2017)

Detection Method of Aggregated Floating Objects on Water Surface Based on Attention Mechanism and YOLOv3

Jiannan Wang[1,2] and Bingcai Chen[1,2(✉)]

[1] School of Computer Science and Technology, Xinjiang Normal University, Urumqi 830054, China
china@dlut.edu.cn
[2] School of Computer Science and Technology, Dalian University of Technology, Dalian 116024, Liaoning, China

Abstract. With the development of smart water conservancy construction, a realistic demand for using computer vision technology to assist in the supervision of floating objects on the water surface has arisen. To address the problem that the current research on water surface floating objects detection mainly focuses on detecting scattered individual floating objects, and an improved YOLOv3 algorithm embedded with the SE attention module is proposed for detecting aggregated water surface floating objects. A self-made dataset containing aggregated floating objects "mixed garbage" and "water pollution" is developed and augmented with four data enhancement methods. The K-means++ algorithm was used to replace the K-means algorithm for clustering the dataset with ground truth box sizes to reduce the negative effects of randomly selecting the initial clustering centers. The localization loss in the loss function of the YOLOv3 model is improved, and GIoU Loss is introduced to improve the localization accuracy. The experimental results show that S-YOLOv3 outperforms other models in the field of water surface object detection on the self-made dataset compared with YOLOv3 and other commonly used models in the field of water surface object detection, and the mAP reaches 83.1%.

Keywords: YOLOv3 · Attention Mechanism · Water Surface Floating Object Detection

1 Introduction

Water surface floating objects pose a serious threat to the safety of the water environment, but the current supervision of water surface floating objects is still based on manual inspection, and the scope of supervision is limited and less efficient. Compared with this, the use of object detection technology for water surface floating objects supervision not only reduces the required manpower but also can monitor the water surface environment in real-time without interruption. Therefore, image-based water surface floating object detection research is gaining attention.

Q. Liang et al. (Eds.): AIC 2022, LNEE 871, pp. 67–74, 2023.
https://doi.org/10.1007/978-981-99-1256-8_9

The algorithm in water surface floating object detection proposed by Senhao Li [1] mainly takes countermeasures against the disturbing factor of water ripples, but its self-made dataset images are taken under good lighting conditions, so there is still some room to improve the robustness of this algorithm against lighting changes. Lieshout C V et al. [2] constructed a plastic floating litter monitoring dataset, and the detection of plastic floating litter was completed by Faster R-CNN [3] and Inception V2 [4], performed two stages of detection from coarse to fine to complete the identification of plastic floating debris. Guojin Li et al. [5] constructed a floating litter dataset using images of the lake on campus and used Faster R-CNN as the base model for more accurate localization of small scattered objects on the water surface using pixel points instead of bounding boxes. However, most of the existing studies focus on detecting sparsely dispersed objects on the water surface, and due to the influence of factors such as water mobility and wind, floating garbage tends to accumulate in the direction of water flow, resulting in a large amount of mixing of multiple types of garbage together, and there is less research on this kind of aggregated floating garbage. Moreover, most of the self-made datasets have a single image acquisition scene and location, some only divide the training set and test set, leading to a lack of convincing generalizability for those models that achieve good detection results on these datasets.

Therefore, this paper researches aggregated water surface floating objects and constructs a water surface floating object detection dataset containing aggregated mixed garbage and water pollution objects. The YOLOv3 [6] model, which is widely used in industry, is used as the base detection model, GIoU is introduced in its localization loss calculation, the anchor frame size is reset using the K-means algorithm, and the SE attention module is embedded into YOLOv3 to obtain the S-YOLOv3 model, to achieve reliable detection of aggregated floating objects.

2   YOLOv3 Model Improvement

2.1   Anchors Presetting Based on K-means++ algorithm

The setting of anchor size has a direct relationship with the detection performance of the model. Theoretically, the closer the anchor size is to the GT frame size in the dataset, the higher the detection performance of the model. Therefore, YOLOv3 uses the K-means clustering algorithm to cluster the GT frame sizes in the dataset, where K is set to 9. The process of IoU-based K-means clustering of anchor frames is as follows ($K = 9$).

(1)   Read the annotation files of the dataset, parse the size information of GT boxes, and initialize the clustering center by selecting the size information of 9 boxes randomly from all ground truth boxes.
(2)   Calculate the distance between each GT frame and the cluster center in turn, which is equal to 1-IoU, where IoU is the intersection ratio of GT frames and the cluster center. A GT box is divided into clusters belonging to the cluster center with the smallest distance from it.
(3)   Recalculate the center of each cluster, and the center of the cluster is equal to the mean of all GT boxes contained in that cluster.
(4)   Repeat steps 2 and 3 until the centers of the clusters no longer change.

In the above process, it can be found that the Initial clustering centers in the K-means algorithm are selected randomly, and this randomness1 is mainly caused by the random selection of the initial clustering centers in the first step, so K-means++ improves K-means in this regard. The idea of K-means++ initializing the clustering centers is that the first clustering center is selected randomly, and the xth clustering center $(x > 1)$ is then selected as far as possible from the first $x-1$ cluster centers, thus ensuring that the cluster centers are as dispersed as possible.

2.2 Localization Loss Function Improvement

The loss function of YOLOv3 consists of localization loss, classification loss, and confidence loss. IoU is equal to the ratio of the intersection area of the prediction bounding box and the GT box to the union area of the two and is used to measure the accuracy of the prediction. However, IoU does not accurately reflect the degree of overlap between the prediction bounding box and the GT box and the distance between them when they do not intersect. H Rezatofighi proposed GIoU [7], which inherits the advantages of IoU while quantifying the distance between the prediction bounding box and the GT frame. Therefore, GIoU is added to the calculation of the YOLOv3 localization loss function as shown in Eq. 2.1, to improve the model detection performance.

$$L_{box} = \lambda_{coord} \sum_{i=0}^{S^2} \sum_{j=0}^{B} 1_{i,j}^{obj} (2 - \omega i \times hi)(1 - GIoU) \qquad (2.1)$$

3   S-YOLOv3 Network

3.1   SE Attention Module

The SE module proposed by hu et al. [8] is an attention mechanism belonging to the channel dimension, and its name is derived from the combination of the initials of the two core operations of the module, Squeeze and Excitation. In terms of the overall architecture, the SE module assigns weights to the features of each channel through the three steps in Fig. 1: first, it compresses all the feature maps of dimension h $\times$ w on the channels, compressing each feature map into one element and obtaining C_2 real numbers; then, it converts the real numbers into corresponding weight values through the excitation operation; and finally, the scale operation applies these weights back to each channel. After these three steps, the features that were originally of the same rank in each channel are given different weights, allowing the introduction of attention into the neural network. As seen in the figure, the SE module does not change the structure of the neural network before and after its processing, so it can be easily embedded in other neural networks with excellent portability.

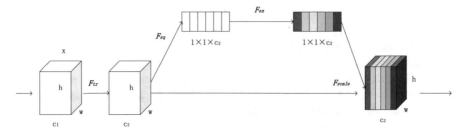

Fig. 1. Structure of SE attention module

3.2 S-YOLOv3 Network Structure

Specifically, the S-YOLOv3 network consists of three main components, the CBL component corresponds to the Convolutional convolutional block in Fig. 2, which consists of a convolutional layer, a batch normalization layer, and an activation function. The Res-unit residual connection component is a residual module with a shortcut structure, which ensures that the complex model after network hierarchy deepening necessarily contains the model before deepening by an addition operation, avoiding model bias caused by network hierarchy deepening. The first convolutional block in each ResN component plays the role of down sampling, and the convolutional blocks in the five ResN components down sample the feature map by a factor of 2 to 5, i.e., 32 times. S-YOLOv3 embeds a SE attention module before each YOLO layer, thus improving the detection accuracy of the model.

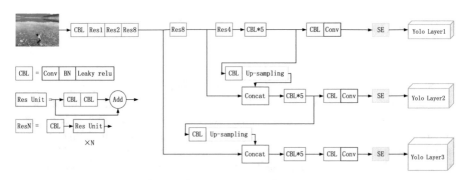

Fig. 2. S-YOLOv3 network structure

4   Experimental Results and Analysis

4.1   Dataset Preparation

Since there is no common dataset such as ImageNet and COCO in the field of water surface floating matter detection, we need to make our dataset. The images in this dataset are all from the Wuxi public data open platform, mainly consisting of images that were intercepted from the river monitoring videos and the real-world images from the water surface pollution reports. The specific steps are as follows. The training set, validation set, and test set are divided in the ratio of 3:1:1 by manually cleaning the data and normalizing the annotation, and the training set is augmented with offline data using horizontal flipping, random brightness, random panning, and adaptive histogram equalization to perform offline data enhancement on the training set to expand the sample size. The dataset has 1752 images and contains two types of targets, namely "mixed garbage" and "water pollution". Aggregated garbage often consists of a large number of targets of multiple types, so it is regarded as a whole - "mixed garbage", and no longer separately identifies the sub-targets in mixed garbage. Water pollution type targets refer to cases where non-water surface objects cause, but rather, a situation where the water body is polluted and has a special color due to phenomena such as water bloom.

4.2 Experimental Environment and Conditions

The experiments were conducted in the Windows 10 system environment with a Graphic Processing Unit of NVIDIA RTX2060 with 6 GB of video memory, CPU model i7-9750H, and a deep learning framework of Pytorch 1.10.2. The initial learning rate was 1.19E−4, which decreased to 9.69E−5 at the 40th training epoch, for a total of 80 training epochs.

4.3 Experiments and Analysis of Results

The first set of experiments compares the detection performance of YOLOv3 before improvement and YOLOv3(1), which adds GIoU to the localization loss calculation. The second set of comparison experiments compares the effects of using the K-means clustering algorithm and the K-means++ clustering algorithm for pre-defined anchors, respectively. Since the K-means clustering algorithm randomly initializes the clustering centers, the clustering results are not stable. 20 K-means clusters were performed on the dataset, and the clustering result with the highest mean IoU value of the final clustering center and ground truth boxes was selected as the configuration of YOLOv3(2). The result of using the K-means++ clustering algorithm is used as the preset anchor size for YOLOv3(3). Figure 3 shows a visualization of the clustering results used in YOLOv3(2). Table 1 shows the anchor sizes used for each model in the second set of comparison experiments when the input image size is 416. Combining Fig. 3 and Table 1, it can be seen that the larger-scale ground truth boxes in the homemade dataset in Sect. 3.1 are more numerous and relatively concentrated. Therefore, the anchor sizes clustered out by K-means will be relatively concentrated, but the K-means++ algorithm will spread out the clustering centers and distribute them evenly across the scales due to the way it initializes the clustering centers. In the third set of comparison experiments, S-YOLOv3 adds the SE attention module to YOLOv3(3).

Table 2 shows the detection results of these three sets of experiments. In the first set of comparison experiments, the mAP of YOLOv3(1) is improved compared with the YOLOv3 before improvement. In the second set of comparison experiments, the YOLOv3(3) model with anchor sizes obtained using the K-means++ clustering algorithm achieves higher mAP. When the ground truth box sizes of the dataset span a large-scale range but are not uniformly distributed across scales, setting the anchor sizes using the K-means++ clustering algorithm is more favorable for YOLOv3. In addition, the addition of the SE attention module makes the model pay more attention to the global information and the relationship between channels in the detection, which further improves the detection capability of the model.

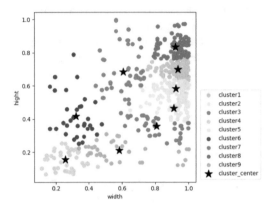

Fig. 3. Results of clustering the self-made dataset using the k-means clustering algorithm

Table 1. Comparison of preset anchor sizes

Clustering algorithm	Small scale	Middle scale	Large scale
K-means clustering	(106,64) (133,173) (243,87)	(252,284) (336,148) (378,192)	(382,347) (384,241) (389,29)
K-means++ clustering	(27,21) (65,16) (77,31)	(108,83) (157,195) (357,221)	(373,346) (378,278) (381,133)

Table 2. Results of ablation experiments

Model	GioU	K-means	K-means++	SE	mAP
YOLOv3					0.709
YOLOv3(1)	√				0.726
YOLOv3(2)	√	√			0.781
YOLOv3(3)	√		√		0.794
S-YOLOv3	√		√	√	0.831

In the field of water surface object detection, the most commonly used detection models besides YOLOv3 are Faster R-CNN and SSD. Therefore, the proposed S-YOLOv3 was compared with Faster R-CNN and SSD on the self-made dataset for experiments.

The experimental results in Table 8 show that S-YOLOv3 achieves 83.1% mAP, which is better than SSD and Faster R-CNN and can meet the needs of water surface aggregated floating object detection.

Figures 4 show the detection results of YOLOv3 and S-YOLOv3 on the self-made dataset, and it can be seen that the improved S-YOLOv3 has improved both in reducing the redundant bounding boxes and improving the confidence of the prediction bounding boxes. For some objects not detected by YOLOv3, S-YOLOv3 can also identify them better. However, S-YOLOv3 also fails to detect the mixed garbage which is similar to the background color and consists of multiple scattered small objects.

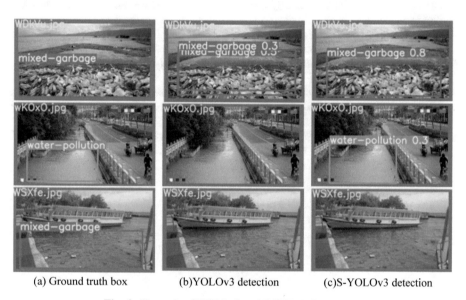

(a) Ground truth box   (b)YOLOv3 detection   (c)S-YOLOv3 detection

Fig. 4. Example of YOLOv3 and S-YOLOv3 detection

5   Conclusions

In this paper, we focus on the detection of aggregated water surface floating objects and improve the YOLOv3 model. To further improve the detection performance of the YOLOv3 model, the SE attention module is introduced into YOLOv3, and the S-YOLOv3 model is proposed by combining experiments to improve the loss function and anchor size of YOLOv3. In addition, the S-YOLOv3 proposed in this paper is compared experimentally with the YOLOv3 model before improvement, and the SSD and Faster R-CNN models commonly used in the field of water surface floating object detection, and the detection performance of S-YOLOv3 is improved significantly, which can meet the needs of surface floating object detection and exceed the effect of the comparison models under the same index.

References

1. Li, S.: Research and Implementation of Small Scale Waters Float Recognition Method Based on Feature Fusion. Chongqing University of Posts and Telecommunications (2019). https://doi.org/10.27675/d.cnki.gcydx.2019.000845
2. van Lieshout, C., van Oeveren, K., van Emmerik, T., et al.: Automated river plastic monitoring using deep learning and cameras. Earth and Space Science **7**(8), e2019EA000960 (2020)
3. Ren, S., He, K., Girshick, R., et al.: Faster R-CNN: towards real-time object detection with region proposal networks. IEEE Trans. Pattern Anal. Mach. Intell. **39**(6), 1137–1149 (2017)
4. Ioffe, S., Szegedy, C.: Batch normalization: Accelerating deep network training by reducing internal covariate shift. International conference on machine learning. PMLR, pp. 448–456 (2015)
5. Li, G., Yao, D., Ai, J., Yi, Z., Lei, L.: Detection and localization of floating objects via improved faster R-CNN. J. Xinyang Normal University (Natural Science Edition) **34**(02), 292–299 (2021)
6. Redmon, J., Farhadi, A.: Yolov3: An Incremental Improvement. arXiv preprint arXiv:1804.02767 (2018)
7. Rezatofighi, H., Tsoi, N., Gwak, J.Y., et al.: Generalized intersection over union: a metric and a loss for bounding box regression. In: Proceedings of the IEEE/CVF Conference on Computer Vision and Pattern Recognition, pp. 658–666 (2019)
8. Hu, J., Shen, L., Sun, G.: Squeeze-and-excitation networks. In: Proceedings of the IEEE Conference on Computer Vision and Pattern Recognition, pp. 7132–7141 (2018)

Research on the Text2SQL Method Based on Schema Linking Enhanced

Jun Tie[1,2], Boer Zhu[1,2(✉)], Chong Sun[1,2], and Ziqi Fan[1,2]

[1] College of Computer Science, South-Central Minzu University,
Wuhan 430074, China
aulaia_zhu@163.com

[2] Hubei Provincial Engineering Research Center for Intelligent Management of
Manufacturing Enterprises, Wuhan 430074, China

Abstract. In recent years, natural language generation of SQL sentences (Text2SQL) has received a lot of attention as an important research direction natural language processing (NLP). Text2SQL makes it easier for users to query complex databases without learning SQL sentences and the underlying database schema. The current mainstream Text2SQL method is the grammar-based IRNet, which attempts to present an efficient method with explanations from the perspective of constructing reasonable grammars and provides a good solution for solving complex nested queries. Still, it makes simple use of external database ontology knowledge, resulting in natural language problems in which words do not correspond well to tables and columns in the database. To address this problem, a new method that considers the entity relationships between natural language problems and data in the database - SLESQL is proposed, which extends some of the functionality of IRNet by using schema linking enhanced Experiments show that SLESQL achieves 6.8% improvement in accuracy over IRNet on the publicly available dataset Spider.

Keywords: Deep learning · IRNet model · Text2SQL · schema linking

1 Introduction

Most of the previous studies have focused on converting natural language problems into SQL query statements that can be executed by existing database software, and now providing users with similar convenient interfaces to query data directly through natural language is the focus of any data opening work that cannot be ignored [1].

The typical Example of Text2SQL is shown in Fig. 1. An effective SQL query sentence is generated by a natural language query sentence, the corresponding database schema (DB Schema), and database content. For example, a query sentence: "Show origin and destination for flights with a price higher than 300." reflects some problems faced by the current Text2SQL method. In fact, "higher"

Industry-University-Research Innovation Fund Project of Science and Technology Development Center of Ministry of Education (No. 2020QT08).

is supposed to refer to the height column (not in table flight) and cannot be extracted directly from the problem. There is also another case: "flights" refers to table flights, and the tag "flights" in natural language problems cannot be mapped directly to a value in a column, table, or database content in a database.

Fig. 1. A Typical Example of Text2SQL

Choosing the correct columns, tables, and values in the data tables involves a major challenge for Text2 SQL - Schema Linkling [4]. Schema linking refers to identifying references of columns, tables, and conditional values in the natural language query sentence. Schema linking involves three difficulties. Firstly, any Text2SQL model must encode the database schema into a representation suitable for decoding into SQL queries, and the decoding process may involve given columns or tables. Secondly, these represent all the information encoded about the schema, such as its column type, foreign key relationship, and the main key used for table connection in the database, as well as external knowledge not in the database and natural language queries, such as the entity type and relationship of each word. Finally, even if the table names, column names, and values in the data tables in the query are different from those encountered during training, the model must identify natural language queries that refer to the values in the table names, column names, and data tables.

Given the above difficulties, we start from the schema linking so that the model can not only extract important information from natural language query sentences but also take into account the relevance of database schema and database content. At the same time, a new named entity recognition method is used to provide more sufficient external knowledge for the model, and a complete candidate strategy is designed to help model selection. Finally, the model generates correct executable SQL query sentences.

2   Related Work

Most current methods use advanced neural network architecture to synthesize SQL queries for given user problems. The SQLOVA method developed by Hwang et al. [5] in 2019 introduces BERT preprocessing model. IRNet [7] used a Transformer encoder and decoder based on an LSTM network, and the general processing flow is shown in the blue part of Fig. 2. In the stage of schema linking, the segmentation method of IRNet is only based on the N-gram method of the string. This will cause the input information to be redundant, and a large number of invalid word segmentations are input, which increases the difficulty of neural network selection. At the same time, IRNet uses a largescale open knowledge map ConceptNet to predict the relationship between the values in the data table in the natural language query sentence and the columns in the database model and puts the fragments in the natural language query sentence into the knowledge map ConceptNet. The results returned by ConceptNet contain two noteworthy information, namely, "relevant terms" and "the same type". This only by looking up the "upper words" to filter the database schema matching column name effect is not ideal, can not accurately understand the relationship between the data table median and the column in the database schema, and because of some spelling errors and ignore the relevant words, thus reducing the accuracy of generating SQL.

We propose a Text2SQL model based on schema linking enhanced (SLESQL) by combining external knowledge and database content. Starting from the schema linking, a new candidate strategy is adopted to make up for the defects of single string comparison (such as N-gram). It not only considers the association between natural language query sentences and database schema, but also considers linking the values in the data table to the database schema so that the model can extract the correct entity type, enhance the reasoning ability of the model, and improve the accuracy of generating SQL sentences.

3   Algorithm Description

The input of IRNet is the natural language query sentence and database schema, and the SLESQL proposed in our work adds the database content (DB Content) as the input on this basis, as shown in the orange section in Fig. 2. In the encoder, SLESQL introduces the named entity recognition model-TENER [8] to enhance the ability of the model to extract the correct entity. After screening and verification of candidate sets, the encoder can more accurately align the database schema with each part of the natural language query sentence, and the process can be seen in Fig. 3.

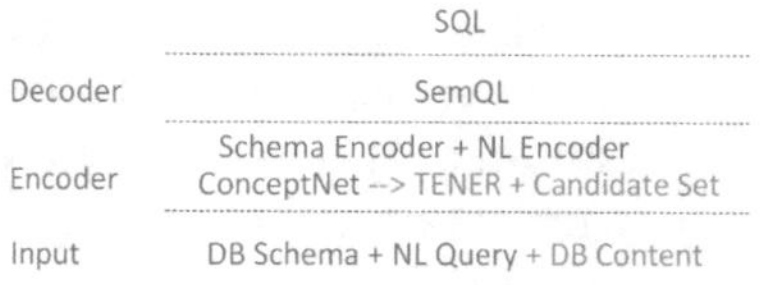

	SQL
Decoder	SemQL
Encoder	Schema Encoder + NL Encoder ConceptNet --> TENER + Candidate Set
Input	DB Schema + NL Query + DB Content

Fig. 2. SLESQL Overall Flow

3.1 Schema Linking Enhanced

The schema linking enhanced phase mainly completes two tasks : (1) Propose an appropriate candidate set. (2) Provide adequate information for predicting values in table names, column names, and data tables. In addition to the database schema and natural language query sentences, the model can predict the correct table name, column name, and value in the data table in the decoder by viewing the database content to provide more information for the neural network.

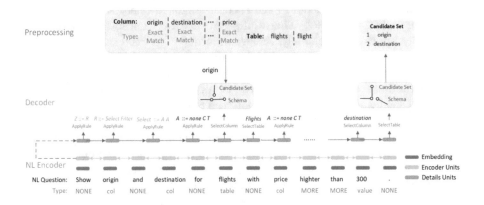

Fig. 3. Schema Linking Enhanced

Candidate Extraction. For a natural language query sentence, we use the named entity recognition (NER) model to extract potential candidates. Since Spider is an English dataset, the TENER model proposed by Yan et al[17] in 2019 has also been applied in our work. This model is an improvement on the original Transformer based on the NER task. The improved Transformer encoder is used to model the character-level features and word-level features. The TENER model is the best indepth learning method on the English dataset CoNLL-2003 so far. we also make an entity analysis of the Question section in Spider [2]. In addition, some simple and effective heuristic methods are used to extract and generate candidates : (1) Content in quotation marks. (2) Terms in capital letters. (3) Single alphabet.

Candidate Generation. After extracting the value from the problem, the candidate needs to be generated. For the value in the data table, the extracted value itself is likely to be the only necessary candidate. Three candidate value generation methods are used in SLESQL : (1) method based on string similarity, (2) heuristic method based on artificial production, and (3) N-gram method based on a string. To evaluate the similarity between the text value extracted from the problem and the value in the database, using the text distance based on word embedding, the model only scans the value of the similarity in the database that exceeds a certain threshold. we further use Damerau-Levenshtein [9] to measure the similarity between each marker, because Damerau-Levenshtein has a good performance between accuracy and running time.

Verify Candidate and Encode. Since candidate generation will generate a large number of potential candidates, it is necessary to verify the candidate set. Three aspects should be considered : (1) the similarity threshold, (2) the number of candidates extracted from natural language problems, and (3) the total number of values in the database. Since the number of candidates directly affects the accuracy of the model, too many candidates make it difficult for the model to select the correct value. Therefore, this article uses database content again to reduce the number of candidate sets. In contrast, Exact Matching is used instead of using similarity to verify the candidate set. We exclude candidates extracted from quotation marks from database validation. In the validation process of the candidate set, the SLESQL model will record the table column position of the candidate in the database.

The work of candidate encoding is similar to that of the table and column encoding. The position of the candidate (table and column) is encoded with the candidate itself. Because of the extra table and column information, not only for the numerical itself, the encoder can also find the location of the numerical attention. Each candidate, together with its location, is separated from other values by using the [SEP] specified by the encoder. Each marker value is further marked as a lexical chunk using the WordPiece segmentation algorithm. Encoder input is a pretrained embedded list, and each lexical block corresponds to it.

3.2 Encoder-Decoder

SLESQL encoder input is information about database schema and candidates extracted from database content. Therefore, SLESQL encoders can also learn the relationship between the tag of natural language problems and the actual values in the database. The nonoverlapping sequence of queries is expressed as $x = [(x_1, T_1), \ldots, (x_n, T_n)]$, where x_1 is the first sequence and is the type of x_1. The encoder takes as input and converts each word into its vector. Then, running bidirectional LSTM for all sequences, the output states of forwarding LSTM and reverse LSTM are connected as the output of natural language query sentences in the encoder. The database model is expressed as a set of different columns and their types assigned in the preprocessing, which is a set of tables.

The decoder receives the encoding of natural language query questions, table names, column names, and data table medians from the encoder as the input, and the output is the synthesized close semantic query language (SemQL). The decoder is composed of LSTM architecture and multiple pointer networks, which are used to select tables, columns, and values. Using a syntax-based decoder (TRANX), LSTM is used to simulate the generation process of SemQL. Formally, the generation process y of SemQL can be formalized as $p(y \mid x, s) = \prod_{i=1}^{T} p(a_i \mid x, s, a < i)$, where is the action taken at the time i, $a < i$ is the action sequence before i, and is the total number of the whole action sequence. The decoder interacts with three types of actions to generate SemQL, including APPLYRULE, SELECTCOLUMN, and SELECTTABLE

[10]. APPLYRULE(r) applies a generation ruler to the current derivation tree of SemQL, where r is the generation rule designed in IRNet. Also, using a Coarse-to-Fine framework [11], the decoding process for SemQL is shown in Fig. 4.

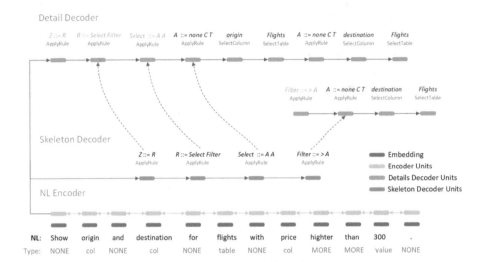

Fig. 4. Encoder-Decoder Processing Flow

4   Experiments and Analysis

The experiment uses the Spider public dataset of Yale University and compares SLESQL with SyntaxSQLNet [3] and IRNet to verify the effectiveness of SLESQL.

4.1   Dataset and Evaluating Indicator

The experiment uses the public dataset Spider, which contains most SQL operators (ORDER BY/GROUP BY/HAVING and nested queries) and is distributed in 200 publicly available databases in 138 fields. Each database has multiple tables, and each database has an average of 5.1 tables. The dataset is divided into a training set, validation set, and test set. The validation set covers 20 different databases that have not appeared in the training set.

To quantitatively evaluate the accuracy of SQL generated by Text2SQL, this paper conducts a comparative test based on the Spider dataset. Exact Matching index is used for evaluation. If the complete statement is correct, the accuracy rate is used as a measure. The specific calculation formula is shown in Formula (1).

$$\text{Exact Matching} = \frac{\#\text{count}}{\#\text{total}} \tag{1}$$

where #*total* is the total number of samples, and #*count* is the consistent number of complete SQL sentences.

4.2   Experiment Setting

The model uses an Adam optimizer, the encoder is 2e–5, the decoder is 1e–3, and the intermediate connection parameter is 1e–4. The learning rate of the encoder is the default parameter for BERT [6] fine-tuning, and all other hyperparameters are set based on experience hyperparameters. After experimental verification, the optimal parameter configuration is as follows: $Batch_{s}ize = 64, Dropout = 0.3$.

4.3   Experimental Results and Analysis

To verify the effectiveness of SLESQL proposed in this paper, SyntaxSQLNet and IRNet models are mainly used as the comparison models of experiments. Spider evaluation index defines difficulty according to the number of SQL components, selection and conditions, so queries containing many SQL keywords will be considered difficult. Spider defines four difficulty levels, simple, medium, Hard and Extra Hard. According to the difficulty level of SQL defined in the Spider dataset, this experiment further studies the performance of SLESQL in different difficulty parts of the test set.

Table 1. SLESQL Exact Match Results in Four Difficulty Levels on the Test Set

Algorithms	Easy	Medium	Hard	Extra Hard
SyntaxSQLNet(BERT)	42.9%	24.9%	21.9%	8.6%
IRNet(BERT)	77.2%	58.7%	48.1%	25.3%
SLESQL	77.6%	60.1%	52.2%	32.1%

As shown in Table 1, SLESQL has higher accuracy than SyntaxSQLNet and IRNet at all difficulty levels. Moreover, the experimental data show that for the more Hard queries, SLESQL is more sufficient to show excellent analytical power and accuracy. For example, compared with IRNet (with BERT to enhance IRNet), SLESQL improves the matching rate of SQL sentences at the Extra Hard level by 6.8%. Experiments show that schema linking enhanced can better solve the problem of containing more database schemas and data tables.

Some problems can be solved by BERT, such as field prediction error and operator misuse, but the improvement for complex nested queries is not obvious. In Fig. 5, we can see that the highest accuracy of SLESQL is 65% in the Spider training set. Moreover, with the increase of training rounds, the accuracy of SQL generated by the model steadily increases. After the convergence in the late training, SLESQL has a higher performance. Similarly, using BERT as the pretraining model, SLESQL performs significantly better than IRNet, indicating that the research work on SLESQL in complex nested queries is effective.

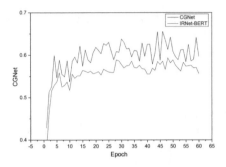

Fig. 5. Encoder-Decoder Processing Flow

5   Summary

In this paper, the method of schema linking enhanced is proposed to link the values in the data table with the database schema, to propose the correct candidate set. The neural network is used to determine the most consistent with the problem intention in these candidate sets as the filling value in the decoder, aiming to solve the mismatch problem between the schema linking and the intermediate representation. The experiment proves the effectiveness of SLESQL in the Spider dataset under complex cross-domain scenarios, which enables the technology to quickly adapt to the enterprise database, simplify the process of data fetching and text analysis, help enterprises freely interact with the database, and effectively activate the knowledge value of the enterprise database.

Acknowledgments. This work is supported and assisted by the Industry-University-Research Innovation Fund Project of Science and Technology Development Center of Ministry of Education (No. 2020QT08), National Ethnic Affairs Commission of the People's Republic of China (Training Program for Young and Middle-aged Talents, MZR20007) and the Hubei Provincial Science and Technology Major Project of China (2020AEA011).

References

1. Pliego, P.I.: Programming an intelligent personal assistant for biomedical database analysis. Universitat Politècnica de Catalunya, Catalonia (2017)
2. Yu, T., Zhang, R., Yang, K.: Spider: a large-scale human-labeled dataset for complex and cross-domain semantic parsing and text-to-SQL task. In: Proceedings of the 2018 Conference on Empirical Methods in Natural Language Processing, pp. 3911–3921. ACL, Brussels (2018)
3. Yu, T., Yasunaga, M., Yang, K.: SyntaxSQLNet: syntax tree networks for complex and cross-domain text-to-SQL task, pp. 1653–1663. ACL, Brussels (2018)
4. Taniguchi, Y., Nakayama, H., Takahiro, K.: an investigation between schema linking and text-to-SQL performance https://arxiv.org/abs/2102.01847
5. Hwang, W., Yim, J., Park, S.: A comprehensive exploration on WikiSQL with table-aware word contextualization https://arxiv.org/abs/1902.01069

6. Devlin, J., Chang, M.W., Lee, K. and Toutanova, K.: BERT: pre-training of deep bidirectional transformers for language understanding. In: Proceedings of the 2019 Conference of the North American Chapter of the Association for Computational Linguistics: Human Language Technologies, pp. 4171–4186. ACL, Minneapolis (2019)
7. Guo J, Zhan Z, Gao Y.: Towards complex text-to-SQL in cross-domain database with intermediate representation, pp. 4524–4535. ACL, Florence (2019)
8. Yan, H., Deng, B., Li, X.: TENER: adapting transformer encoder for named entity recognition https://arxiv.org/abs/1911.04474
9. Damerau, F.J.: A technique for computer detection and correction of spelling errors. Commun. ACM **7**(3), 171–176 (1964)
10. Yin, P., Neubig, G.: A syntactic neural model for general-purpose code generation. In: Proceedings of the 55th Annual Meeting of the Association for Computational Linguistics, pp. 440–450. ACL, Vancouver (2017)
11. Dong, L., Lapata, M.: Coarse-to-fine decoding for neural semantic parsing. In: Proceedings of the 56th Annual Meeting of the Association for Computational Linguistics, pp. 731–742. ACL, Melbourne (2018)

A Multi-task Learning Model for Emotion-Cause Extraction Based on Emotion Classification

Lu Liu[1,2,3(✉)], Jun Qin[1,2,3], Kai Meng[1,2,3], Jing Liu[1,2,3], and Zejin Zhang[1,2,3]

[1] School of Computer Science, South-Central Minzu University,
Wuhan 430000, China
549631871@qq.com

[2] Hubei Provincial Engineering Research Center for Intelligent Management of
Manufacturing Enterprises, Wuhan 430000, China

[3] Hubei Provincial Engineering Research Center of Agricultural Blockchain and
Intelligent Management, Wuhan 430000, China

Abstract. Emotion-Cause joint extraction is to extracting both the emotion and its corresponding cause from the given text, which has a wide range of application scenarios. Previous work only considered emotion extraction, cause extraction and emotion-cause relation classification tasks. This paper proposes a multi-task learning model based on emotion classification to perform emotion-cause joint extraction in a unified model, which obtains semantic features of different granularity based on Bert and attention mechanism. We also introduce focus loss function to deal with the sample imbalance problem. Experimental results show that the emotion classification subtask can effectively extract the emotional features of documents. Compared with the existing models, the experimental results show that our model outperforms the state-of-the-art model on emotion-cause joint extraction.

Keywords: Emotion Classification · Emotion-Cause Extraction · Multi-task Learning

1 Introduction

The massive texts produced by Internet media contain rich information such as people's opinions and emotions, which has great social and commercial value. Therefore, sentiment analysis has become a hot research direction in the field of natural language processing. Sentiment analysis focuses on mining people's views, feelings, evaluations, attitudes and emotions about an entity in the objective world, which are generally divided into three categories: positive, neutral and negative. On this basis, a more fine-grained emotional classification task is extended to further subdivide human emotions. The American psychologist

The National Ethnic Affairs Commission of the People's Republic of China (Training Program for Young and Middle-aged Talents, MZR20007).

Q. Liang et al. (Eds.): AIC 2022, LNEE 871, pp. 84–94, 2023.
https://doi.org/10.1007/978-981-99-1256-8_11

Plutchik identified eight basic emotions: grief, fear, surprise, acceptance, ecstasy, rage, vigilance and hatred, and psychologists suggested many more categories. However, only knowing the emotions or emotions expressed by people is not enough to meet the needs of practical application. It is more meaningful and valuable to dig out the real causes of emotions.

The task of emotional reason extraction aims to extract the corresponding reasons of emotion expressed in a given text, which was first proposed by Sophia [1]. In 2010 Based on the balanced Chinese corpus, this study artificially constructed a data set and corresponding linguistic rules, and found that most of the emotion-expressing texts in the data set had corresponding causal text cues. Lin Gui [2]. pointed out that there are three challenges in the field of emotional reason extraction: I. There is no open and accessible data set; II. The extraction of emotional reasons lacks some formal definitions; III. The lack of good public data sets makes the research progress in this field slow.

Most of the emotion cause extraction tasks are based on the pre-marked emotion and then the corresponding cause extraction, which greatly limits the specific application scenarios. In 2019, Rui Xia [3] proposed a deep learning framework for the joint extraction of emotion and reason, realizing the end-to-end simultaneous extraction of emotion and reason. Based on previous studies, this paper finds that Emotion Classification (EC) is closely related to emotional reason pair extraction task. Because sentiment classification task goal is to identify emotional categories in clause and thus can promote the emotion and reason extraction task in emotional clause identification, so as to improve the extraction of emotional reasons for performance as a result, we built contains emotion classification and emotional reasons for extraction of ECPE-EC multitasking learning model, compared with existing methods, This model has the best comprehensive performance in emotion-cause pair extraction. The main contributions of this paper are as follows:

(1) A bert-based emotion-reason joint extraction multi-task learning model is proposed by introducing emotion classification tasks.
(2) In order to fully capture semantic features in documents, feedforward neural networks and attention mechanisms at different levels are used to capture contextual semantic features.
(3) Experimental results show that the proposed model significantly improves the performance of emotion-reason pair extraction task, and has the best comprehensive performance compared with existing models.

2   Related Work

After the task of emotion cause mining was proposed by Sophia [1] in 2010, Sophia [4] identified seven groups of language clues through the analysis of corpus, and generated language rules for detecting emotional causes. Neviarouskaya [5] explored the causes of implicit emotion from the phrase level. The causal relationship of eight emotions is analyzed by combining semantic features and dependency syntax. Gao [6] adopted the rule-based method to detect emotional causes on Chinese micro-blogs, and obtained the fine-grained corresponding emotional causes from the event results, subject behavior and object. Li Yiwei [7]. Regarded emotional cause extraction as a sequence labeling problem, and significantly improved the accuracy of emotional cause extraction by using the context features of the text. As Chinese microblogs contain rich text data, Lin Gui [8] constructed data sets of emotional reason extraction task based on Chinese microblogs.

As Attention Model was proposed and proved to be effective in many fields, it was also introduced into emotional reason extraction task. Xiangju Li [9] constructed an emotion-reason clause extraction model combining BiLSTM and CNN based on collaborative attention mechanism. XiangjuLi [10] applied attention mechanism to extract emotional reasons from different perspectives. These methods mainly explore the semantic relationship between emotion clauses and cause clauses, and need to extract the corresponding emotional reasons under the premise of emotion labeling, which makes the extraction cost of emotional reasons high and limits the application scenarios. In 2019, Rui Xia [11] regarded emotion and reason extraction as the joint extraction of two sub-tasks, realizing the simultaneous end-to-end extraction of emotion and reason. Subsequently, Zixiang Ding et al. [12] used two-dimensional Transformer to model the relationship between emotion clauses and cause clauses. Penghui Wei [13] ranked candidate emotion and cause pairs based on a ranking method, and then selected emotion and cause pairs based on the maximum probability. Hao Tang [14] proposed emotion detection and emotion-reason pair multilevel attention model. Most of these methods adopt a two-step framework: firstly, emotion and cause are extracted, and then emotional cause pairs are paired and filtered, making the extraction results of emotion-cause pairs depend on the extraction results of emotion and cause, and the correlation features between subtasks are not fully utilized. Although Sixing Wu [15] proposed a unified multi-task emotion-reason pair extraction model, which could obtain training samples for relationship classification without relying on the extraction results of emotion and reason, they failed to further study the degree of feature association between similar sub-tasks.

3   Model

3.1   Problem Description

The objective of emotion-cause Pair extraction task is to extract clause pairs containing emotions and their causes from a given document D, known as emotion-cause pairs [11]. $D = C_1, C_2, ..., C_n$. Every clause C_i in the document refers to the sequence of M characters separated by Chinese punctuation marks, that is, $C_i = w_1, w_2, ..., w_m$. These clauses contain the emotional clause C_e and the cause clause C_c.

3.2   Model Framework

As shown in Fig. 1, the multi-task learning model of extraction (ECPE) and emotion classification (EC) based on emotional reasons is ECPE-EC. On the left is the task model that implements end-to-end extraction of emotional reasons, and on the right is the emotion classification task model that implements clauses. They jointly use BERT to encode clauses in documents and share basic text features. Emotional reason to be due to emotional clause and clause, and emotional clause is to describe the mood of clause, therefore, in order to further study the correlation of two tasks, sentiment classification task of extracting provide important task for emotional reasons emotional characteristics, can catch clause characteristics of emotional expression, emotion clause in help model identification document. Thus improving the performance of the task for emotional reasons.

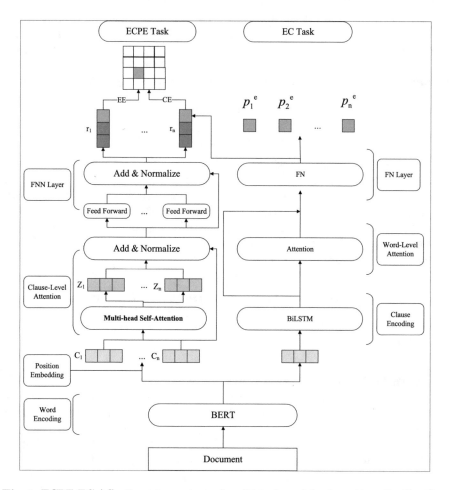

Fig. 1. ECPE-EC Affective cause pairs and multi-task models of emotion classification.

Performance on Emotional Categorization Tasks

The task of emotional classification is to identify the emotional categories expressed by each clause in a given document, which can be divided into six emotional categories (happiness, sadness, fear, anger, Disgust, Surprise, and other). There are also clauses that do not express any emotional categories. D of each clause on a given document, first by BERT of coding sequence of text, every word semantic expression, again through the Bi-LSTM coding clause word level of context information, due to the influence of different word to express sentiment has different, so the application level of attention mechanism, to learn the different weight of each word. Highlight the characteristics of certain keywords to better distinguish the categories of emotions expressed in the clauses.

Encode clause information by using BERT, the vector of C_i is obtained to represent $C_i R^d$ (i is the serial number of the clause in the document and D is the word vector dimension). Then Bi-LSTM is used to encode clause information, and the output of time step in two directions is spliced to obtain the context information h_{ij} of clause at the word level (j is the time step), as shown in formula (1)–(2).

$$c_i = BERT(C_i), \tag{1}$$

$$h_{ij} = [\overleftarrow{(LSTM(c_i))}, \overrightarrow{(LSTM(c_i))}], \tag{2}$$

Using the word level, highlight related to emotional words, through the hidden layer output hf_i and Bi-LSTM each time step the output of the dot product similarity calculation, through softmax function calculating the probability distribution of similarity scores, to get to a focus on different words, multiplied to obtain the final clause information h_i^e, softmax function is used to determine the probability of each category and obtain the emotional label y_i^e of the clause based on the maximum probability.

$$y_i^e = softmax(linear(\sum_j^m softmax(h_{ij} hf_i^T) h_{ij})), \tag{3}$$

where linear is the linear layer.

Draw Tasks Based on Transformer for Emotional Reasons

The objective of emotional reason pair extraction task is to extract emotional clauses and cause clauses from a given document. Firstly, for the given document D, the encoding information C_i of each clause in the document is obtained through BERT coding (i is the clause number). Then the relative position feature RP between clauses is obtained by using the position feature extraction network in Sect. 3.3, and the attention information between clauses is calculated by the self-attention mechanism in Transformer. Finally, by combining the feature of clauses and the emotion feature of emotion classification task, we can judge whether each clause in the document is the corresponding emotion cause pair.

Assume that the linear layer is used to directly integrate the text feature r_i of the i clause and r_j of the j clause, as well as the emotional feature p_i^e

of the clause. Then softmax function is used to calculate the probability of different categories of clauses and obtain the final label $y_i j$ to achieve end-to-end extraction of emotional causes. Reduce the propagation error in the two-step method, as shown in Formula (4)–(5).

$$f_{ij} = linear([r_i; r_j; p_i^e]), \tag{4}$$

$$y_{ij} = softmax(f_{ij}), \tag{5}$$

where linear is the linear layer of fusion features.

3.3 Position Feature Coding

Statistical analysis of the data set found that the relative distances between emotion clause C_e and cause clause C_c in each document were close in most cases, as shown in Table 1. The Relative Position (RP) of emotion-cause clause pairs less than 2 is more than 85%, and the Relative distance between emotion-cause clauses is less than 5%, indicating that there is a strong Relative Position relationship between emotion-description clauses and emotion-cause clauses.

Table 1. Relative distance statistics

RP	0	1	2	>2
Value	507	1302	209	87
Percentage	24.09%	61.85%	9.93%	4.13%

We use position embedding (PE) matrix to represent the relative position relationship between emotion and cause clauses. Firstly, a random matrix is used to represent the Position embedding feature. $P_V = PE(P)$ represents the Position Vector (PV). Then, the relative position features of the emotion clause and the cause clause are adjusted by linear network and nonlinear function ReLU. During the model training, the value of the position matrix PE is dynamically adjusted by back propagation mechanism, and the position feature coding P_f is finally obtained.

3.4 Focal Loss Function

The modulation coefficient $(1 - y')^\gamma$ (γ is the focus parameter) is added to the basic cross entropy function to reduce the proportion of easily distinguishable samples, so that the neural network model can pay more attention to the samples with no obvious distinction in the learning process. Finally, combined with the characteristics of the above formula, the loss function with weight parameters can be adjusted for different types of samples, as shown in formula (6).

$$L_{fl} = \begin{cases} -\alpha(1 - y')^\gamma log y', & y = 1 \\ (1 - \alpha)y''^\gamma log(1 - y'), & y = 0 \end{cases} \tag{6}$$

3.5   Training Objectives

In this multi-task model, emotion classification and emotion-cause pairs are jointly trained to extract two subtasks using the focus loss function in Sect. 3.5, as shown in Formula (7).

$$L = \lambda_e L_{fl}^e + \lambda_p L_{fl}^p + \lambda_2 \tag{7}$$

where λ_e and λ_p are the weight coefficients of the loss of two subtasks respectively, is the regularization term of L_2.

4   Experimental Analysis

4.1   Dataset Analysis

This paper uses a data set constructed based on emotion-reason extraction task [17]. This dataset consists of different documents, each of which contains multiple clauses, in which each document is marked with one or more pairs of emotional reasons, as well as corresponding emotions expressed, as shown in Table 2. There were 1,945 documents in the whole data set, including 1,746 documents containing 1 pair of emotional cause pairs, 177 documents containing 2 pairs, and 22 documents greater than 2. The data set labeled six emotions: happiness, sadness, fear, anger, Disgust, and surprise. Each emotion clause corresponds to one or more cause clauses.

Table 2. Dataset information statistics

Item(pair)	Num	Item(emotion)	Num
All	1945	happiness	551
1	1746	sadness	568
2	177	fear	402
≥ 3	22	other	564

4.2   Experimental Parameter Setting

This chapter builds the model based on neural network framework Pytorch1.7, uses Google's open source Chinese pre-training language model BERT, and fine-tunes the implementation scheme of third-party library pytorch-pretrained-BERT. The learning rate of BERT fine tuning is set to 1E–5. Adam optimizer [18] was used to update the neural network model parameters, and training was carried out on G200eR2 graphics card equipment. The dataset was randomly divided for 10 times in a ratio of 9:1 to obtain the training set and test set. The experimental result in this section is the average of ten experiments.

The learning rate of the model was set to 1E–4, batch size was set to 5, and weight decay was set to 1E–5. The model adopted an early termination strategy, the number of training rounds (EPOCH) was 40, and the test was performed every 200 steps. Other detailed parameters of the model are shown in Table 3.

Table 3. Dataset information statistics

Item	Value
Bi-LSTM layers	1
hidden layers of Bi-LSTM	1024
Position embedding size	200
γ	1.5
Dropout Value	0.5

4.3   Experimental Results and Analysis

In order to verify the effectiveness of the model, the neural network-based models in recent years are selected for comparison. The specific information is as follows.

E2EECPE [19]: A neural network model for end-to-end extraction of emotional reasons

MTNECP [15]: A multi-task learning network model. Emotion clause extraction and emotion reason clause extraction are combined with emotion reason extraction for joint learning of extraction task to improve model performance.

TDGC [20]: A neural network model based on directed graph structure, which transforms the emotional reason pair extraction task into a problem of finding whether there are edges between the neutron sentences in a document.

RANKCP [13]: An end-to-end neural network model based on ranking, which uses ranking mechanism to screen candidate emotional cause pairs.

Ecpe-2d [12]: is an end-to-end model based on Transformer two-dimensional cross-paths.

Table 4. Emotional reasons for data set model performance comparison

Model	Emotion Extraction			Cause Extraction			Emotion-Cause Pair Extraction		
	P	R	F1	P	R	F1	P	R	F1
E2EECPE	0.8595	0.7915	0.8238	0.7062	0.6030	0.6503	0.6478	0.6105	0.6280
MTNECP	0.8595	0.7915	0.8238	0.7062	0.6030	0.6503	0.6478	0.6105	0.6280
RANKCP	0.8662	0.8393	0.8520	0.7400	0.6378	0.6844	0.6828	0.5894	0.6321
TDGC	0.8548	0.8703	0.8406	0.6824	0.6927	0.6743	0.6610	0.6698	0.6546
ECPE-2D	0.8716	0.8244	0.8474	0.7562	0.6471	0.6974	0.7374	0.6307	0.6799
ECPE-EC	**0.8627**	**0.9221**	**0.8910**	**0.7336**	**0.6934**	**0.7123**	**0.7292**	**0.6544**	**0.6889**

The experimental results are shown in Table 4. ECPE-EC is a joint learning model proposed in this paper based on emotion-cause pairs and emotion classification tasks. In order to conduct a comprehensive performance comparison with the baseline model, the extraction of emotion clauses and cause clauses in ECPE-EC were extracted separately during the training process as intermediate products of the model. It can be seen that our model has the best performance on recall rate R and F1 of extraction task for final emotional reasons. Because of the feature of emotion classification task, it has the highest performance on all task indexes of emotion clause extraction. The recall rate of reason clause extraction is also the highest. In summary, the multi-task model proposed in this paper has obvious advantages. In order to test the relationship between each sub-task, we compared the experimental results of multi-task joint training and independent training, as shown in Table 5. "Joint" refers to the joint training of ECPE and EC, and "independent" refers to the independent training of ECPE and EC.

Table 5. Comparative experimental results of multi-task learning based on emotion categorization task

Method	Emotion-Cause Pair Extraction			Emotion Classification		
	P	R	F1	P	R	F1
Joint	**0.7004**	**0.7146**	**0.7061**	0.6435	0.7200	0.6736
Independent	0.6114	0.6437	0.6250	**0.7511**	**0.8035**	**0.7710**

It can be seen that multi-task joint learning based on emotion classification can effectively improve the performance of emotion-reason pair extraction. The results showed that the affective categorization task positively moderated the affective cause-pair task. However, for The EC task, the emotion classification task index was better when trained separately, indicating that joint learning has a negative effect on the emotion classification task, which is also a prominent problem in multi-task learning, that is, the performance of one task needs to be sacrificed to promote the performance of another task. In the ECPE-EC multi-task learning model proposed in this paper, the performance of emotion classification task should be sacrificed to improve the performance of emotion reason pair extraction task. This is due to the difference in learning efficiency of different sub-tasks in the process of model training.

5   Summarize

This paper proposes a bert-based emotion-reason pair extraction framework for multi-task learning. Emotion classification was regarded as a subtask of emotion-cause extraction, and the performance of emotion clause extraction was improved by sharing the emotion characteristics of the emotion classification task, thus improving the extraction effect of the target task – emotion-cause pair. In order to fully capture semantic features in documents, feedforward neural networks and attention mechanisms at different levels are used to capture contextual semantic features. The focus loss function is introduced to solve the problem of data set label imbalance. Compared with the existing multi-task model of emotion-cause pair extraction, the proposed model has the best comprehensive performance, and the F1 value of emotion-cause pair extraction is improved by 1.72% points.

Although the experimental results show that the proposed model can effectively improve the F1 value of emotion-reason pair extraction, the accuracy of emotion-reason pair extraction is slightly reduced in the task of reason clause extraction, which affects the accuracy of emotion-reason pair extraction. In the future, we will further explore the correlation between the sub-tasks of emotion-reason pair extraction, and try to use the graph model to further improve the extraction performance of emotion-reason pair.

Acknowledgments. This work is supported and assisted by the National Ethnic Affairs Commission of the People's Republic of China (Training Program for Young and Middle-aged Talents, MZR20007), the Industry-University-Research Innovation Fund Project of Science and Technology Development Center of Ministry of Education (No. 2020QT08), Hubei Provincinal Science and Technology Major Project of China (2020AEA011), the Fundamental Research Funds for the Central Universities, South-Central Minzu University (No. CZQ20012), the Graduate Innovation Fund, South-Central Minzu University (No. 3212021yjshq008).

References

1. Lee, S.Y.M., Chen, Y., Li, S., et al.: Emotion cause events: Corpus construction and analysis. In: ELRA. Proceedings of the Seventh International Conference on Language Resources and Evaluation (LREC 2010), Valletta: European Language Resources Association (ELRA), vol. 2010, pp. 1121–1128 (2012)
2. Gui, L., Wu, D., Xu, R., et al.: Event-driven emotion cause extraction with corpus construction. In: Proceedings of the 2016 Conference on Empirical Methods in Natural Language Processing (EMNLP), Austin, vol. 2016, 1639–1649. ACL (2016)
3. Xia, R., Ding, Z.: Emotion-cause pair extraction: a new task to emotion analysis in texts. In: Proceedings of the 57th Annual Meeting of the Association for Computational Linguistics (2019)
4. Lee, S.Y.M., Chen, Y., Huang, C-R.: A text-driven rule-based system for emotion cause detection. In: Proceedings of the NAACL HLT 2010 Workshop on Computational Approaches to Analysis and Generation of Emotion in Text, Los Angeles, vol. 2010, pp. 45–53. ACL (2010)
5. Neviarouskaya, A., Aono, M.: Extracting causes of emotions from text. In: Proceedings of the Sixth International Joint Conference on Natural Language Processing, vol. 2013, pp. 932–936. ACL (2013)
6. Gao, K., Xu, H., Wang, J.: A rule-based approach to emotion cause detection for Chinese micro-blogs. Expert Syst. Appl. **42**(9), 4517–4528 (2015)
7. Lee, Y.W., Lee, S.S., Huang, J.R.: Identification of emotional causes based on sequence annotation model. J. Chin. Inf. Sci. **27**(5), 93–100 (2013)
8. Gui, L., Yuan, L., Xu, R., Liu, B., Lu, Q., Zhou, Yu.: Emotion cause detection with linguistic construction in Chinese Weibo text. In: Zong, C., Nie, J.-Y., Zhao, D., Feng, Y. (eds.) NLPCC 2014. CCIS, vol. 496, pp. 457–464. Springer, Heidelberg (2014). https://doi.org/10.1007/978-3-662-45924-9_42

9. Li, X., Song, K., Feng, S., et al.: A co-attention neural network model for emotion cause analysis with emotional context awareness. In: Proceedings of the 2018 Conference on Empirical Methods in Natural Language Processing, Brussels, vol. 2018, pp. 4752–4757. ACL (2018)

10. Li, X., Feng, S., Wang, D., et al.: Context-aware emotion cause analysis with multi-attention-based neural network. Knowl.-Based Syst. **2019**(174), 205–218 (2019)

11. Xia, R., Ding, Z.J.: Emotion-cause pair extraction: a new task to emotion analysis in texts. In: Proceedings of the 57th Annual Meeting of the Association for Computational Linguistics. Florence, Association for Computational Linguistics (2019)

12. Ding, Z., Xia, R., Yu, J.: ECPE-2D: Emotion-cause pair extraction based on joint two-dimensional representation, interaction and prediction. In: Proceedings of the 58th Annual Meeting of the Association for Computational Linguistics, vol. 2020, pp. 3161–3170. Association for Computational Linguistics (2020)

13. Wei, P., Zhao, J., Mao, W.: Effective inter-clause modeling for end-to-end emotion-cause pair extraction. In: Proceedings of the 58th Annual Meeting of the Association for Computational Linguistics, vol. 2020, pp. 3171–3181. Association for Computational Linguistics (2020)

14. Tang, H., Ji, D., Zhou, Q.J.N.: Joint multi-level attentional model for emotion detection and emotion-cause pair extraction. Neurocomputing **2020**(409), 329–340 (2020)

15. Wu, S., Chen, F., Wu, F., et al.: A multi-task learning neural network for emotion-cause pair extraction. In: ECAI 2020 - 24th European Conference on Artificial Intelligence. Santiago de Compostela, vol. 2020, pp. 2212–2219. ECAI (2020)

16. Lin, T.Y., Goyal, P., Girshick, R., et al.: Focal loss for dense object detection. In: 2017 IEEE International Conference on Computer Vision (ICCV), Honolulu, vol. 2017, pp. 2999–3007. IEEE (2017)

17. Hen, Y., Lee, S.YM., Li, S., et al.: Emotion cause detection with linguistic constructions. In: Proceedings of the 23rd International Conference on Computational Linguistics. Beijing: Coling 2010 Organizing Committee, vol. 2010, pp. 179–187. ACL (2010)

18. Kingma, D., Ba, J.: Adam: a method for stochastic optimization. arXiv preprint arXiv:1412.6980, vol. 2014 (2014)

19. Haolin, S., Chen, Z., Qiuchi, L., et al.: An end-to-end multi-task learning to link framework for emotion-cause pair extraction. arXiv preprint arXiv:2002.10710 (2020)

20. Fan, C., Yuan, C., Du, J., et al.: Transition-based directed graph construction for emotion-cause pair extraction. In: Proceedings of the 58th Annual Meeting of the Association for Computational Linguistics, vol. 2020, pp. 3707–3717. Association for Computational Linguistics (2020)

Research on Helmet Detection Algorithm Based on Improved YOLOv5s

Xiangshu Peng[1], Zhiming Ma[1]([✉]), Ping Wang[2], Yaoxian Huang[3], and Lixia Zhang[1]

[1] School of Computer Science and Technology, Xinjiang Normal University, Urumchi 830054, China
2298882664@qq.com
[2] School of Electricity and Computer, Jilin JianZhu University, Changchun 130000, China
[3] School of Computing, Harbin Finance University, Harbin 150000, China

Abstract. The construction environment is complex and dangerous, and it is difficult to achieve all-round, whole-process and real-time helmet detection. In order to ensure the safety of personnel, this paper proposes a helmet detection algorithm based on improved YOLO v5s. First, replace the backbone feature extraction network of YOLOv5s with the end-side neural network architecture GhostNet, which greatly reduces the amount of network parameters. Introduce the lightweight module attention mechanism ECA-Net in the C3 module to improve the feature extraction ability, and finally use the CIOU as the loss function to improve the positioning accuracy. The average accuracy (mAP) of the improved model on the SHWD dataset reaches 93.86%, and the processing speed (FPS) reaches 49. Compared with the original YOLOv5s, the amount of parameters is reduced by 13.33% without reducing the mAP, and the size of the model is reduced. 26.6%, processing speed increased by 22.5%. The experimental results show that it can effectively reduce the amount and size of model parameters and meet the real-time detection requirements of embedded devices.

Keywords: Safety helmet detection · YOLOv5s algorithm · GhostNet · ECA-Net

1 Introduction

Proper wearing of safety caps is an important guarantee for the life and property of construction workers. It mainly protects the head against falling, striking and collision of high-altitude objects, and has the functions of buffering, dampening and dispersing shocks. Due to the complex environment of the construction site and the high risk factor, it is difficult for the supervisors to achieve all-round, whole-process and real-time safety supervision and management, and there are problems such as high labor management costs, low supervision efficiency, and poor timeliness. Regarding the wearing detection of helmets, there are mainly traditional machine learning methods and methods based on deep learning. The detection of traditional methods needs to rely on manual labor and cannot meet the current detection needs.

© The Author(s), under exclusive license to Springer Nature Singapore Pte Ltd. 2023
Q. Liang et al. (Eds.): AIC 2022, LNEE 871, pp. 95–102, 2023.
https://doi.org/10.1007/978-981-99-1256-8_12

Detection algorithms based on deep learning can effectively solve the above problems. Wu Di et al. [6] optimized the network structure based on the YOLOv3 model in three aspects: target dimension clustering, multi-scale detection, and dense connection, and proposed a helmet wearing detection method based on the OpenPose algorithm. Zhong Zhifeng et al. [9] replaced the YOLOv4 backbone network with the Mobilenetv3 structure and proposed a lightweight target detection algorithm (ML-YOLO). In this paper, GhostNet is used to replace the backbone feature extraction network of YOLOv5s to reduce the amount of network parameters, and a lightweight module attention mechanism ECA-Net is introduced into the backbone network C3 module to improve the feature extraction ability. Real-time detection requirements of type equipment.

2 Related Algorithms

2.1 YOLOv5 Algorithm Principle

The YOLOv5 target detection algorithm is a lightweight detection model based on the Python framework. Its network model is divided into 4 parts, including the input end, the backbone network (Backbone), the Neck module and the output end. For the input, Mosaic data enhancement, adaptive anchor box calculation, and adaptive image scaling are used. Backbone is the benchmark network for YOLOv5, consisting of Focus, Conv, C3 and Spatial Pyramid Pooling (SPP) modules [8]. Among them, the Focus module first slices the input image to obtain four images. After splicing the images, the number of channels is expanded to 4 times the original number. Finally, the obtained new image is subjected to convolution operation, and finally a $32 \times 320 \times 320$ image is obtained. Feature map. Conv module is the convolution operation unit of YOLOv5, which consists of convolutional layer BN layer and SiLU activation function. The C3 module consists of multiple Bottleneck modules, which is a residual structure. The input features of the SSP module are first convolved by the Conv module and then subjected to a maximum pooling operation with convolution kernel sizes of [8] respectively, the results are concatenated by Concat, and then the results are output through the Conv operation. The structure of FPN + PAN is used in Neck; GIOU_Loss is used as the loss function of the target detection frame in the output end, and the NMS non-maximum suppression mechanism is introduced at the same time. YOLOv5 is constantly updated iteratively, and it contains 4 versions: YOLOv5s, YOLOv5m, YOLOv5l, YOLO5x [9]. The main difference between the different versions is that the depth and width of the feature map are different. The YOLOv5s network is the network with the smallest depth and the width of the feature map in the YOLOv5 series, and it is also the version with the least amount of computation. Its model parameters are about 7.5M, which has low performance requirements and is easy to deploy. It is good to meet the needs of real-time rapid detection of helmets.

2.2 YOLOv5s Algorithm Principle

Aiming at the problems of complex background, strong interference, small and few objects to be detected, and high requirements for detection speed faced by safety helmet detection under on-site monitoring of construction sites. In this paper, the lightweight

YOLOv5s algorithm is used for improvement. The YOLOv5s network mainly uses the C3 structure [5], and the structure is shown in Fig. 1. The C3 structure is divided into two parts, the first part performs the Bottleneck operation, which is a classic residual structure [6]. After a 1×1 and 3×3 convolution operation, the convolution result is added to the input. The other part is reduced by 1×1 convolution to reduce the number of channels by half; the last two are combined for output.

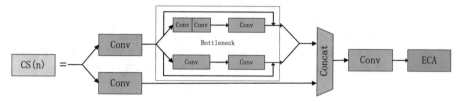

Fig. 1. C3 structure

There are many practical application scenarios of helmet detection, and the detection effect is greatly affected by the environment. In order to eliminate the influence of the background on the detection performance, the mosaic (Mosaic) data enhancement algorithm is used, which is a very important data enhancement method in the YOLOv5s algorithm [4]. Randomly select 4 pictures, first cut them randomly, then splicing them clockwise into one picture, and finally adjust to the set input size, and pass them into the model as new samples. This fusion of different detection backgrounds enriches the contextual information of images, increases the number of small targets, effectively enhances the robustness of the model, and achieves a balance between targets of different scales [7] (Fig. 2).

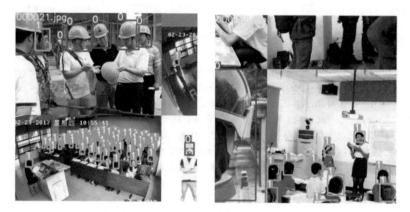

Fig. 2. Mosaic data enhancement renderings

YOLOv5s loss function includes classification loss, localization loss (error between prediction box and GT box) and confidence loss; the total loss function: classification loss + localization loss + confidence loss. The DIOU [5] is proposed by combining the distance between the center point of the ground-truth box and the predicted box and the IOU. Considering the overlapping area, center point distance, and aspect ratio, the CIOU is proposed. YOLOv5s uses CIOU Loss as the loss of bounding box regression. CIOU combines the center point distance, overlap rate, scale and penalty between the predicted box and the real box to make the target box regression more stable. The CIOU calculation formula is as in formula (1), where: $\rho^2(b, b^{gt})$ is the Euclidean distance between the center point of the real box and the center point of the prediction box; c is the diagonal distance of the smallest area that can contain both the prediction box and the real box; v is the similarity of the aspect ratio of the real box and the prediction box. α is the penalty factor. The calculation formulas of α and v are as formulas (2) and (3). w^{gt}, h^{gt}, w, h are the width and height of the ground-truth box and the predicted box, respectively. Therefore, the loss function of CIOU can be defined as Eq. (4).

$$CIOU = IOU - \frac{\rho^2(b, b^{gt})}{c^2} - \alpha v \tag{1}$$

$$\alpha = \frac{v}{1 - IOU + v} \tag{2}$$

$$v = \frac{4}{\pi^2}\left(arctan\frac{w^{gt}}{h^{gt}} - arctan\frac{w}{h}\right)^2 \tag{3}$$

$$LOSS_{CIOU} = 1 - IOU + \frac{\rho^2(b, b^{gt})}{c^2} + \alpha v \tag{4}$$

3   Algorithm Improvement

3.1   GhostNet

GhostNet [6] is a new end-to-side neural network structure proposed by Huawei Noah's Ark Lab. The overall structure is modeled after the structure of MobileNet-v3, but the effect is better than MobileNet. Using Ghost Module as a basic component, it aims to generate more feature maps through cheap operations. Based on a set of original feature maps, a series of linear transformations are applied to generate many features that can extract the required information from the original features at a small cost. Picture. The Ghost module is plug-and-play, and the Ghost bottleneck is obtained by stacking the Ghost modules, and then a lightweight neural network, GhostNet, is built. Based on the Ghost module, the size and channel size of the output feature map of the convolution are not changed, so that the entire calculation amount is reduced. And the number of parameters is greatly reduced. GhostNet is mainly a lightweight network designed for mobile devices. While reducing the amount of calculation and improving the running speed, the accuracy is reduced less, and it is suitable for any convolutional network. Ghost bottleneck is shown in Fig. 3

3.2   ECA-Net

ECA-Net is a very lightweight plug-and-play block based on SE-Net extension, which can tune up the performance of various CNNs, ECA-Net proposes an efficient correlation channel (ECA) module. The module consists of one-dimensional convolutions

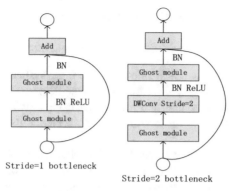

Fig. 3. Ghost bottleneck

determined by nonlinear adaptation, and adopts grouped convolutions to improve the CNN architecture, where high-dimensional (low-dimensional) channels include long (short) convolutions with a fixed number of groups, channel dimension C and volume The kernel size k is proportional to. A nonlinear function is used, and the number of convolution kernels is set to the k power of 2, as shown in formula (5), and the convolution size is shown in formula (6). In the formula |t⁻| |odd represents the nearest odd number, and γ and b are 2 and 1, respectively.

$$C = \emptyset(K) = 2^{(\gamma * K - b)} \tag{5}$$

$$K = \varphi(c) = \left| \frac{\log_2}{\gamma} + \frac{b}{\gamma} \right|_{odd} \tag{6}$$

The structure of ECA-Net is shown in Fig. 4. (1) Global avg pooling generates feature maps of size 1*1*C1*1*C1*1*C; (2) Calculate the adaptive kernel_size; (3) Apply kernel_size to one-dimensional convolution to get the weight of each channel.

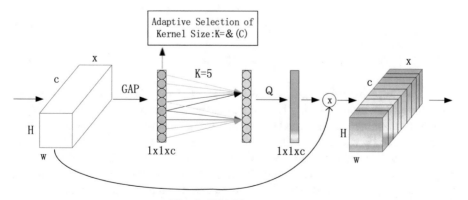

Fig. 4. ECA-Net structure

4   Experimental Results and Analysis

4.1   Experimental Environment Configuration and Data Introduction

The experimental environment is Windows 10 operating system, AMD R7 5800H processor, NVIDIA RTX 3070 graphics card, through the Pytorch deep learning framework to achieve model building, training and verification, using CUDA 11.1 computing architecture, and adding cudnn to the environment to accelerate the computer's Calculate ability.

This experiment selects the safety helmet public dataset SHWD (Safety Helmet Wearing-Dataset), It includes 7581 images with 9044 items of human safety helmet wearing (positive) and 111514 items of normal head (unworn or negative). Divide the dataset and validation set with a ratio of 8:2, and convert the format from XML to txt format, the image resolution size is 640 × 640, the training batch is set to 8, and the IoU threshold is set to 0.5, and all reference models are trained for 300 epochs according to this parameter.

4.2   Evaluation Indicators

In order to verify that the improvement points of the network have an effect on the performance improvement of the model, this paper conducts comparative experiments, using Precision, Recall, Single Category Precision (AP), Average Precision (mAP), Compare the performance of each model with the amount of parameters, calculation amount, and model size. The specific formula is as follows:

$$Iou = \frac{A \cap B}{A \cup B} \tag{7}$$

$$Precision = \frac{TP}{TP+FP} \tag{8}$$

$$Recall = \frac{TP}{TP+FN} \tag{9}$$

$$AP = \frac{TP+TN}{TP+TN+FP} \tag{10}$$

$$mAP = \frac{\sum_{i=1}^{n} AP_i}{n} \tag{11}$$

In the above formula (7), A and B are the prediction frame and the real frame, respectively. In formula (8–10), TP refers to the number of people who wear a helmet and detects the correct number, and FN refers to the number of people who wear a helmet but the detection is wrong. Quantity, TN refers to the number of people who did not wear a helmet and detected the correct number at the same time, FP refers to the number of people who did not wear a helmet but detected errors, n in formula (11) represents the number of categories, i represents the i-th category, AP_i represents the precision of the ith class.

4.3 Analysis of Experimental Results

In order to verify the reliability of the model, under the same hardware conditions and helmet data set, different detection algorithms are used to conduct comparative experiments. The recall rate (recall), average precision (mAP), parameter amount, calculation amount, and model size are evaluated. Common indicators of training model performance and reliability, this paper also uses the above evaluation indicators to evaluate the performance of the helmet wearing detection model. The experimental results are shown in Table 1.

Table 1. Comparative experimental results

Detect Model	Recall	mAP	Parameter quantity	FPS	Model size
YOLOv3	91.80%	92.62%	58.7M	27	236M
YOLOv3-SPP	92.75%	93.13%	20.8M	31	240M
YOLOv5s	85.23%	93.84%	7.5M	40	27M
YOLOv5s + GhostNet	96.00%	93.47%	5.6M	47	12.3M
YOLOv5s + GhostNet + ECA-Net	96.32%	93.86%	6.5M	49	10.7M

It can be seen from Table 1 that the size of the original YOLOv5s model is 27M and the parameter amount is 7.5M. The backbone network is replaced by the end-side neural network architecture GhostNet, and the lightweight ECA-Net attention mechanism is added. After the model is reduced to 12.3 MB, After adding the ECA-Net attention mechanism, the model is reduced to 10.7M, the average accuracy (mAP) reaches 93.86%, and the processing speed (FPS) reaches 49. Compared with the original YOLOv5s, the amount of parameters is reduced by 13.33% without reducing the mAP., the model size is reduced by 26.6%, and the processing speed is increased by 22.5%. Compared with the YOLOv3 and YOLOv3-spp models, the number of parameters is reduced by 88.78% and 68.75%, the model size is reduced by 95.46% and 95.54%, and the mAP is increased by 1.24% and 0.73%. %, to sum up, the improved model ensures a high mAP while greatly reducing the parameter quantity and model size, and the FPS can also meet the real-time monitoring requirements.

In order to more intuitively compare the difference between the improved model and the original model, some of the detection results are visualized. Figure 5 shows the detection effect of the original model of YOLOv5s. It can be seen that the detection effect is not good for dense small targets and occlusions. It is obvious that the small targets behind are seriously missed. Figure 6 shows the detection effect of the improved model. This experimental model For dense small targets and high detection accuracy in the case of occlusion, the above-mentioned missed detection problem is effectively solved, and the desired effect is achieved. In order to make the detection effect of the improved model more intuitive, some detection results are visualized. Figure 5 shows the detection effect of the improved model.

Fig. 5. Improved model detection effect

5   Conclusion

In order to detect the wearing situation of helmets in real time and accurately, in view of the problems of the existing helmet wearing detection algorithm with many parameters, complex network, large amount of calculation, and unfavorable deployment in mobile devices, this paper makes lightweight improvements based on the YOLOv5s model. For the first time, the end-side neural network architecture GhostNet replaces the backbone feature extraction network of YOLOv5s, and adds a lightweight module attention mechanism ECA-Net. The improved model can effectively improve the average detection accuracy rate while meeting real-time monitoring needs and ensuring personnel lives.

References

1. Pang, X., Wang, L.: Recognition method of helmet based on color and shape. Proceedings of the 2017 Smart Grid Information Construction Seminar (2017)
2. Yang, Y., Li, D.: Improved YOLOv5 lightweight helmet wearing detection algorithm. Computer Engineering and Applications: 1–8 [21 Mar 2022]
3. Zhang, J., Qu, P.: Helmet wearing detection method based on improved YOLOv5. Computer Application: 1–11 [21 Mar 2022]
4. Zhao, R., Liu, H.: Helmet detection algorithm based on improved YOLOv5s. Journal of Beijing University of Aeronautics and Astronautics: 1–16 [21 Mar 2022]. https://doi.org/10.13700/j. bh.1001-5965.2021.0595
5. Zheng, Z.H., Wang, P., Liu, W., et al.: Distance - IoU loss: faster and better learning for bounding box regression. In: Proceedings of the AAAI Conference on Artificial Intelligence. New York, NY, USA: IAAA, 2020: 12993-13000
6. Han, K., Wang, Y., Tian, Q., et al.: GhostNet: More Features From Cheap Operations. In: 2020 IEEE/CVF Conference on Computer Vision and Pattern Recognition (CVPR). IEEE (2020)
7. Cao, Z.H., Shao, M.F., Xu, L., et al.: MaskHunter: real-time object detection of face masks during the COVID-19 pandemic. IET Image Processing **14**(16), 20–32 (2020)
8. Zhong, Z., Xia, Y., Zhou, D., Yan, Y.: Lightweight target detection algorithm based on improved YOLOv4. Computer Application: 1–8 [21 Mar 2022]
9. He, K., Zhang, X., Ren, S., et al.: Deep residual learning for image recognition. In: Proceedings of the IEEE Conference on Computer Vision and Pattern Recognition, pp. 770–778 (2016)

Label Embedding Based Scoring Method for Secondary School Essays

Chao Song, Ge Ren, YinZhong Song, JunJie Liu, and Yong Yang[✉]

College of Computer Science and Technology, Xinjiang Normal University, Urumqi 830054, China
68523593@qq.com

Abstract. Automatic essay scoring techniques can automatically evaluate and score essays, and they have become one of the hot issues in the application of natural language processing techniques in education. Current automatic essay scoring methods often use large pre-trained models to obtain semantic features, which do not perform well in the field of automatic essay scoring because the training corpus does not match the content domain of the essay, and the extraction of features for long essays is not effective. We propose a label embedding-based method for scoring secondary school essays, using a modified bidirectional long- and short-term memory network and a BERT model to extract domain features and abstract features of essays, while using a gating mechanism to adjust the influence of both types of features on essay scoring, and finally automatic scoring of essays through feature fusion. The experimental results show that the proposed model performs significantly better on the essay auto-scoring dataset of the Kaggle ASAP competition, with an average QWK value of 81.22%, which verifies the effectiveness of the proposed algorithm in the essay auto-scoring task.

Keywords: Automatic essay scoring · pre-trained embeddings · semantic enhancement · feature fusion · natural language processing

1 Introduction

Automated Essay Scoring (AES) is a technology that uses linguistic, statistical and natural language processing techniques to automatically evaluate and score written essays. AES has a wide range of applications in the education field, helping markers to improve the accuracy and reduce the workload of markers, and helping markers to eliminate the influence of subjective factors in the marking process. At present, there is still a gap between the accuracy of automatic marking of essays using computer technology and that of manual marking, which is one of the problems that need to be solved in this field.

Essays can be divided into lower grade essays and upper grade essays depending on the age of the writer. Unlike adult writers, the vocabulary used in the lower grades is relatively basic and the structure of the essay is simple, even though the content of the essay is sometimes long, the sentences are still mainly simple and the grammatical structure used is more standardized. Therefore, there are significant differences in semantic

© The Author(s), under exclusive license to Springer Nature Singapore Pte Ltd. 2023
Q. Liang et al. (Eds.): AIC 2022, LNEE 871, pp. 103–118, 2023.
https://doi.org/10.1007/978-981-99-1256-8_13

expression, grammar and syntax between the compositions in the early grades and the texts on the Internet.

The main approach of early AES techniques was to extract the grammatical [1], syntactic [2], and shallow semantic features of compositions by means of feature engineering [3], and to use traditional machine learning methods such as logistic regression for automatic scoring of compositions. Such methods cannot extract the deep semantic features of compositions and have poor generalization ability. In recent years, deep learning methods have been widely applied in the field of AES, such as CNN [4], LSTM [5], etc., and their performance has been improved to a certain extent, but essay scoring is a complex task that requires the marker to deeply understand the main idea of the essay and evaluate the ideas and wording expressed in the essay, and obviously using a single neural network cannot capture the semantic meaning expressed in the essay comprehensively, so The model performance of neural networks is therefore somewhat constrained. Pre-trained models have achieved SOTA performance in many natural language processing tasks, but the performance of pre-trained language models represented by BERT models applied to AES domain is not satisfactory, mainly because the training corpus of pre-trained language models such as BERT is mostly from BooksCorpus and English Wikipedia, which is different from the samples in AES tasks in terms of semantic expressions, logical structures, and language styles are very different from those in the AES task. Second, the corpus used in the pre-training model has a looser line structure, the semantic expressions are more random, and the logical structure is much less clear than that of the lower grade compositions. In addition, the maximum input length of the BERT model is 512, and the text exceeding this length will be truncated, thus causing the loss of semantic information. Using a single-level semantic embedding alone will not only not improve the performance of AES tasks, but will also make it difficult to fine-tune the parameters of the pre-trained model effectively due to the insufficient amount of data and the "biased" information in the pre-trained corpus, resulting in lower performance [6].

To address these problems, we propose a label embedding-based automatic scoring method (LEM) for lower grade essays. For the pre-training model based on large corpus data for training, we use the abstract feature extraction (AFE) unit of the feature fusion layer to extract the deep semantic features of the composition text. Also, we employ a Domain Feature Extraction (DFE) unit consisting of a modified bidirectional long and short-term memory network to adapt to the domain characteristics of lower grade students' compositions and to solve the long text modeling problem. In addition, we incorporate a gating mechanism to adjust the weights of the extracted features from the AFE module and the DFE module, enabling the deep semantic features to be beneficially adjusted to obtain domain features that are more relevant to the composition scores. Since the parallel use of maximum pooling and average pooling can fuse features at different levels [7], we also use a dual pooling approach to further improve the performance of the model. Our contributions are as follows:

(1) We use the AFE module to extract abstract features from composition texts and the DFE module to extract AES domain features, and the DFE module is a useful complement to the AFE module.

(2) We propose a simple and effective method to improve the performance of the pre-trained model on automatic essay scoring tasks by category-labeled text (e.g., "excellent", "good", "fair ", "poor", "poor") to enhance contextual representation learning while not changing the original encoder network structure.

(3) We employ a gating mechanism to adjust the weights of the features extracted by the pre-trained and self-trained models, and employ a maximum and average pooling strategy to fuse different features for automatic scoring of essays.

(4) Our proposed LEM model achieves SOTA performance on the open dataset of the Kaggle ASAP competition, and the experimental results show that the LEM model can significantly improve the performance of automatic scoring of essays.

2 Related Work

Research on automatic essay scoring started in the 1960s, and early studies mainly used artificial features and traditional machine learning methods. In recent years, neural network models and pre-trained language models have achieved better performance in the field of AES. In general, the two most important aspects of AES technical-level development are feature extraction and model construction. The selection of features has gradually transitioned from a single shallow text feature to a trend toward diverse and discriminative features that reflect the linguistic, structural, and content aspects of compositions. At the same time, the feature extraction methods gradually transitioned from basic statistical methods to hybrid methods combining new NLP technologies. We will introduce the development of AES from three aspects: feature engineering, neural network model, and pre-trained language model.

(1) Automatic scoring of essays based on feature engineering. Constructing artificial features mainly considers the features of composition such as topic, content, structure and language. Ming-Yang Liu and Bing Qin et al. [8] proposed a heuristic approach to automatically identify prose as well as metaphorical rhetorical devices, the shortcoming of this approach is that it cannot cope with the existence of rewriting in quotations, which affects the recognition accuracy. Cai Li et al. proposed a method to calculate the difficulty coefficient of word use [9], using the sum of the difficulty coefficients of all words used in the text as the writing level characteristic of the composition. Ming Zhou et al. [10] extracted fine-grained features such as structure, vocabulary and syntax of chapters to construct an automatic scoring model of chapter structure based on linear regression. Yang Zhengxiang and Liu Jie [11] used Markov chain model and sentence ranking correlation algorithm, incorporated Word2Vec and synonym word forest for semantic expansion, and analyzed the inter-sentence coherence and paragraph logic rationality of compositions.

(2) Neural network-based automatic scoring of essays, which uses an end-to-end approach to train the extracted features. Dong et al. [7] used a hierarchical convolutional neural network model based on an attention mechanism to automatically learn features from two levels, sentence structure and text structure, and score essays. The SkipFlow neural network model proposed by Tay et al. [12] can better Liu et al. [13] proposed a two-stage learning framework that first uses a deep neural network

to obtain the semantic representation of the composition and then passes it to the XGboost classifier to predict the composition score. Zhou Xianbing [14] et al. constructed an essay scoring model based on multi-level semantic features from the perspective of hierarchical semantics and achieved better performance.

(3) Pre-training based automatic scoring of essays, which first trains a model on a large dataset with annotations and later performs the model as a pre-trained model for downstream tasks. Rodriguez et al. [15] applied BERT and XLNet models to the field of automatic scoring of essays and achieved good results. Li et al. [16] proposed a cross-topic knowledge transfer deep neural network model and achieved SOTA performance in cross-topic essay scoring. Ormerod [17] proposed an efficient language model based on Transformer structure to address the problem that large-scale language models are difficult to train, which can predict essay scores accurately and efficiently.

3   Method

Our proposed LEM model can be mainly divided into a feature extraction module and a feature fusion module, and the model diagram is shown in Fig. 1. The feature extraction layer contains the AFE module and the DFE module. First, in order to better extract deep semantic features, we freeze the weights of the first eleven layers of the BERT model and only fine-tune the weights of the last layer. Secondly, to better adapt to long texts and reduce "bias" we use a bidirectional long and short-term memory network to extract text features. In order to make the network interact more thoroughly with the text embedding inherent in the BERT model, we also classify and label the scores of different essays and incorporate them into the training sample. Next, in order to adjust the contribution of different features to the essay scores, the feature vectors extracted by the DFE module are multiplied by a linear layer with the fused vectors to adjust the

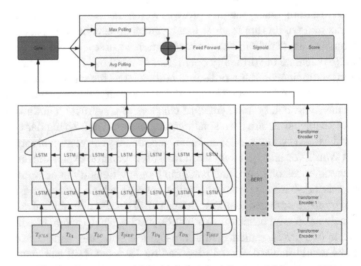

Fig. 1. Our proposed fused label embedding model framework

feature weights extracted by the DFE module and the AFE; in the feature fusion stage, we use both average pooling and maximum pooling strategies to fuse the text features; finally, we fuse the fused vectors The final score of the essay is obtained by passing the fused vectors through the linear layer and Sigmoid activation function.

3.1 Feature Extraction Module

The semantic content of an essay and its expressive effect play an important role in scoring the essay [18]. The pre-trained model is trained from a large-scale corpus containing rich semantic information, which enables the neural network model to acquire text structure and underlying semantic features, thus enabling the model to understand the information embedded in the composition at a deeper level [11]. The pre-trained BERT model can learn rich external knowledge and deep semantic expressions from a large amount of unstructured text data, thus obtaining deep features of the text data [22].

The self-attention mechanism in the BERT model connects any two words in a sentence by computing the semantic similarity and semantic features of each word in the sentence with other words to obtain better distance dependence, and this section uses the multi-headed self-attention proposed by Ashish [19]. For giving a specific Query, Key, Value all have $Q \in R^{(n_1 \times d_1)}$, the $K \in R^{(n_1 \times d_1)}$, the $V \in R^{(n_1 \times d_1)}$, we use the dot product operation to compute the attention parameter. The formula is as follows.

$$Attention(Q, K, V) = softmax\left(\frac{QK^T}{d_1}\right)V \tag{1}$$

where d_1 is the number of neurons in the hidden layer.

The multi-headed attention mechanism maps the input vector X to Query, Key, and Value by linearly varying. in our task, key = value. Subsequently, the model learns the semantic features between words by the attention mechanism, set i be the number of attention heads in the attention mechanism. For the first i attention head, let the parameter matrix $W_i^Q \in R^{(n_1 \times \frac{d_1}{i})}$, $W_i^K \in R^{(n_1 \times \frac{d_1}{i})}$, $W_i^V \in R^{(n_1 \times \frac{d_1}{i})}$, we use the dot product operation to calculate the semantic features between them. The formula is as follows.

$$M_i = Attention(QW_i^Q, KW_i^K, VW_i^V) \tag{2}$$

The vector representation obtained from the multiple attention mechanism is stitched to obtain the final feature representation.

$$H^s = concat(M_1, M_2, M_3)W_o \tag{3}$$

where W_o is the parameter matrix, and *concat* denotes the connection operation.

The use of pre-trained word vectors in the composition scoring domain is not effective in improving the performance of the model due to insufficient data volume and the "bias" of pre-trained word vectors [19]. In order to better extract features from the training corpus in the composition domain and adapt to long texts, we used a modified two-way long and short-term memory network. Inspired by the BERT model that fuses textual representations of samples with location vectors [11], we classified different essay scores into five categories of labeled texts. Scores of 80% or more of the total composition score were labeled as "excellent"; scores of 60%–80% of the total composition score were labeled as "good"; scores of 40%–60% of the total composition score were labeled as "fair"; and scores of 80%–80% of the total composition score were labeled as "good". "Fair"; 20%–40% of the total composition score is marked as "Poor"; 0%-20% of the total composition score is marked as "Poor". Next, the marked text is stitched with the original sample as input. The formula is as follows.

$$
Label \begin{cases} \text{"Excellent"} & 0.8 \leq \dfrac{Scoring_{current}}{Scoring_{sum}} \leq 1 \\[2mm] \text{"Good"} & 0.6 \leq \dfrac{Scoring_{current}}{Scoring_{sum}} < 0.8 \\[2mm] \text{"Fair"} & 0.4 \leq \dfrac{Scoring_{current}}{Scoring_{sum}} < 0.6 \\[2mm] \text{"Poor"} & 0.2 \leq \dfrac{Scoring_{current}}{Scoring_{sum}} < 0.4 \\[2mm] \text{"Bad"} & 0 \leq \dfrac{Scoring_{current}}{Scoring_{sum}} < 0.2 \end{cases} \tag{4}
$$

$$
T_i = concat(Ori_i, Label_i) \tag{5}
$$

where for the first i sample, the $Label$ is the corresponding labeled text for different essay scores, and T_i is the semantic representation of the essay.

3.2 Feature Fusion Module

To better mitigate the semantic bias caused by the word vectors output by the pre-trained model, we used a gating mechanism to adjust the weights of AFEs and modules. The formula is as follows.

$$
B_i = Sigmoid(W \cdot D + b) \tag{6}
$$

$$
X^{feature} = Concat(X_{sh}, X_{dp}) \tag{7}
$$

$$
S_i = B_i \odot X^{feature} \tag{8}
$$

where D is the random embedding representation of the composition, b is the bias term, X_{sh}, X_{dp} are the shallow features and deep features of the composition, respectively. $\odot$ is the dot product operation.

In order to capture the salient features and average semantics in the composition, we employ both average pooling and maximum pooling strategies to fuse the text features The formula is as follows.

$$
P_{max} = Pooling\_max(S_1' + S_2' + \cdots + S_L') \tag{9}
$$

$$
P_{avg} = Pooling\_average(S_1' + S_2' + \cdots + S_L') \tag{10}
$$

$$P_W = Concat(P_{max} + P_{avg}) \tag{11}$$

of which P_{max} and P_{avg} are the eigenvectors of the two pooling layers, and L is the composition length, and P_W is the feature vector after fusion.

3.3 Essay Scoring Module

In summary, we used both BERT and a modified bidirectional long and short-term memory network to extract features, a gating mechanism to adjust the feature weights of different modules, and dual pooling for feature fusion to finally obtain a vector representation of the essay using *Sigmoid* activation function to get the score of the essay. The formula is as follows.

$$P_i = Relu(W \cdot P_w + b) \tag{12}$$

$$Score = Sigmoid(W' \cdot P_i + b') \tag{13}$$

of which W and W' are the parameter matrices, the b and b' are the bias terms, and *Relu* and *Signoid* is the activation function, and *Score* is the final score of the essay.

For the scoring process, we use the mean square error as the loss function. The formula is as follows.

$$MSE = \frac{1}{n} \sum_{i=1}^{n} (y_i - \hat{y}_i)^2 \tag{14}$$

where n is the number of essay samples, and each essay sample x_i is the true score of y_i and the model's prediction of x_i The predicted value of the score is y_i.

4 Experiments and Analysis of Results

This section first introduces the dataset and the evaluation method, followed by the experimental parameters, then a detailed comparison of the performance of our proposed model and the baseline model, and finally an ablation experiment to analyze the performance of the different modules in detail.

4.1 Data Sets and Evaluation Metrics

We use publicly available datasets from the Kaggle ASAP (Automated Student Assessment Prize) competition, which is widely used in the field of automated essay scoring. The ASAP dataset is divided into eight subsets based on the subject matter of the essays, and each subset contains one essay writing requirement and multiple essays on related topics, all written by students in grades 7–10. The details of the datasets are shown in Table 1.

Table 1. ASAP Data set information

Essay Data Subset	Number of essays	Average length	Score range	Text Length
D1	1783	350	2–12	600
D2	1800	350	1–6	600
D3	1726	150	0–3	300
D4	1772	150	0–3	300
D5	1805	150	0–4	300
D6	1800	150	0–4	400
D7	1569	250	0–30	500
D8	723	650	0–60	800

To maintain consistency with the baseline model, we use the quadratic weighted kappa coefficient (QWK) as an evaluation index. QWK is a consistency evaluation index to assess whether the model results are consistent with the actual results. Assuming that the papers are classified into N levels, the QWK is calculated as follows.

$$QWK = 1 - \frac{\sum W_{i,j} O_{i,j}}{\sum W_{i,j} E_{i,j}} \tag{14}$$

$$W_{i,j} = \frac{(i-j)^2}{(N-i)^2} \tag{15}$$

where O is a n histogram matrix of order, and $O_{i,j}$ denotes an expert score of i and the model score is j the number of essays, and $W_{i,j}$ denotes the quadratic weighting matrix based on the difference between expert and model scores, and $E_{i,j}$ denotes the number of essays with an expert score of i and the probability that the model score is j the product of the probabilities of $E_{i,j}$ and $O_{i,j}$ are normalized.

4.2 Experimental Configuration

The test set for the Kaggle ASAP competition is not yet publicly available, and similar to references [4, 20, 21], we used a 5-fold cross-validation approach to evaluate the proposed model in our experiments, with 60% of the training data, 20% of the validation set, and 20% of the test set in each fold.

In the training process, the input to the AFE module is the BERT word vector [11] and the input to the DFE module is the random initialization vector. The number of layers of the two-way long and short-term memory network is 2, and the output space dimension is 64. The number of neurons in the fully connected layer is 100. The optimization function is Nadam, the decay rate is set to 0.1, and the learning rate is set to 0.00005. an early stop mechanism is used during the training period to prevent overfitting. In addition, depending on the length of the dataset, different maximum text lengths were used for the input to the AFE module, up to 512.

4.3 Analysis of Results

To verify the validity of the proposed method, we compared it with the following baseline methods.

CNN, LSTM [5] Methods: Convolutional neural networks or long and short-term memory networks alone were used to extract essay features and to score the essays.

SkipFlow LSTM [6]: The SkipFlow mechanism is added to the LSTM network, which uses the semantic relationships between the LSTM hidden layers as auxiliary features for essay scoring.

CNN + LSTM [5]: Using the integrated learning method, the prediction results of 10 CNN models and 10 LSTM models are averaged and used as the final prediction results.

CNN + LSTM + ATT [1]: A hybrid neural network using CNN and LSTM with an attention mechanism.

Topic + BiLSTM + ATT [22]: The semantic representation of the essay and the semantic representation of the prompt are obtained using a bidirectional long and short-term memory network and an attention mechanism, respectively, and then the topic relevance of the essay is obtained using vector multiplication, and finally the topic relevance is incorporated into the semantic vector of the essay for essay scoring.

BERT [3]: Use BERT model to extract text features for automatic scoring of essays.

BERT + XLNet [17]: Using an integrated learning approach, the prediction results of six different BERT models and six different XLNet models are averaged and used as the final prediction results.

Stacking [24]: Firstly, different encoding methods are used to obtain the lexical vectors of the text. Secondly, features of the seven aspects of the text are fully extracted.Finally, a model based on the stacking method is proposed.

Electra + Mobile-BERT [20]: Applying the efficient language models Electra [23] and Mobile-BERT [24] to automatic scoring of essays, while using integrated learning to further improve scoring performance.

LEM: Our proposed automatic essay scoring model based on label embedding.

Table 2. Performance comparison between our method and other baseline methods

模型	D1	D2	D3	D4	D5	D6	D7	D8	Avg QWK (%)
CNN[*]	79.70	63.40	64.60	76.70	74.60	75.70	74.60	68.70	72.25
LSTM[*]	77.50	68.70	68.30	79.50	81.80	81.30	80.50	59.40	74.63
SkipFlow LSTM[*]	83.20	68.40	69.50	78.80	81.50	81.00	80.00	69.70	76.51
CNN + LSTM[*]	82.10	68.80	69.40	80.50	80.70	81.90	80.80	64.40	76.08
CNN + LSTM + ATT[*]	82.20	68.20	67.20	81.40	80.30	81.10	80.10	70.50	76.38

(*continued*)

Table 2. (*continued*)

模型	D1	D2	D3	D4	D5	D6	D7	D8	Avg QWK (%)
Topic + BiLSTM + ATT[*]	82.70	69.60	69.10	81.60	81.10	82.30	80.90	70.70	77.30
BERT[*]	79.20	67.99	71.52	80.08	80.59	80.53	78.51	59.58	74.75
BERT + XLNet[*]	80.78	69.67	70.31	81.90	80.82	81.45	80.67	60.46	75.78
Stacking[*]	**86.30**	**71.90**	69.00	79.1	80.40	80.50	78.20	73.70	77.40
Electra + Mobile-BERT[*]	83.10	67.90	69.00	82.50	81.70	82.20	84.10	74.80	78.20
LEM	85.52	78.13	**74.00**	**85.20**	**81.40**	**83.22**	**84.58**	**77.66**	**81.22**

A comparison of our proposed label embedding-based scoring method for secondary school essays with previous work is presented in Table 2, and the experimental results show that:

(1) The low performance of training separate CNN or LSTM networks in an end-to-end form indicates that separate neural network models are not good at extracting semantic features of longer documents and using them for essay scoring.LSTM has a small performance improvement after adding the SkipFlow mechanism to model the semantic relationships between hidden layers. Compared to LSTM, the performance improvement is only significant for D1 and D8 subsets, while the performance is comparable for other subsets. The overall performance of BERT alone is 0.12% higher than the LSTM model, but still does not surpass the effect of SkipFlow LSTM, indicating that large pre-trained language models do not work well for topic-specific compositions.

(2) The hybrid model with CNN and LSTM can effectively improve the overall performance of essay scoring, with improvements in all six subsets compared to the single model; the average performance of the hybrid model with Electra + Mobile-BERT improves 3.45% over the single BERT model; and, the model performance is further improved by using the attention mechanism. This indicates that the performance of the hybrid model is better than the single network model in most of the data sets, and the attention mechanism can effectively capture the semantics of the composition, which is crucial for the performance improvement of the model.

(3) For long text datasets, the performance advantage of the pre-trained model over the traditional neural network is not significant, and for D1, D2 and D7 datasets, Electra + Mobile-BERT with SkipFlow LSTM model does not achieve a great improvement. This indicates that since the long text is truncated in the pretrained model, which leads to the loss of semantic information, simply using the semantic embedding of the pretrained model will not improve the performance of the AES task, but will make it difficult to effectively fine-tune the parameters of the upstream pretrained model due to the insufficient data samples and the "biased" information of the pretrained corpus. This results in lower performance.

Our proposed method achieves SOTA performance on all data subsets with an average quadratic weighted kappa value of 81.22%, which is 3.11% better than the Electra + Mobile-BERT method, especially on the longer composition datasets D1, D2, D7, and D8, and 10.23% better on the D2 dataset. Compared with other deep pre-trained language models, our proposed model has fewer parameters, is faster to train, and achieves better performance on longer composition datasets. The experimental results show that the combination of self-training and pre-training approaches helps the model to better understand the overall writing level of compositions.

4.4 Feature Extraction Layer Analysis

In order to verify the impact of AFE module and DFE module on the model performance, we compare the performance of the two modules in detail. The experimental results are shown in Fig. 2, where red represents using only the AFE module, yellow represents using only the DFE module, and green represents using both modules.

The experimental results show that, except for the D5 dataset, the performance of extracting text features using the DFE module alone is significantly better than that of using the AFE module alone. This is mainly due to the fact that the self-training embedding approach uses a random initialization method, while the pre-training embedding approach includes a certain semantic representation, so the pre-training embedding approach can better adapt the model parameters to the representation of the downstream composition corpus in the task. However, the difference between the pre-trained corpus and the task text makes the pre-trained model also carry the "bias" of its own corpus. This "bias" does not improve the performance of the model in the task, but causes distortions in the model's understanding of the written expressions of the compositions in the training set. We note that using both AFE and DFE to extract essay features on the

Fig. 2. Model performance with different embeddings

D2, D7, and D8 datasets significantly outperforms using one module alone. This further indicates that using a combination of self-training and pre-training not only enables the embedded word vectors to have deep semantic expressions, but also facilitates the incorporation of semantic information beyond the maximum text length processed by the pre-trained model into the model. At the same time, using the self-trained embedded word vectors as biases can effectively correct the "biased" information in the pre-trained model.

To verify the effectiveness of the label embedding method, we compare the performance of label embedding for different modules in detail. The experimental results are shown in Fig. 3. Blue represents no label embedding; orange represents label embedding for AFE module only; gray represents label embedding for DFE module only; and yellow represents label embedding for both modules at the same time.

The experimental results show that the label embedding approach in the AFE module only outperforms the other two label embedding schemes in most cases. We believe that the reason for this phenomenon is due to the fact that the Next Sentence Prediction (NSP) task, which is used in BERT to learn sentence-level representations, is to concatenate two natural language sentences with a [SEP] label. When we splice a labeled text with a training sample, the [SEP] token combines a non-natural language sequence with a natural language sentence. This difference may cause a bias between pre-training and fine-tuning in BERT, resulting in a degradation of performance. Therefore, simply adding a tagged text as a prefix to the AFE module can be more effective in providing information gain, resulting in more consistent improvements.

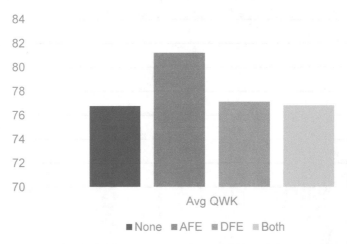

Fig. 3. Performance comparison of different label embedding

When the BERT model is fine-tuned for downstream tasks, it is often necessary to freeze some parameters to obtain better performance if the fine-tuned dataset differs significantly from the original training dataset. To verify the effect of freezing the weights of different layers of network parameters in the BERT model on the performance of the essay scoring task, we gradually freeze each layer of BERT in order to obtain the best performance. As shown in Fig. 4, all_train indicates that all parameters of the BERT model are open for training; embedding indicates that only the parameters of the embedding layer of the BERT model are frozen; emb + 4 indicates that the first four layers of the embedding layer and encoder are frozen; emb + 8 indicates that the first eight layers of the embedding layer and encoder are frozen; emb + 10 indicates that the first eight layers of the embedding layer and encoder are frozen; emb + 10 indicates that the parameters of the BERT model are frozen. Emb + 8 means freeze the first eight layers of embedding layer and encoder; emb + 10 means freeze the first ten layers of embedding layer and encoder; emb + 11 means freeze the first eleven layers of embedding layer and encoder; emb + 12 means freeze the first twelve layers of embedding layer and encoder; onlyBert_emb + 11 means use only freeze the first eleven layers of embedding layer and encoder. Emb + 11 means freeze the first eleven layers of embedding and encoder; emb + 12 means freeze the first twelve layers of embedding and encoder; onlyBert_emb + 11 means use only the BERT model that freezes the first eleven layers of embedding and encoder.

As shown in Fig. 4, the experimental performance is poor if the parameters of the DFE module are not frozen. This is mainly because the DFE module uses the BERT_Base version, which has nearly 110 million parameters, and our dataset has fewer samples involved in training, which not only fails to fine-tune the parameters of the model well, but also destroys the data distribution in the network layer and exacerbates the bias of the parameters in the pre-trained model for the semantic representation. Secondly, when freezing the shallower layers, the model performance fails to reach the best, which is due to the fact that the word embedding representations obtained by the DFE module, whose training corpus are all from BooksCorpus and English Wikipedia, have large differences in semantic expressions, logical structures, and linguistic styles from the samples in our AES task, and thus the performance is not satisfactory. The experimental results show that the model performs best when the embedding layer and the first eleven layers of encoder of the AFE module are frozen, and the performance decreases when the parameters are

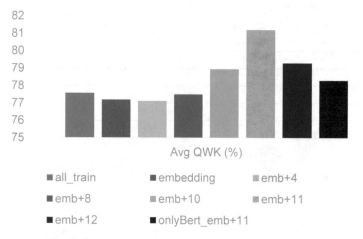

Fig. 4. Freeze model performance of different layer parameters

all frozen instead. This is due to the fact that freezing the weight parameters of the final encoder layer prevents the upstream pre-trained model parameters from fitting the AES task in a targeted manner, making the model performance poor. Therefore, the method of freezing the first 11 layers of the embedding layer and encoder of the BERT model can effectively solve the problem of semantic bias due to insufficient task data samples and poor relevance of the training corpus.

4.5   Feature Fusion Layer Analysis

Our proposed model includes two modules, AFE and DFE, to extract the semantic features of compositions, and both contribute differently to the classification performance of the model, and we experimentally compare the model performance with and without the gating mechanism. As shown in Fig. 5, blue is without the gating mechanism and red is with the gating mechanism. The experimental results show that the gating mechanism can be used as a way to adjust the parameter weights of the AFE module and the DFE module. When the AFE module is combined with the DFE module, it is essentially a repetitive splicing of two sentence vectors that imply similar semantics. In the prediction process, sometimes the deep semantic knowledge possessed by the AFE module is required, and sometimes the bias brought by the DFE module for the weights of the pre-trained model is required. Therefore, the gating mechanism can be used to adjust the semantic knowledge weights at different levels. The gating mechanism not only enables the model to extract multiple semantic features, but also eliminates the redundant information brought by the combination of the two approaches to the sentence itself, thus improving the performance of the model.

The pooling layer can speed up the computation to prevent overfitting by using maximum pooling to obtain the most significant feature expression of each word vector in the sentence and average pooling to obtain the average semantic expression of each word vector in the sentence [5]. We experimentally compare the performance of the two pooling methods in automatic scoring of essays. As shown in Fig. 6, blue represents the scheme without pooling; orange represents the scheme with average pooling only; gray represents the scheme with maximum pooling only; and yellow represents the scheme with double pooling. The experimental results show that the performance of the model without pooling degrades significantly, and using average pooling or maximum pooling alone can improve the model performance. In this task, maximum pooling outperforms

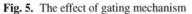
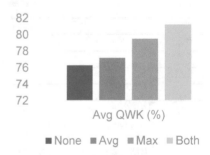

Fig. 5. The effect of gating mechanism   **Fig. 6.** The impact of different pooling schemes

average pooling, and the combined use of the two pooling strategies can improve the model performance to some extent.

5   Conclusion

We use a combination of pre-training and self-training modules for feature extraction, which can add semantic bias to the "bias" of the pre-training model. Also, in order to better extract features in the composition domain, we enhance the training samples by classifying and labeling the composition scores. Second, we adopt a gating mechanism to adjust the weights according to the contribution of different modules to the classification performance of the model; in addition, we use a dual-pooling strategy to obtain different semantic expressions of essays. Experimental results on the open dataset of automatic essay scoring in the Kaggle competition show that our proposed label embedding-based secondary school essay scoring model can significantly improve the performance of the automatic essay scoring task, achieving the best known performance so far.

There are multiple types of essays, and the semantic features of different types of essays vary greatly, making it difficult to solve the automatic scoring problem for all essay types with a unified model. In the next work, we will further explore more domain features of essays into the neural network model. In addition, for problems such as the long text length of essays, how to perform long text processing on pre-trained models is also a direction worthy of attention.

Acknowledgment. This work was supported by XinJiang Uygur Autonomous Region Natural Science Foundation Project (No. 2021D01B72, 2022D01A99), the Natural Science Foundation of China (No.62167008, 62066044), and National Natural Science Foundation young investigator grant program (No. 62006130).

References

1. Ghasemi, N., Aliannejadi, M., Hiemstra, D.: BERT for Target Apps Selection: Analyzing the Diversity and Performance of BERT in Unified Mobile Search (2021)
2. Stefanick, M.L., Anderson, G.L., Margolis, K.L., et al.: Effects of conjugated equine estrogens on breast cancer and mammography screening in postmenopausal women with hysterectomy. JAMA **295**(14), 1647 (2006)
3. Rodriguez, P.U., Jafari, A., Ormerod, C.M.: Language models and Automated Essay Scoring (2019)
4. Chen, M., Li, X.: Relevance-based automated essay scoring via hierarchical recurrent model. In: 2018 International Conference on Asian Language Processing (IALP) (2018)
5. Alikaniotis, D., Yannakoudakis, H., Rei, M.: Automatic text scoring using neural networks. arXiv e-prints (2016)
6. Dong, F., Yue, Z., Jie, Y.: Attention-based recurrent convolutional neural network for automatic essay scoring. In: Proceedings of the 21st Conference on Computational Natural Language Learning (CoNLL 2017) (2017)
7. Dong, F., Zhang, Y.: Automatic Features for Essay Scoring -An Empirical Study (2018)
8. Taghipour, K., Ng, H.T.: A neural approach to automated essay scoring. The 2016 Conference on Empirical Methods in Natural Language Processing (EMNLP 2016) (2016)

9. Zhou, X., Yang, L., Fan, X., et al.: Self-training vs Pre-trained Embeddings for Automatic Essay Scoring (2021)
10. Yin, W., Schütze, H., Xiang, B., et al.: ABCNN: Attention-Based Convolutional Neural Network for Modeling Sentence Pairs. Computer Science (2015)
11. Devlin, J., Chang, M.W., Lee, K., et al.: BERT: Pre-training of Deep Bidirectional Transformers for Language Understanding (2018)
12. Tay, Y., Phan, M.C., Tuan, L.A., et al.: SkipFlow: Incorporating Neural Coherence Features for End-to-End Automatic Text Scoring (2017)
13. Ormerod, C.M., Malhotra, A., Jafari, A.: Automated Essay Scoring Using Efficient Transformer-Based Language Models (2021)
14. Xia, L., Mc, A., Jyn, C.: SEDNN: shared and enhanced deep neural network model for cross-prompt automated essay scoring. Knowledge-Based Systems **210**, 106491 (2020)
15. Rodriguez, P.U., Jafari, A., Ormerod, C.M.: Language Models and Automated Essay Scoring. arXiv preprint arXiv:1909.09482 (2019)
16. Liu, J., Yang, X., Zhao, L.: Automated Essay Scoring based on Two-Stage Learning (2019)
17. Yi, T., Phan, M.C., Tuan, L.A., et al.: SkipFlow: Incorporating Neural Coherence Features for End-to-End Automatic Text Scoring (2017)
18. Fei, D., Yue, Z.: Automatic features for essay scoring – an empirical study. In: Conference on Empirical Methods in Natural Language Processing (2016)
19. Vaswani, A., Shazeer, N., Parmar, N., et al.: Attention Is All You Need. arXiv. arXiv (2017)
20. Page, E.B.: Grading essays by computer: progress report. In: Proceedings of the invitational Conference on Testing Problems (1967)
21. Shen, D., Wang, G., Wang, W., et al.: On the Use of Word Embeddings Alone to Represent Natural Language Sequences (2018)
22. Zupanc, K., Bosnic, Z.: Automated essay evaluation with semantic analysis. Knowledge-Based Systems **120**(Mar.15), 118–132 (2017)
23. Clark, K., Luong, M.T., Le, Q.V., et al.: ELECTRA: Pre-training Text Encoders as Discriminators Rather Than Generators (2020)
24. Li, C., Lin, L., Mao, W., Xiong, L., Lin, Y.: An automated essay scoring model based on stacking method. In: 2022 IEEE 2nd International Conference on Software Engineering and Artificial Intelligence (SEAI), pp. 248–252 (2022)

Research on Wheat Ears Detection Method Based on Improved YOLOv5

Hong Wang[1,2,3], Mengjuan Shi[1,2,3], Shasha Tian[1,2,3(✉)], Yong Xie[1,2,3], and Yudi Fang[1,2,3]

[1] College of Computer Science, South -Central Minzu University,
Wuhan 430074, China
`947330234@qq.com, shashatian77@mail.scuec.edu.cn`
[2] Hubei Provincial Engineering Research Center for Intelligent Management of Manufacturing Enterprises, Wuhan 430074, China
[3] Hubei Provincial Engineering Research Center of Agricultural Blockchain and Intelligent Management, Wuhan 430074, China

Abstract. Wheat ears detection and counting play a crucial role in wheat yield prediction and breeding. In this paper, a deep neural network wheat ears detection method Wheat-YOLOv5 based on improved YOLOv5 is proposed for the problem of low accuracy of traditional wheat ears detection methods. Fusion of the ECANet attention module with the Backbone part of the YOLOv5s network to improve network feature extraction. Using SPConv convolution to replace the original ordinary convolution in the neck layer to improve the model's ability to cope with complex scenes of wheat ears. Using αEIoU Loss instead of GIoU Loss as the target bounding box regression loss function to improve the accuracy of wheat ears localization. The detection average accuracy AP value of the improved algorithm reaches 94.30% and F1 value reaches 91.50%, which has certain recognition accuracy and robustness and can effectively improve the detection of wheat ears in actual agricultural scenes.

Keywords: Wheat ears detection · YOLOv5 · ECANet Attention Modules · Detection layer · αEIoU

1 Introduction

Yield prediction is one of the important aspects of wheat production management, which can provide an important reference for wheat breeding, etc. Early methods of wheat yield prediction mainly include manual field judgment prediction, etc., which have low accuracy and high workload, in addition, in the actual field environment, the growth of wheat is complex and diverse, and the ears of wheat overlap and shade each other, etc., so the detection effect of traditional methods on wheat ears is not ideal. In recent years, machine learning has become an indispensable key technology for China's agriculture to enter into intelligence, and it is

H. Wang—National Ethnic Affairs Commission of the People's Republic of China (Training Program for Young and Middle-aged Talents, MZR20007).

Q. Liang et al. (Eds.): AIC 2022, LNEE 871, pp. 119–129, 2023.
https://doi.org/10.1007/978-981-99-1256-8_14

the core of artificial intelligence. Fan Mengyang et al. [1] used the SVM learning method to accurately extract the wheat ears contour, refine the binary image of wheat ears to obtain the wheat ears skeleton, and infer the yield of wheat ears by calculating the number of wheat ears skeleton and the number of effective intersection points of wheat ears skeleton with an accuracy of 93.1%; the related method has achieved some success, but the calculation process is tedious and less robust. With the rapid development of deep learning, the development of convolutional neural network has brought a new direction for wheat detection, among which the deep learning detection models represented by Faster R-CNN algorithm [2] and YOLO algorithm [3–5] have made significant breakthroughs with their high accuracy and high robustness, etc. Meanwhile, many deep learning-based image analysis techniques have been applied to wheat image localization and detection tasks with good results. Zhang Quanbing et al. [6] introduced a channel attention mechanism and a spatial attention mechanism in the encoding and decoding regions of the original feature extraction network, respectively, in order to improve the detection performance of the network for obscured wheat ears and generate more accurate detection frames for smaller ears that are difficult to detect, in response to the problems such as false detection or missed detection in the detection results. In addition, many methodological research pairs were self-constructed datasets and validated on a limited dataset with little variation in genotype, size, etc. of wheat ears with a single limitation, but in practice, the complex diversity of wheat ears such as differences in genotype makes the task of ears detection more complex. In May 2020, David et al. made public the global wheat ears detection dataset GWHD_2020 [7], which contains 4700 high-resolution RGB images and 193,634 marked wheat ears from different regions of different countries in the world, with a wide range of genotypes, different sizes, overlapping shading, and other complex and diverse scenarios. GWHD_2020 successfully attracted the attention of the computer vision and agricultural science community, further pinning wheat ears diversity as well as labeling reliability for improvement, re-checking and labeling the dataset, supplementing it with 1722 images from four new countries, adding 81,553 wheat ears to strengthen the quantity and quality of the dataset, constituting the dataset GWHD_2021 [8], which is richer and more complex and diverse, and how to effectively improve the detection performance of the network under complex situations such as wheat ears genetic diversity and overlapping occlusion are yet to be solved.

2   Problems with YOLOv5 in the Detection of Wheat Ears

The Yolov5s model has good performance in target detection in terms of accuracy and other indicators, but for the problem solved in this paper, the following defects still exist.

(1) The backbone network in the algorithm has more Bottleneck structures, and the convolution kernel in the convolution operation contains a large number of parameters, leading to a large number of parameters in the recognition

model, which increases the deployment cost of the model. Due to the different genotypes of wheat ears, the smaller wheat ears are more dependent on the shallow features, and the feature extraction is easy to cause certain information loss for the feature extraction of small target objects after passing a large number of conventional convolutions.

(2) The FPN [9] combined with PANet structure is used in Neck network, although the CSP structure [10] is fused in PANet, and the deep feature map and shallow feature map are stitched to fuse different levels of feature information, but the redundancy within the features is not considered, which has some impact on the detection of wheat ears with overlapping occlusion.

(3) In the process of wheat ears detection, when the prediction frame is inside the target frame and the prediction frame size is the same, the boundary frame regression loss function GIoU completely degrades to the IoU loss function, which will lead to the situation of missed detection and false detection and cannot realize the high accuracy of wheat ears localization.

3   Wheat-YOLOv5 Detection Algorithm

In this paper, we take GWHD_2021 wheat ears as the research object, consider the complex scenario of wheat ears, and improve the model to improve the detection effect of wheat ears for the problems of YOLOv5 in the process of wheat ears detection as follows.

(1) Incorporating ECANet attention mechanisms [11] in backbone to enhance the semantic information of the network, reduce noise information, enhance feature propagation and generate more accurate detection frames for dense wheat ears with overlapping occlusions.

(2) Replace the original neck layer ordinary convolution with SPConv [12] to eliminate the redundancy problem existing in the input feature map of wheat ears within the same layer and improve the detection of obscured overlapping wheat ears.

(3) Combining the ideas of EIoU [13]and Alpha IoU [14], αEIoU is proposed as the loss function of bounding box regression to improve the original loss function GIoU [15]and reduce the missed and false detection of wheat ears. Therefore, this paper uses a neural network based on improved YOLOv5 for detection and analysis, aiming to achieve efficient and accurate wheat ears detection.

3.1   Fusion ECA Effective Channel Attention Mechanism

Due to the large variety of wheat ears size and shapes in different regions, the features of different sizes are easily ignored, and the attention mechanism is an important method to improve the target detection performance. In this paper, after feature extraction of images by backbone network, we add a channel attention mechanism ECANet after part of the C3 layer of the backbone network and use a local cross-channel interaction strategy without dimensionality reduction and an adaptive selection of one-dimensional convolutional kernel size, which helps the model to locate the target of interest more accurately and achieve performance improvement.

As shown in Fig. 1, this paper incorporates the ECANet attention module in the backbone network of YOLOv5s to weight and adjust the feature weights of different channels to enhance the network's ability to extract cross-channel information features and improve the accuracy of detection. For the input wheat ears feature map X, all channels are subjected to an averaging pooling operation and then learned by a one-dimensional convolution with shareable weights to translate neurons in such a way that each channel interacts with its k neighboring elements to achieve cross-channel interaction. The information on different sizes of wheat ears obtained in the feature extraction network is fused at 3 different scales, and the attention mechanism is used in the channel dimension for adaptive learning of weights to improve the performance of the model for wheat ears detection.

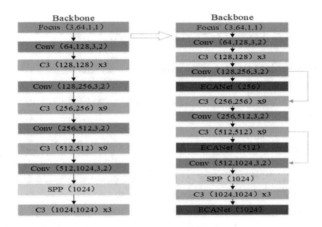

Fig. 1. Converged ECANet module

3.2   Introduction of Separation-Based Convolution

In the process of wheat detection, conventional convolution directly convolves the input feature map to obtain the output feature map when performing the operation, but many features within the same layer have similar but not identical presentation patterns, and these similar features are difficult to determine whether they are redundant or contain important detail information.

In this paper, the SPConv convolution module is used to rethink the feature redundancy problem in conventional convolution, replacing the conventional convolution in the neck layer with the separation-based convolution operation SPConv, reducing the number of parameters while effectively eliminating the redundancy between the feature maps of each layer, and improving the model AP while reducing the model redundancy.

3.3   Improved Loss Function

The YOLOv5 loss function is defined as shown in Formula 1, and its loss function consists of three components, which are confidence loss l_{obj}, classification loss l_{cls}, and position loss l_{box}.

$$Loss = l_{obj} + l_{cls} + l_{box} \tag{1}$$

where the target box position error l_{box} is defined using the GIoU loss function as shown in Formula 2.

$$L_{GIoU} = 1 - IoU + \frac{C - (A \cup B)}{C} \tag{2}$$

YOLOv5s uses GIoU Loss as the bounding box regression loss function to measure the distance between the predicted box and the real box, but when the predicted box is inside the real box and of the same size, resulting in the GIoU value being the same, then GIoU Loss completely degrades to IoU Loss, which cannot distinguish the wheat detection regression, and such a scenario is more common in the actual wheat detection task.

Therefore, this paper combines the ideas of EIoU and Alpha IoU, considers the aspect ratio, overlap area, and distance to the center point between the predicted frame and the real frame, as well as the problem that the gradient is difficult to converge in the actual training process, introduces the power transformation in the existing IoU Loss, idempotents the loss function, and obtains more accurate bounding box regression by focusing more on high IoU targets, and proposes EIoU Loss is proposed as the regression loss function for the target detection task, and the adjustment factor (α is taken as 3 in this paper) is used to make the model more flexible for the regression process of complex wheat detection, reduce the situation of wheat miss detection and false detection, and improve the detection performance of the model for wheat ears. The equation used is as in Formula 3.

$$L_{\alpha EIoU} = 1 - IoU^{\alpha} + \frac{\rho^{2\alpha}(b, b_{gt})}{C^{2\alpha}} + \frac{\rho^{2\alpha}(w, w_{gt})}{C_w^{2\alpha}} + \frac{\rho^{2\alpha}(h, h_{gt})}{C_h^{2\alpha}} \tag{3}$$

where IoU is the ratio of intersection and concatenation between the prediction frame and the real frame, the prediction frame centroid is denoted by b, the target frame centroid is denoted by b_{gt}, and ρ represents the Euclidean distance between the two centroids of the prediction frame and the real frame, C represents the diagonal distance of the smallest outside frame that can contain both the prediction frame and the real frame, C_w and C_h are the width and height of the smallest outside frame, w and h represent the width and height of the prediction frame, and w_{gt} and h_{gt} represent the width and height of the real frame.

4   Experimental Results and Analysis

4.1   Experimental Dataset

The experimental dataset in this paper is derived from the global wheat ears detection dataset GHWD_2021. 6515 high-resolution RGB images with different genotypes from 11 different countries and regions, consisting of 1024×1024 pixel images, are converted to image annotation format by scripting and are also processed for manual verification. 6375 images are selected in total. The global wheat ears dataset has different genotypes, and there will be some differences in the color and size characteristics of wheat ears, etc. Some visualization results of the above scenarios are shown in Fig. 2.

(a) overlap   (b) different growth stages   (c) different sizes   (d) dim

Fig. 2. Some pictures of wheat ears

For algorithm training, 4590 data sets were selected as the training set, 510 as the validation set, and 1275 as the test set. Mosaic data blending enhancement was also performed on the high-resolution wheat images to ensure that the wheat images at different resolutions were used as the dataset for this study. 4 wheat images were randomly selected for each image, randomly scaled and distributed then stitched into one image.

4.2   Experimental Environment and Parameters

The operating system for this experiment is Linux Ubuntu 20.04.1 LTS, the GPU is Tesla K80 with 11 GB video memory, the CPU is Intel(R) Xeon(R) CPU E5-2609 v4 @ 1.70 GHz, and the framework is Pytorch.

The processed data set is fed into the improved neural network for training. In the training phase, the input image size is set to 640 × 640, the batch size is set to 16, the initial learning rate value is 0.01, the momentum and decay weights are set to 0.937 and 0.0005, respectively, to prevent overfitting, and the epoch is set to 100 rounds. The iterative loss curves of the Wheat-YOLOv5 algorithm during the training of the model are shown in Fig. 3, from which it can be seen that the loss of the model decreases rapidly in the first 25 epochs, after which there is a small oscillation and the loss gradually plateaus, and the loss almost ceases to change after 100 training epochs, and the algorithm has reached convergence.

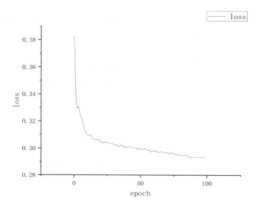

Fig. 3. Loss decline curve during training

4.3   Experimental Analysis

4.3.1   Effect of Improved Methods on Model Performance

In order to verify the effectiveness associated with the improved methods in this paper, ablation experiments were set up to evaluate the effectiveness of each improvement point for this algorithm, and the original YOLOv5 algorithm, the algorithm YOLOv5-ECA after adding the attention module ECANet to the backbone network, the algorithm YOLOv5-αIoU using EIoU as the loss function, the algorithm YOLOv5-SP after introducing SPConv, and the algorithm Wheat-YOLOv5 after fusion of all the improvements proposed in this paper were used for comparison, and the experimental results are shown in Table 1.

Table 1. Comparison of detection performance of different improvement strategies(%)

Algorithms	P	R	F1	AP
YOLOv5	92.40	87.10	89.60	92.30
YOLOv5-ECA	93.00	87.00	89.90	92.80
YOLOv5-αEIoU	92.40	88.10	90.20	93.10
YOLOv5-SP	92.80	89.20	90.96	93.90
Wheat-YOLOv5	**93.30**	**89.70**	**91.50**	**94.30**

Comprehensive results of all experiments can be seen that the improved algorithm Wheat-YOLOv5 proposed in this paper is better than the original YOLOv5 algorithm, and the fusion experimental effect of the improvement points proposed in this paper is generally higher than the effect of each ablation experiment, in which the accuracy P of Wheat-YOLOv5 is 93.3%, the recall R is 89.7%, the F1 value is 91.5%, and the AP value is 94.3%, which has a certain improvement effect, indicating that the improved algorithm in this paper, which is beneficial to the detection of complex and diverse wheat ears, reduces the error detection as well as the leakage detection situation, and effectively improves the detection accuracy, and the changes of the average accuracy of detection during the training of some different models are shown specifically in Fig. 4.

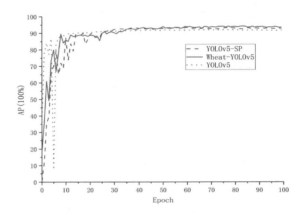

Fig. 4. Variation of AP during the training of some different models

4.3.2   Comparison of the Improved Algorithm with Other Target Detection Algorithms

In order to further verify the effectiveness of the proposed algorithm in this paper, the proposed algorithm Wheat-YOLOv5 is compared with other algorithms widely used in target detection for experiments, the compared algorithms are deep learning based YOLOv3, YOLOv4, and Faster R-CNN networks, each model is trained with 100 epochs, and the experimental results are shown in Table 2.

Table 2. Comparison of the improved algorithm with other target detection algorithms(%)

Algorithms	P	R	F1	AP
YOLOv5	92.40	87.10	89.60	92.30
YOLOv3	81.70	91.50	86.32	91.30
YOLOv4	82.30	91.80	86.79	91.80
Faster R-CNN	81.90	88.60	85.11	89.20
Wheat-YOLOv5	**93.30**	**89.70**	**91.50**	**94.30**

As can be seen from Table 2, compared with other target detection algorithms, the algorithm proposed in this paper has a higher accuracy with an AP value of 94.30%, which is a 3% improvement compared with YOLOv3, and a 2.5% improvement compared with YOLOv4; and compared with the Faster R-CNN algorithm, there is also a certain improvement, which verifies the effectiveness of the improved strategy proposed in this paper, the applicable to wheat ears detection.

4.3.3   Detection Results

In order to more intuitively verify the effectiveness of this paper on the detection of wheat sheaves in complex environments, YOLOv5 and Wheat-YOLOv5 are used to compare the detection. Some of the test results are shown in Fig. 5 below, with the original image on the left, the original YOLOv5 algorithm results in the middle, and the Wheat-YOLOv5 results on the right, with the word "wheat" above the wheat ears and the confidence level. Figure 5(a) shows the detection under the fuzzy scene, and it can be seen that the original YOLOv5 model has incomplete detection and missed detection for the fuzzy boundary of the wheat ears, while the improved model can detect the wheat ears completely.Figure 5(b) shows the wheat ears in near background color, the wheat ears and the background color are highly similar, and the comparison shows that the improved model detects the wheat ears missed by the original model.Figure 5(c) shows the mutually occluded and overlapping wheat ears, and it can be seen that the original model missed some of the mutually occluded and overlapping ears. From the above comparison of the detection under various wheat scenarios, it is clear that the improved YOLOv5 model is more effective in detecting wheat in complex environments.

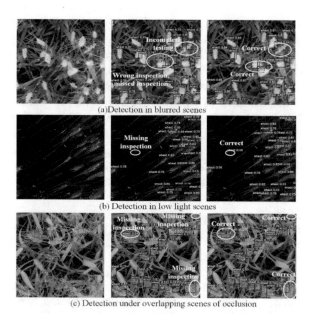

(a)Detection in blurred scenes

(b) Detection in low light scenes

(c) Detection under overlapping scenes of occlusion

Fig. 5. Comparison of detection results in some different scenarios

To better observe the feature information possessed by different network layers of the Wheat-YOLOv5 module for complex wheat ears, the wheat ears feature information is displayed using a heat map. From Fig. 6, it can be found that as the number of network layers increases, the extracted semantic features of wheat ears become stronger and stronger, which can better express the image information that humans can understand, verifying that the improved YOLOv5 model has a stronger feature extraction ability for wheat ears in complex and diverse environments, and to a certain extent, it can remove the influence of interference information such as complex background and enhance the robustness of the model.

    (a) Original image   (b) Wheat-YOLOv5   (c) Wheat-YOLOv5   (d) Wheat-YOLOv5
                        Conv layer   BN layer   Relu layer

Fig. 6. Heat map of different layers of Wheat-YOLOv5

5   Summary

This paper addresses the problem that the complex and diverse global wheat ears dataset GWHD_2021 is difficult to detect with high accuracy in a general deep learning target detection model, and proposes a wheat ears detection model Wheat-YOLOv5 with improved YOLOv5, which integrates the ECANet network into the YOLOv5 backbone network to enhance the feature extraction effect for different sizes of wheat ears; introduces SPConv eliminates the redundancy of feature maps and strengthens the model's ability to cope with obscured overlapping wheat ears, and uses EIoU as the loss function of border regression to improve the localization accuracy of wheat ears detection bounding boxes. Experiments show that the detection method proposed in this paper can locate wheat ears more accurately, and the method provides useful help for wheat breeding as well as yield prediction.

Acknowledgments. This work is supported and assisted by the National Ethnic Affairs Commission of the People's Republic of China (Training Program for Young and Middle-aged Talents, MZR20007), the Industry-University-Research Innovation Fund Project of Science and Technology Development Center of Ministry of Education (No. 2020QT08), Hubei Provincial Science and Technology Major Project of China (2020AEA011), Wuhan Science and Technology Program Applied Basic Frontier Project (2020020601012267) and the Graduate Innovation Fund, South-Central Minzu University (No. 3212022sycxjj335).

References

1. Mengyang, F.: A machine vision-based method for counting wheat ears in a large field environment. China Agricultural University (2015). (in Chinese)
2. Ren, S., He, K., Girshick, R., et al.: Faster R-CNN: towards real-time object detection with region proposal networks. IEEE Trans. Pattern Anal. Mach. Intell. **39**(6), 1137–1149 (2017)
3. Redmon, J., Farhadi, A.: YOLO9000: better, faster, stronger. In: IEEE Conference on Computer Vision & Pattern Recognition, pp. 6517–6525 (2017)
4. Redmon, J., Farhadi, A.: YOLOv3: an incremental improvement. arXiv e-prints, (2018)
5. Bochkovskiy, A., Wang, C.Y., Liao, H.: YOLOv4: optimal speed and accuracy of object detection (2020)
6. Quanbing, Z.: Attention mechanism pyramid network based wheat spike detection method. Anhui University (2021). (in Chinese)
7. David, E., Madec, S., et al.: Global wheat head detection (GWHD) dataset: a large and diverse dataset of high-resolution RGB-labelled images to develop and benchmark wheat head detection methods. Plant Phenomics (2020)
8. Etienne, D., Mario, S., et al.: Global wheat head detection 2021: an improved dataset for benchmarking wheat head detection methods. Plant Phenomics (2021)
9. Lin, T.Y., Dollar, P., Girshick, R., et al.: Feature pyramid networks for object detection. In: 2017 IEEE Conference on Computer Vision and Pattern Recognition (CVPR), IEEE (2017)
10. Wang, C.Y., Liao, H., Wu, Y.H., et al.: CSPNet: a new backbone that can enhance learning capability of CNN. In: 2020 IEEE/CVF Conference on Computer Vision and Pattern Recognition Workshops (CVPRW), IEEE (2020)
11. Wang, Q., Wu, B., Zhu, P., et al.: ECA-Net: efficient channel attention for deep convolutional neural networks. In: 2020 IEEE/CVF Conference on Computer Vision and Pattern Recognition (CVPR), IEEE (2020)
12. Zhang, Q., Jiang, Z., Lu, Q., et al.: Split to be slim: an overlooked redundancy in vanilla convolution (2020)
13. Zhang, Y.F., Ren, W., Zhang, Z., et al.: Focal and efficient IOU loss for accurate bounding box regression (2021)
14. He, J., Erfani, S., Ma, X., et al.: Alpha-IoU: a family of power intersection over union losses for bounding box regression (2021)
15. Rezatofighi, H., Tsoi, N., Gwak, J.Y., et al.: Generalized intersection over union: a metric and a loss for bounding box regression. IEEE (2019)

Aircraft Target Detection Algorithm Based on Improved YOLOv5s

Lixia Zhang, Zhiming Ma[✉], Xiangshu Peng, and Menglin Qi

School of Computer Science and Technology, Xinjiang Normal University, Urumchi, China
1412820596@qq.com

Abstract. Aiming at the characteristics of multi-scale, diversity and complex background of aircraft targets, in order to improve the average detection accuracy of YOLOv5s algorithm, an improved aircraft target detection algorithm based on YOLOv5s model is proposed, Firstly, replace conv module in the backbone network of YOLOv5s with RepVGGBlock to reduce the number of parameters; Secondly, a small target detection head is added to enhance the recognition ability of small targets; Finally, GAM_Attention mechanism is introduced in front of each detection head to improve the detection accuracy. The research shows that the RG-YOLOv5s (Repvggblock GAM_Attention you only look once) algorithm proposed in this paper improves the average accuracy by about 1% and reaches 97.3% when IOU = 0.5, which is more suitable for the detection of aircraft remote sensing targets.

Keywords: RepVGGBlock · YOLOv5s · GAM_Attention

1 Introduction

Remote sensing image technology has great development space in both military and civil fields. It has great application value in military war. Because of the low resolution of aircraft image, the detection process is more difficult. In the field of deep learning, target detection algorithms mainly include RCNN series [1–3], SSD and YOLO [4]. Researchers at home and abroad have done a lot of research in the field of target detection and achieved phased results. Zhong Zhifeng et al. Proposed a lightweight target detection algorithm based on improved YOLOv4 [5], which improved the detection speed, but reduced the detection accuracy; Chen et al. Proposed the research on traffic sign recognition based on the improved YOLOv4 model [6], which improved the detection accuracy, but reduced the detection speed of the model. Therefore, this paper proposes a network structure RG-YOLOv5. The improved algorithm is mainly reflected in the following aspects:

1) Replace conv module in the backbone network of YOLOv5s with RepVGGBlock to reduce the number of parameters.
2) Add a small target detection head to enhance the recognition ability of small targets.
3) GAM_Attention mechanism is introduced in front of each detection head to improve the detection accuracy.

Q. Liang et al. (Eds.): AIC 2022, LNEE 871, pp. 130–136, 2023.
https://doi.org/10.1007/978-981-99-1256-8_15

2 Related Algorithms

2.1 YOLOv5 Algorithm

YOLOv5s algorithm is a simple and efficient target detection model, which reduces the training threshold. Only a conventional gtx-1660ti can be used to train YOLOv5. The main body of the algorithm is composed of input terminal, backbone network, neck feature enhancement module and head detection module. Compared with YOLOv3 and YOLOv4 [7], YOLOv5 puts forward some improvement ideas in the model training stage, mainly including mosaic data enhancement, adaptive anchor frame calculation and adaptive image scaling.The structure diagram of YOLOv5 [8-9] is shown in Fig. 1:

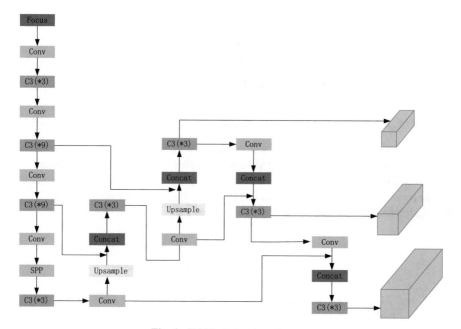

Fig. 1. YOlOv5 structure diagram

2.2 RepVGG

RepVGG is a simple VGG structure, which uses 3 × 3 convolution, BN layer and relu activation function in large quantities. It uses parameterization to improve performance. It is characterized by using multi branch network during training and fusing multi branches into single branches during reasoning. During training, add parallel 1 × 1 convolution branches and identity mapping branches to each 3 × 3 convolution layer to form a RepVGGBlock. Learn from the practice of Resnet, the difference is that Resnet adds a branch every two or three layers, and RepVGGBlock adds a branch every layer. The structure of RepVGGBlock is shown in Fig. 2:

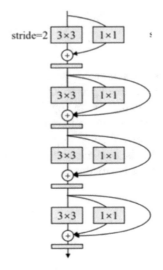

Fig. 2. RepVGGBlock structure diagram

2.3 GAM_Attention

GAM_ Attention mechanism can enlarge the global dimension interaction features while reducing information diffusion. The sequential channel spatial attention mechanism is adopted and the CBAM sub module is redesigned [10]. The whole process is shown in Fig. 3 and in formulas 1 and 2. Given the input characteristic mapping $F_1 \in R^{C \times H \times W}$, the intermediate state F_2 and output F_3 are defined as:

$$F_2 = M_c(F_1) \otimes F_1 \tag{1}$$

$$F_3 = M_s(F_2) \otimes F_2 \tag{2}$$

M_c and M_s are channel attention map and spatial attention map respectively; $\otimes$ Represents multiplication by element.

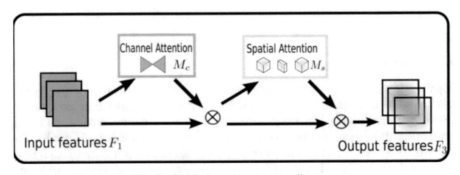

Fig. 3. GAM_Attention structure diagram

3 Algorithm Improvement

The improvement of RG-YOLOv5s algorithm is mainly in the following three aspects: first, replace conv module in YOLOv5s backbone network with RepVGGBlock to reduce the number of parameters; Second, add a small target detection head to enhance the recognition ability of small targets; Third, the GAM_Attention mechanism is introduced in front of each detection head to improve the detection accuracy. The improved RG-YOLOv5s network structure is shown in the Fig. 4:

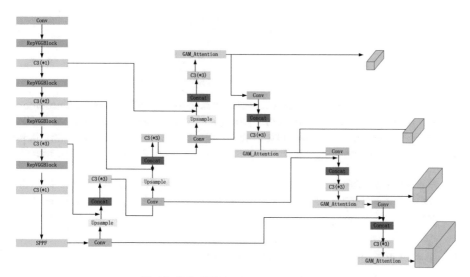

Fig. 4. RG_YOLOv5s structure diagram

After the image passes through the backbone feature extraction network, the feature maps of the second, fourth and sixth layers are extracted respectively and input into the neck network for feature fusion between different scales. The decoding algorithm of YOLOv5s is carried out through four YOLO detection heads to generate the final prediction.

4 Experimental Results and Analysis

4.1 Experimental Environment Configuration and Data Introduction

Because there are few aircraft images in a single remote sensing data set, this experiment selects 446 aircraft images in the RSOD remote sensing data set marked by Wuhan University and 784 aircraft images in the Dior optical data set marked by Han Junwei research group of Western Polytechnic University, a total of 1230 images. The ratio of training set and test set is 7:3. The operating system used in the experiment is ubantu16 4. Processor model: Intel (R) Xeon (R) CPU e5–2630 V4 @ 2.20GHz, graphics card model: NVIDIA Tesla K80, using NVIDIA cuda10 2 acceleration toolbox.

In order to comprehensively evaluate the performance of the model and prove the effectiveness of the algorithm proposed in this paper, the following evaluation indexes are selected.

1) Average precision and mean average precision (mAP): are the average values of accuracy of all categories. The calculation formula of sum is as follows:

$$\Pr ecision = \frac{TP}{TP + FP} \tag{3}$$

$$AP = \frac{\sum \Pr ecision}{images} \tag{4}$$

$$AP = \frac{\sum \Pr ecision}{images} \tag{5}$$

where *TP* is the number of positive samples with correct prediction, *FP* is the number of positive samples with wrong prediction, and *N* is the number of total categories. *TP* and *FP* are determined according to the *IOU* (intersection over union) threshold.

2) Parameters: reducing parameters can reduce the amount of calculation and training time.

3) Frame per second: represents the number of pictures that can be detected by the model per second. This index can evaluate the detection speed of the proposed model.

4.2 Ablation Experiment and Result Analysis

In order to verify the optimization effect of each module used in RG-YOLOv5s on YOLOv5s, ablation experiments were carried out in this paper. As shown in Table 1:

Table 1. Model test results

Algorithm	Backbone	Neck	Aircraft(mAP%)	Parameters	FPS
YOLOv5s	Focus + Conv	FPN + PAN	96.4	7063542	166
YOLOv5s	RepVGGBlock	FPN + PAN	96.7	5441922	160
YOLOv5s	RepVGGBlock	FPN + PAN + detect(small)	97.1	5947028	151
RG-YOLOv5s(Ours)	RepVGGBlock	FPN + PAN + GAM + detect(small)	97.3	6383764	142

It can be seen from Table 1 that the accuracy of YOLOv5s + repvggblock is slightly improved and the amount of parameters is reduced compared with YOLOv5s. After adding a small target detection head to YOLOv5s + repvggblock, the accuracy is further improved while adding a few parameters. The algorithm RG-YOLOv5s proposed in this paper improves the accuracy after introducing gam attention mechanism in the neck part.The detection effect of RG-YOLOv5s algorithm is shown in Fig. 5:

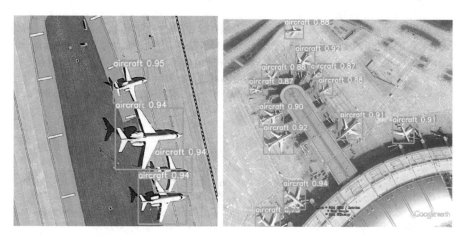

Fig. 5. RG_YOLOv5s detection effect

4.3 RG-YOLOv5 Compared with Other Algorithms

Comparative tests are conducted on RSOD and Dior data sets respectively. The experimental results are shown in Table 2. The experimental results show that the RG-YOLOv5s backbone network proposed in this paper adopts RepVGGBlock module, and GAM_Attention is introduced into the feature fusion module,Adds a small target detection head, enhances the receptive field, greatly improves the detection accuracy, and can meet the real-time requirements of remote sensing target detection. From the perspective of map and FPS, RG-YOLOv5s has superior performance.

Table 2. RG-YOLOv5s compared with other algorithms

Algorithm	Backbone	mAP/%	FPS
SSD	Resnet50	85.68	281
Faster RCNN	VGG16	94.71	207
YOLOv3spp	Darknet53	88.09	201
YOLOv5s	Focus + Conv	96.5	166
RG-YOLOv5s(ours)	RepVGGBlock	97.4	142

5   Conclusion

In this paper, RepVGGBlock is used to replace the conv module in the YOLOv5s backbone network, which reduces the amount of parameters, adds a small target detection head in the head, and improves the recognition rate of small target detection.In the neck part, GAM_Attention mechanism is introduced to improve the detection accuracy, and the mAP value of the improved model reaches 97.3%.However, the algorithm in this paper also has shortcomings. Although the model size is greatly improved compared with the original algorithm, there is still room for improvement. In view of this problem, the backbone network and neck enhancement network will be improved in the later stage, which will be the focus of further research in this paper.

References

1. Han, C., Gao, G.Y., Zhang, Y.: Real-time small traffic sign detection with revised faster-RCNN. Multimedia Tools and Appl. **78**(10), 13263–13278 (2019)
2. Hung, G.L., Sahimi, M.S.B., Samma, H., et al.: Faster R-CNN deep learning model for pedestrian detection from drone images. SN Computer Science **1**(4), 17–23 (2020)
3. Ren, S.Q., He, K.M., Girshick, R., et al.: Faster R-CNN; towards real-time object detection with region proposal network. IEEE Trans. Pattern Anal. Mach. Intell. **39**(6), 1137–1149 (2017)
4. Zhao, J., Han, R., Sun, J., et al.: Research on the ETC vehicle detection algorithm based on improved YOLOv3. Computer and Digital Engineering **50**(01), 90–94 + 139 (2022)
5. Zhong, Z., Xia, Y., Zhou, D., et al.: Lightweight object detection algorithm based on improved YOLOv4. Computer application: 1–8 [2021–10–22]. http://kns.cnki.net/kcms/detail/51.1307.TP.20210929.1334.012.html
6. Chen, M., Yu, S.: Study on Traffic Sign Identification Based on improved YOLOV4 Model. Microelectronics and Computers: 1–10 [2021–10–27]. http://kns.cnki.net/kcms/detail/61.1123.TN.20210923.0115.001.html
7. Du, S., Zhang, P., Zhang, B., et al.: Weak and occluded vehicle detection in complex infrared environment based on improved YOLOv4. IEEE Access **9**, 25671–25680 (2021)
8. Matthew, C., et al.: Rapid DNA origami nanostructure detection and classification using the YOLOv5 deep convolutional neural network. Sci. Rep. **12**(1), 3871 (2022)
9. Upesh, N., Hossein, E.: Comparing YOLOv3, YOLOv4 and YOLOv5 for Autonomous Landing Spot Detection in Faulty UAVs. Sensors **22**(2), 464 (2022)
10. Yuteng, X., et al.: TReC: transferred resnet and CBAM for detecting brain diseases&13. Front. Neuroinform. **15**, 781551 (2021)

A Survey in Virtual Image Generation Based on Generative Adversarial Networks

Xiaojun Zhou[1(✉)], Yunna Wei[1], Gang Xing[1], Yanan Feng[1], and Li Song[2]

[1] MIGU Co., Ltd, Beijing, China
zhouxiaojun@migu.cn

[2] Institute of Image Communication and Network Engineering Shanghai, Jiao Tong University Shanghai, Shanghai, China

Abstract. With the rapid development of the metaverse, this is the inevitable trend content form from 2D to 3D, from real to virtual to virtual-real combination. The demand for virtual scenes will surge in the coming period. How to quickly generate a large number of high-quality scene graphs has sparked extensive academic attention. GAN being the most popular generation algorithm, and we discuss the prevalent architectures to solve the problems of training collapse and uncontrollable images during training. Besides, this paper focuses on the development of GAN and its variants in the field of image synthesis. A comprehensive review of recent advances in virtual image generation in the last two years is presented, with a summary of the classification of algorithms and an introduction to techniques related to multimodal generation. Finally, some suggestions for future research directions are proposed.

Keywords: Metaverse · GAN · Virtual image synthesis · Multimodal generation

1 Introduction

GANS has received a lot of academic attention since it was proposed, and has been applied to many fields such as computer vision [1, 2], medicine [3], and natural semantic processing [4]. As shown in Fig. 1, GAN is one of the research hotspots in recent years, and it also occupies a large proportion in the field of virtual image generation. In computer vision, in addition to the image generation introduced in this paper, image restoration, super-resolution reconstruction, image captioning, image segmentation, image denoising and other applications are also relatively active application research. The traditional GAN consists of two parts, the generator(G) and the discriminator(D), which can be vividly expressed as the story of a thief and a policeman. The thief (G) is responsible for learning the data probability distribution from the input noise to fake images in an attempt to fool the police (D), and the discriminator is responsible for improving the discrimination ability and trying to distinguish real samples from fake ones. The two sets of adversarial networks are trained to improve their capabilities and eventually form a Nash equilibrium [5] that outputs a perfect fake picture.

Virtual image generation is an interesting and important task because it has many practical applications, such as art generation, image editing, virtual reality, video games,

Fig. 1. It shows the most frequent keywords in CVPR conferences from 2020 to 2022. The "GAN","visual" and "generation" keywords are often have been used.

and computer-aided design [6]. We divide virtual image generation into two categories according to the input. When the input and output are both images, we consider this as a direct method for generating images from images, which aims to use pairs of images as training sets [1, 7] and learn the mapping from the input image to the output image.The development of various GAN variants [8–14] has gradually eliminated the limitation of data [15, 16]. Although the current image generation effect has been able to reach the point of being fake, it cannot generate images according to the required semantic information. So researchers turned their attention to multimodal generation techniques. It can be seen as an extension of conditional GANs, which can generate target images from text describing images. Besides text, one or more of layout, scene graph, semantic mask, mouse traces can also be supported as input [17].

In this paper, we first introduce the classification of recent variants of GAN in terms of GAN architecture, and focus on the field of virtual image generation. The development history of representative algorithms and problem optimization are described in terms of direct and step-by-step classification. Finally, problems and researchable directions are pointed out.

2   GAN Preliminaries

In this section, we review the network architecture and core concepts of GANS and its representative variants.

The original GAN architecture consists of a single generator(G) and discriminator(D),as shown in Fig. 2 (a). It can be composed of various types of neural network structures, first MLP, and later there are deep fully convolutional networks to enhance the results [10]. Letting θ and φ be the learnable parameters in G and D respectively, GAN training can be generalized as a minimax problem. Reduce the gap between the generated data distribution $\rho(G(z : \theta))$ and the real data distribution $\rho(x)$; the discriminator does the opposite.

$$\min_{\Phi} \max_{\theta} V(\theta, \phi) \tag{1}$$

As an unsupervised learning algorithm, GAN can generate uncontrollable pictures only by random noise z, and often produce unexpected effects. CGAN [8] solves the problem that traditional GAN algorithms cannot control the generated data by adding category labels C on G to guide data generation with prior knowledge. The architecture is shown in Fig. 2(b). C can be another image (image to image conversion), text (text to image synthesis) or categorical labels (label to image synthesis) The proposal of CGAN is particularly crucial and many subsequent algorithms have been improved based on this idea. [1, 7, 9, 18, 19].

Suppose we want to control certain features of the input to change the thickness or skew angle of the font in MNIST. However, the feature is very messy and we cannot control it clearly in practice. The purpose of Infogan [9] is to clarify and regularize these disorganized features. It uses mutual information to constrain C and optimise the model objective function. Assuming that C is interpretable with respect to the generated data g, then c and g should be highly correlated and their mutual information large. And if it is unconstrained, the mutual information is close to zero. Borrowing ideas from CGAN [8], ACGAN [19] reconstructs auxiliary information by adding an auxiliary encoder network,such as image categories. But when discriminating the output of the network, ACGAN feeds back the label as the target and submits it to the discriminating network. The architecture of Infogan and ACGAN is shown in Fig. 2(c).

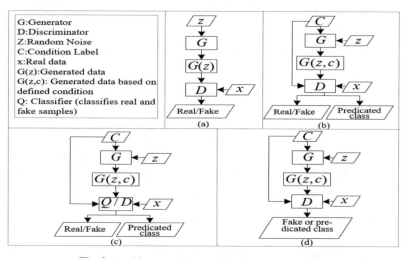

Fig. 2. Architecture diagram of GAN and its variants

GAN has the problem of training collapse [20], and insufficient feature learning is an important reason. This is especially evident on small sample data. BAGAN [12] (as shown in Fig. 2(d))performs data augmentation by introducing an auto-encoder. The problem of insufficient GAN features is addressed by learning common data features for both few-shot and multi-shot categories. In addition, when the two output losses of ACGAN are generated in small sample data, since the discriminator is biased towards the real data with more occurrences, it will generate a data imbalance problem that the small sample category data generates less. BAGAN merges the outputs into one class, which resolves the potential conflict brought by the two parts of the output.

3   General Approaches of Virtual Scene Generation

3.1   Direct Method

In terms of applications, direct methods are defined as methods to generate virtual image using image transformation, such as style transfer, image transformation, animation, and image dehazing. Divided from the algorithm structure, direct methods are defined as a class of methods that use a set of G/D to generate images,as shown in Fig. 3.

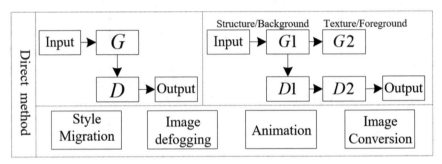

Fig. 3. The method of direct of virtual image synthesis and related applications

Early GANs all follow a generator and a discriminator structure [7, 9, 12, 13, 19, 21]. The same is true for srargan [23] for multi-domain transformation. Traditional multi-domain image translation requires mutual feature extraction between different domains, which leads to the need to learn C*(C-1) generators for transformations containing C domains. Therefore, the versatility is not strong when performing image conversion in different domains. StarGAN solves this problem, being able to train multiple datasets with different domains simultaneously in a single G. The role of D is not only to determine whether the input image is true, but also to determine the domain to which the input image belongs. G accepts an image and target domain label C, and generates a fake image.

Cyclegan [15] is a ring structure with two sets of generators and discriminators. It eliminates the limitation that the pix2pix data field must be one-to-one, and does not need to have a strict correspondence. In order to make up for the lack of information caused by unpaired images, CycleGAN adds an opposite mapping $F : Y- > X$ to the mapping G on the basis of GAN, and proposes a cycle consistency loss function to ensure that $F(G(X)) \approx X$ (and vice versa). The two parallel G/D network structures of Cyclegan have no difference in processing images. There are also two-layer GANS algorithms responsible for processing the structure/texture, foreground/background of the image, respectively [0][0][0].

3.2 Staged Method

Divided in terms of application, the staged method is defined as generating virtual images using multimodal information such as semantics and text. Secondary editing using latent space is supported, such as text-generated images,image editing, and image latent space editing. In terms of algorithmic structure,staged methods are defined as a class of methods that use multiple sets of G/D combinations or pre-trained G to generate images,as shown in Fig. 4.

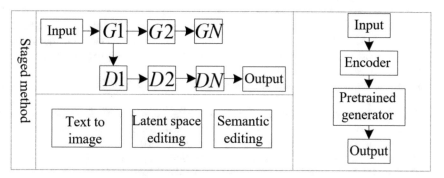

Fig. 4. The method of direct of virtual image synthesis and related applications

Multiple groups can be composed of the same network structure side by side, sharing weights with each other, and also use different network structures in cyclic, parallel or sequential [14, 15, 27–29]. StackGAN is a typical representative of the step-by-step method, which consists of two parts. The Stage-I network generates 64x64 low-resolution thumbnails of basic colors and sketches, with Stage-II adding more detail to generate 256x256 photorealistic images with text guidance. Pixel2style2pixel (pSp) [30] relies on a pre-trained StyleGAN to expand the style vector generated based on the encoder network into a latent space. The emergence of Psp breaks the traditional image editing process of flipping first and then editing [31–35], but uses the encoder to directly solve the image conversion task, which simplifies the training process.

4 Summary

This paper reviews the development of the basic architecture of Generative Adversarial Networks and its variants. The related algorithms in the field of virtual image generation are divided into direct method and staged method. For direct methods, generalization models for multi-domain adaptation(DA) will be the focus of research for some time. For the staged method, how to better control the characteristics of the latent space to generate anthropomorphic images quickly and accurately is a worthy research direction.

References

1. Wang, T.C., Liu, M.Y., Zhu, J.Y., Tao, A., Kautz, J., Catanzaro, B.: High-resolution image synthesis and semantic manipulation with conditional gans. In: Proceedings of the IEEE conference on computer vision and pattern recognition, pp. 8798–8807 (2018)

2. Shamsolmoali, P., et al.: Image synthesis with adversarial networks: a comprehensive survey and case studies. Information Fusion **72**, 126–146 (2021)
3. Armanious, K., et al.: MedGAN: medical image translation using GANs. Comput. Med. Imaging Graph. **79**, 101684 (2020)
4. Huang, J., Johanes, M., Kim, F.C., Doumpioti, C., Holz, G.C.: On GANs, NLP and architecture: combining human and machine intelligences for the generation and evaluation of meaningful designs. Technology| Architecture+ Design **5**(2), pp.207–224 (2021)
5. Goodfellow, I.J., et al.: Generative Adversarial Networks. arXiv e-prints. arXiv preprint arXiv: 1406.2661, 1406 (2014)
6. Liu, M.Y., Huang, X., Yu, J., Wang, T.C., Mallya, A.: Generative adversarial networks for image and video synthesis: algorithms and applications. Proc. IEEE **109**(5), 839–862 (2021)
7. Isola, P., Zhu, J.Y., Zhou, T., Efros, A.A.: Image-to-image translation with conditional adversarial networks. In: Proceedings of the IEEE conference on computer vision and pattern recognition, pp. 1125–1134 (2017)
8. Mirza, M., Osindero, S.: Conditional Generative Adversarial Nets. arXiv preprint arXiv:1411. 1784 (2014)
9. Chen, X., Duan, Y., Houthooft, R., Schulman, J., Sutskever, I., Abbeel, P.: Infogan: Interpretable representation learning by information maximizing generative adversarial nets. Advances in neural information processing systems, **29** (2016)
10. Mandal, B., Puhan, N.B., Verma, A.: Deep convolutional generative adversarial network-based food recognition using partially labeled data. IEEE Sensors Letters **3**(2), 1–4 (2018)
11. Yi, Z., Zhang, H., Tan, P., Gong, M.: Dualgan: Unsupervised dual learning for image-to-image translation. In: Proceedings of the IEEE international conference on computer vision, pp. 2849–2857 (2017)
12. Mariani, G., Scheidegger, F., Istrate, R., Bekas, C., Malossi, C.: Bagan: Data Augmentation with Balancing Gan. arXiv preprint arXiv:1803.09655 (2018)
13. Nowozin, S., Cseke, B., Tomioka, R.: f-gan: Training generative neural samplers using variational divergence minimization. Advances in Neural Information Processing Systems **29** (2016)
14. Zhang, H., Xu, T., Li, H., Zhang, S., Wang, X., Huang, X., Metaxas, D.N.: Stackgan: Text to photo-realistic image synthesis with stacked generative adversarial networks. In Proceedings of the IEEE International Conference on Computer Vision, pp. 5907–5915 (2017)
15. Zhu, J.Y., Park, T., Isola, P., Efros, A.A.: Unpaired image-to-image translation using cycle-consistent adversarial networks. In: Proceedings of the IEEE International Conference on Computer Vision, pp. 2223–2232 (2017)
16. Park, T., Efros, A.A., Zhang, R., Zhu, J.Y.: Contrastive learning for unpaired image-to-image translation. In: European Conference on Computer Vision pp. 319–345. Springer, Cham (2020). https://doi.org/10.1007/978-3-030-58545-7_19
17. Frolov, S., Hinz, T., Raue, F., Hees, J., Dengel, A.: Adversarial text-to-image synthesis: a review. Neural Netw. **144**, 187–209 (2021)
18. Brock, A., Donahue, J., Simonyan, K.: Large Scale GAN Training for High Fidelity Natural Image Synthesis. arXiv preprint arXiv:1809.11096 (2018)
19. Odena, A., Olah, C., Shlens, J.: Conditional image synthesis with auxiliary classifier gans. In International Conference on Machine Learning, pp. 2642–2651. PMLR (2017)
20. Jabbar, A., Li, X., Omar, B.: A survey on generative adversarial networks: variants, applications, and training. ACM Computing Surveys (CSUR) **54**(8), 1–49 (2021)
21. Radford, A., Metz, L., Chintala, S.: Unsupervised Representation Learning with Deep Convolutional Generative Adversarial Networks. arXiv preprint arXiv:1511.06434 (2015)
22. Salimans, T., Goodfellow, I., Zaremba, W., Cheung, V., Radford, A., Chen, X.: Improved techniques for training gans. Advances in Neural Information Processing Systems **29** (2016)

23. Choi, Y., Choi, M., Kim, M., Ha, J.W., Kim, S., Choo, J.: Stargan: Unified generative adversarial networks for multi-domain image-to-image translation. In: Proceedings of the IEEE Conference on Computer Vision and Pattern Recognition, pp. 8789–8797 (2018)
24. Abdal, R., Zhu, P., Mitra, N.J., Wonka, P.: Labels4free: unsupervised segmentation using stylegan. In: Proceedings of the IEEE/CVF International Conference on Computer Vision, pp. 13970–13979 (2021)
25. Alaluf, Y., et al.: Third Time's the Charm? Image and Video Editing with StyleGAN3. arXiv preprint arXiv:2201.13433 (2022)
26. Zhu, P., Abdal, R., Femiani, J., Wonka, P.:. Barbershop: Gan-based Image Compositing Using Segmentation Masks. arXiv preprint arXiv:2106.01505 (2021)
27. Huang, X., Li, Y., Poursaeed, O., Hopcroft, J., Belongie, S.: Stacked generative adversarial networks. In: Proceedings of the IEEE Conference on Computer Vision and Pattern Recognition, pp. 5077–5086 (2017)
28. Im, D.J., Kim, C.D., Jiang, H., Memisevic, R.: Generating Images with Recurrent Adversarial Networks. arXiv preprint arXiv:1602.05110 (2016)
29. Wang, X., Gupta, A.: Generative image modeling using style and structure adversarial networks. In: European Conference on Computer Vision, pp. 318–335. Springer, Cham (2016). https://doi.org/10.1007/978-3-319-46493-0_20
30. Richardson, E., et al.: Encoding in style: a stylegan encoder for image-to-image translation. In : Proceedings of the IEEE/CVF Conference on Computer Vision and Pattern Recognition, pp. 2287–2296 (2021)
31. Yang, C., Shen, Y., Zhou, B.: Semantic hierarchy emerges in deep generative representations for scene synthesis. Int. J. Comput. Vision **129**(5), 1451–1466 (2021)
32. Tewari, A., et al.: Stylerig: Rigging stylegan for 3d control over portrait images. In: Proceedings of the IEEE/CVF Conference on Computer Vision and Pattern Recognition, pp. 6142–6151 (2020)
33. Härkönen, E., Hertzmann, A., Lehtinen, J., Paris, S.: Ganspace: discovering interpretable gan controls. Adv. Neural. Inf. Process. Syst. **33**, 9841–9850 (2020)
34. .Shen, Y., Gu, J., Tang, X., Zhou, B.: Interpreting the latent space of gans for semantic face editing. In: Proceedings of the IEEE/CVF Conference on Computer Vision and Pattern Recognition, pp. 9243–9252 (2020)
35. Collins, E., Bala, R., Price, B., Susstrunk, S.: Editing in style: uncovering the local semantics of gans. In: Proceedings of the IEEE/CVF Conference on Computer Vision and Pattern Recognition, pp. 5771–5780 (2020)

Research on the Service for the Disabled in Interior Design and Product Design Based on Artificial Intelligence

Chaoran Bi[1]([✉]), Hai Wang[2], and Weiyue Cao[1]

[1] College of Media Design, Tianjin Modern Vocational Technology College, Tianjin 300350,
China
806281351@qq.com

[2] Teaching Center for Experimental Electronic Information, College of Electronic Information
and Optical Engineering, Nankai University, Tianjin 300350, China

Abstract. The development of modern science and technology has brought us a far-reaching and significant impact. Artificial intelligence is an important product of this era. With the continuous development of society, the status of social vulnerable groups has been paid more and more attention. As a country with a large population, China has the highest number of disabled people in the world. The existing products and services for the disabled can not meet the needs of users. The application of artificial intelligence technology in interior design and product design practice is still in the exploratory stage, and there is a large space for development. Through the user research of observation method and experience prototype method, this paper deeply excavates the communication needs and expectations of Chinese disabled people. Through user research and design, a set of smart home technology service system is proposed, which provides a reasonable and feasible solution for the daily communication of the disabled.

Keywords: Artificial intelligence · People with disabilities · Smart home · Interior design · Product design

1 Introduction

As a branch of computer science, artificial intelligence technology, together with genetic engineering technology and nanoscience and technology, is called the three major scientific and technological achievements in the 21st century [1–3]. It is an advanced technology type that is currently focused on in various fields。In recent years, with the expansion of the application scope of artificial intelligence technology and the continuous improvement of technical application conditions, It gradually appears in the interior design work, In addition, it has also provided strong programs and technical support for the improvement and help of the living and working conditions of some disabled people [4]. Under the background of actively promoting innovation and development in the field of interior design and product design, more and more designers take artificial intelligence technology as an important entry point for technology and method innovation, By optimizing design tools, methods and ideas, we can make full use of artificial intelligence technology to provide more considerate and better services for different groups of people [5, 6].

2 Application Analysis of Artificial Intelligence in Interior Design and Product Design

The development of artificial intelligence promotes the progress of the smart home industry. The continuous growth in the number of smart home enterprises is a strong proof. From the market sales of smart home enterprises, the overall development of the smart home industry continues to grow upward. "Smart home" is the main form of artificial intelligence in daily life, Smart home is bound to have more advanced perception and perception of the preferences of the users of the house. Analyze users' preferences and habits through data [7].

Smart home essentially belongs to household appliances, interior design and some related product design. But it is more intelligent. Compared with traditional household appliances, it looks like another brain. Get along with them for a long time, it will remember your preferences and habits, and put everything in order. This is a major advantage of smart home. Smart home serves people, security must be ensured while providing services, including ensuring the security of its own smart home system, the security of the whole residence, information security, network security, communication security, etc. [8].

The application of artificial intelligence technology in interior design is mainly reflected in the presentation of interior design effect and the layout of smart home. Product design is a series of technical work from the formulation of new product design specification to the design of product samples.The main idea of the product service system is to derive physical products in a sustainable way from the content of enterprise production, that is, the material products exported from the traditional manufacturing industry, Foreseeable factors or outcomes that evolve into a post industrial society, The secondary foreseeable factors and the manifestation of achievements are no longer the products understood in the traditional manufacturing industry, it is a service that can extend the product life cycle, that is, product service system, which links intangible services with tangible products, become an integrated system. The value of this process is that it can enhance the relevant technical systems in the system composed of different participants, technical products, services, business models and driving forces such as sustainable development and dematerialization. Its task is to provide consumers with products and services that meet their needs, and to meet the ecological, economic and social requirements in the whole product life cycle [9].

3 Analysis of the Disabled and Special Groups

The disabled are divided into the disabled in the broad sense and the disabled in the narrow sense.

In a broad sense, people with disabilities can be understood as those with weak physical functions, such as injured children, pregnant women, the elderly and patients.

In the narrow sense, people with disabilities can be understood as visual, hearing and language disabilities, physical disabilities, intellectual disabilities and mental disabilities. The United Nations World Health Organization divides disability into three categories, namely: "(1) functional and morphological disability refers to the defect or abnormality of human structure or function due to accidental injury and disease sequelae; [10] (2) Disability refers to the structural defect and dysfunction of the human body, which makes the human body lose the ability it should have; (3) The term "handicap" refers to the situation that the disabled participate in normal social activities due to the defects or abnormalities of the body's shape and function. Artificial intelligence in smart home system and product design mainly solves the problem that disabled people will encounter limited activity space in life. The range of daily activities of the physically disabled is obvious, they mainly focus on their own family, a large proportion of the disabled are physically handicapped,and they don't want to add trouble to others, seldom go out every week, about once or twice a week. At the same time, the unreasonable design of public activity space and the limited resources of barrier free public facilities have also led to the reluctance of the physically disabled to go out for activities. It also leads to the limited activity space for the physically disabled [11, 12].

4 The Application of Artificial Intelligence in Residence

Artificial intelligence is applied in many aspects. It is changing our living habits and also better serving the disabled. The application of artificial intelligence in interior design enables the disabled to have a "professional housekeeper" to provide services 24 h a day. Because each user is an independent individual, they face different problems [13]. No product can meet the needs of all users. The intelligent home system can provide personalized services according to the needs of each family. It changes our special living habits and brings many conveniences to life. With the further development of science and technology, this simple and fast model will be more recognized and widely used. While the intelligent home system serves people to improve their life experience, its security is another major advantage. Safety hazards can be eliminated from the source. Compared with traditional home furnishings, it shows strong advantages in fire prevention and theft prevention [14, 15].

4.1 Application Research of Artificial Intelligence in Interior Space Design

The application of artificial intelligence in all aspects of bedroom and product design is analyzed from four aspects that are closely related to public life: bedroom, living room, kitchen and bathroom.

4.1.1 Application Analysis of Bedroom Space

The bedroom is an important place for daily rest. In the design of ordinary people's living and use, the space should be transparent, bright, quiet and private. Artificial intelligence elements can be added through remote control and mobile app, Voice or gesture can meet the special needs of people with disabilities who are unable to move, it can also better adapt to the needs of different scenarios, normal sleep, rest and reading. The artificial intelligence system can adjust the indoor temperature and brightness according to different needs. For example, the curtain is automatically closed and the light is turned off during the afternoon nap, and the night light is automatically turned on according to the induction at night. When setting the wake-up time, the curtain can be automatically opened to reduce human limb operation, and more simulate personal life clock and living habits to meet the needs of users.

4.1.2 Application Analysis and Design of Bathroom Space

The toilet is an important and dangerous space for the disabled, and a living and bathing space for ordinary users. In the face of the special situation of the disabled, attention should be paid to the installation of reasonable handrails and anti-skid floor tiles in the basic interior design stage. The height of the toilet should match the height of the user's body. Artificial intelligence can control the water outlet, water stop and water temperature through waterproof remote control or voice recognition, prepare in advance according to the user's habits, such as opening the toilet cover or hot water before bathing. It can avoid the embarrassment of forgetting to turn on the water heater when hot water is needed, and also avoid the energy waste caused by the water heater working all the time when no one is using it.

4.1.3 Application Analysis of Living Room Space

The living room is the most frequently used space for ordinary people and some disabled people who do not need to stay in bed. The functions of the living room are more diversified, which need to meet more needs such as hospitality, movie viewing, reading, family gatherings, etc. These requirements require different conditions. It is laborious to arrange the conditions according to the needs during use. At this time, the importance of artificial intelligence is highlighted. For example, you can close windows, lights and curtains with one button or voice control while watching the movie. Automatically put the screen down or turn on the TV. After watching the movie, it will automatically return to the normal state and start automatic cleaning. As a user, all you need to do is simply follow two instructions.

4.1.4 Application Analysis of Household Appliances

Artificial intelligence is not only applied to intelligent systems in home space, but also widely used in household appliances. There are many kinds of intelligent household appliances, ranging from refrigerators, televisions, washing machines and air conditioners to rice cookers, induction cookers, desk lamps, night lights and bedside lamps. The functions of these intelligent appliances are different, but the human-computer interaction mode after connecting the network to realize intelligence is the same. The refrigerator temperature can be saved and the rice can be cooked through the intelligent terminal.

Key control can even achieve voice control. Compared with the cumbersome operation methods in the past, the application of artificial intelligence in the home appliance industry has created a new interaction mode and improved the control efficiency, which also plays a significant role in improving the user experience.

5 Application of Artificial Intelligence in Product Design for the Disabled

5.1 Proposed Framework for Face Detection and Recognition in CCTV Images

The maximum and minimum thresholds are selected initially. If the pixel's value is greater than the specified threshold, then one is assigned to the pixel, and if the value of the pixel is less than the threshold, then it is set to 0. Another case is when the value is the same as the threshold; it remains the same. Lastly, the edges are added to the original image to get the final enhanced image. Thus, detection and extraction of facial features become easy and increase the efficiency of the overall system [16].

5.2 Intelligent Monitoring

Intelligent monitoring is mainly through artificial intelligence technology. And Internet technology to design a comprehensive artificial intelligence system that can dynamically monitor, track and alarm indoor and outdoor conditions. With the support of this system, the owner can view the status of the indoor space through the system anytime and anywhere, grasp the indoor security status in time, and strengthen the safety management of the indoor space. At present, the more commonly used intelligent monitoring is embedded in the video server. This monitoring device integrates the intelligent behavior recognition algorithm, which can accurately identify and judge the human behavior in the picture scene, and send out alarm prompts under appropriate conditions (Fig. 1).

Fig. 1. Intelligent monitoring

At present, with the continuous improvement of functions and the continuous enrichment of products, intelligent monitoring equipment has gradually become a popular artificial intelligence technology product in interior design.

5.3 Intelligent Voice Prompt

The intelligent voice prompt function is mainly applicable to the environmental space that provides various services for the public, such as banks. When designing such spaces, designers and users will consider the personalized needs of the public in terms of services as much as possible. For this, artificial intelligence technology can be used to reduce the complexity of relevant operations. For example, some artificial intelligence voice prompt systems will be designed in the bank hall. Through the prompt function, customers can find the corresponding business counterpart faster and reduce the difficulty of customer business handling. When this technology is applied to the indoor smart home, it will be more helpful to some disabled people, such as those with visual impairment, blindness or mobility difficulties. Voice prompts can be more convenient for interaction.

5.4 Intelligent Security System

With the help of artificial intelligence technology, the intelligent security system has built residential alarm system, perimeter alarm system, electronic patrol system, closed-circuit television monitoring system, visitor intercom system and other systems with intelligent security protection functions for indoor space security and protection, so as to enhance the intelligence and security of indoor space protection.in terms of performance, the intelligent security system not only has the function of intelligent security, but also has the characteristics of low cost and strong function.

The above is to analyze the service design for the disabled from the field of smart home interior design and product design.

6 Conclusion

The application of artificial intelligence in interior design and product design will become more and more extensive with the development of technology. As an important technical tool, it has the advantages of expanding the design scheme, improving the service value in design and innovation ability in the service application to the disabled in interior design and product design. Smart home is the concrete manifestation of this phenomenon, and will be more and more widely used in more design fields to better serve mankind.

References

1. Mekonnen, A.K., Tangdiongga, E., Koonen, A.M.J.: High-capacity dynamic indoor all-optical-wireless communication system backed up with millimeter-wave radio techniques. J. Lightwave Technol. **36**(19), 4460–4467 (2018)
2. Jovanovic, M., Schmitz, M.: Explainability as a user requirement for artificial intelligence systems. Computer **55**(2), 90–94 (2022)

3. Perez Escoda, A.,Pedrero Esteban, A.: Framing ethical considerations on artificial intelligence bias applied to voice interfaces. In: ACM International Conference Proceeding Series, vol. 10, no. 21, pp. 250–255 (2020)
4. Ji, Q.: The design of the lightweight smart home system and interaction experience of products for middle-aged and elderly users in smart cities. Comput. Intell. Neurosci. (1), 279–351 (2022)
5. Yan, W., Wang, Z., Wang, H., Wang, W., Li, J., Gui, X.: Survey on recent smart gateways for smart home: systems, technologies, and challenges. Trans. Emerg. Telecommun. Technol. **33**(6), 40–67 (2022)
6. Sadikoglu-Asan, H.: User-Home relationship' regarding user experience of smart home products. Intell. Build. Int. **14**(1), 114–130 (2022)
7. Zhou, S., Van Le, D., Tan, R.: Configuration-adaptive wireless visual sensing system with deep reinforcement learning. IEEE Trans. Mob. Comput. (2022)
8. Bennett, K.C.: Prioritizing access as a social justice concern: advocating for ableism studies and disability justice in technical and professional communication. IEEE Trans. Prof. Commun. **65**(1), 226–240 (2022)
9. Zhang, N., Wang, J.: Application of ergonomics in the living bathroom environment. In: IEEE 11th International Conference on Computer-Aided Industrial Design and Conceptual Design, no. 1, pp. 651–654 (2010)
10. Li, L.: Research on the furniture arrangement in residential interior space. Agro Food Ind. Hi-Tech **28**(1), 2919–2921 (2017)
11. Matsumoto, S.: How do household characteristics affect appliance usage application of conditional demand analysis to Japanese household data. Energy Policy **94**, 214–223 (2016)
12. Liu, Y., Xiao, F.: Intelligent monitoring system of residential environment based on cloud computing and internet of things. IEEE Access **9**, 58378–58389 (2021)
13. Ceaparu, M., Toma, S.-A., Segarceanu, S., Suciu, G., Gavat, I.: Multifactor voice-based authentication system. J. Eng. Sci. Technol. Rev. 131–136 (2020)
14. Vijayakumar, P., Chang, V.: Achieving privacy-preserving DSSE for intelligent IoT healthcare system. IEEE Trans. Ind. Inform. **18**(3), 2010–2020 (2022)
15. Sreenivasulu, G., Anithaashri, T.P.: Implementation of IoT-based intelligent patient healthcare monitoring system using KNN algorithm. In: Pandian, A.P., Fernando, X., Haoxiang, W. (eds.) Computer Networks, Big Data and IoT, pp. 215–222. Springer, Singapore (2022). https://doi.org/10.1007/978-981-19-0898-9_16
16. Hayat-Hassan, S., Afsah-Abid: A real-time framework for human face detection and recognition in CCTV images. Math. Probl. Eng. **03**, 12 (2022)

Application and Prospect of Artificial Intelligence Image Analysis Technology in Natural Resources Survey

Yuehong Wang and Hao Luo[✉]

Tianjin Natural Resources Investigation and Registration Center, Tianjin 300201, China
40623249@qq.com

Abstract. With the rapid development of artificial intelligence, the application of Computer vision and Machine learning in the field of photogrammetry and remote sensing continues to enrich. Cognitive reasoning based on spatiotemporal big data has gradually deepened, which greatly promotes the development of remote sensing and surveying and mapping geographic information technology. The natural resource survey and monitoring technology system presents the characteristics of intelligence, spatialization, ubiquity and multi-source, which promotes the development of natural resource survey and monitoring business.

Keywords: Machine learning · Artificial intelligence · Remote Sensing Image Processing · CNN · FCN

1 Introduction

In the survey and monitoring of natural resources, remote sensing technology plays a important role. Compared with the traditional research work of land use and coverage change, remote sensing technology has the advantages of fast, accurate and real-time, and has been widely used in large-scale land use change monitoring. Remote sensing image refers to the image data obtained through an imaging sensor mounted on an aviation or space vehicle that contains the detailed object information of the observation scene. Remote sensing image classification aims to infer the correct object category by analyzing the object information contained in remote sensing images, which can provide theoretical basis and scientific guidance for practical application in natural resource survey such as environmental monitoring, geological exploration, precision agriculture and urban planning.

Remote sensing image classification usually extracts its features, and then use a classifier to divide its categories. In terms of feature extraction, it is mainly divided into manual design and deep learning. Based on manually designed feature extraction, artificial feature descriptors are usually used to extract the visual features of remote sensing images. Among them, Yang and Newsam [1] proposed to use the word bag model to encode the scale-invariant feature [2] of the local image blocks, and Zhao [3] introduced the Fisher kernel [4] to encode the manual features according to the

© The Author(s), under exclusive license to Springer Nature Singapore Pte Ltd. 2023
Q. Liang et al. (Eds.): AIC 2022, LNEE 871, pp. 151–158, 2023.
https://doi.org/10.1007/978-981-99-1256-8_18

gradient vector. Negrel et al. proposed the use of a second-order feature coding method to aggregate square deep network-based remote sensing image classification studies to the gradient histogram [5]. To effectively express high-resolution remote sensing images, Cheng et al. have proposed using pre-trained component detectors to capture the visual element [6].

In recent years, the spatial resolution of remote sensing images reaches submeters, which greatly improves the visual differentiation of ground objects, so that the morphology and the interval between them can be clearly captured. At the same time, due to the improvement of spatial resolution, in the case of constant geographical coverage, the remote sensing image information is also increased, this is the change of the traditional remote sensing image classification algorithm, can not make full use of and express complex abstract ground features in high resolution remote sensing image, resulting in the classification process, both in form and type accuracy and the actual difference, as shown in Fig. 1.

Fig. 1. High-resolution remote sensing image

Facing the massive data in high-scoring remote sensing images, the remote sensing image processing technology based on deep learning has become a research hot spot. Deep learning simulates the multi-layer nested neural structure of the human brain, extracts the features of the data layer by layer, and automatically learns the intrinsic laws and feature representation of the data.

2   Deep Learning

Deep learning, as a branch of machine learning, is a very important way to realize artificial intelligence. Compared with the traditional machine learning, deep learning has a deeper network structure, and the extracted features are highly abstract. In 2012, Alex designed AlexNet that won the ImageNet International Image Classification Competition, and defeated the shallow structure of machine learning methods such as support vector machine by an absolute advantage, which completely triggered the research boom of deep learning. The field of deep learning is huge, and its core algorithms include: CNN

(Convolutional Neural Network), RNN (Recurrent Neural Network) and FCN (Fully Convolutional Networks). Today, CNN are almost the dominant approach for image recognition and detection tasks. Since remote sensing image data is the basis of earth observation research, it is feasible for to apply deep learning algorithms in the field of remote sensing [7].

2.1 Convolutional Neural Network

Convolutional neural network is one of the important models in deep learning, through the convolutional layer and pooling layer staggered setting, as well as local feeling field and weight sharing two characteristics, makes CNN can efficiently extract abstract features in remote sensing image, general convolutional neural network structure includes convolutional layer, pooling layer and fully connected layer. The LeNet-5 model as an example is shown in Fig. 2. LeNet-5 consists of three convolutional layers, two pooling layers, and two fully connected layers.

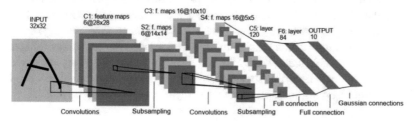

Fig. 2. LeNet-5 structure diagram

The convolution layer is used to extract features. Through the convolution layer, the network can convolve the input original image or the output feature map of the upper layer with the convolution kernel, and the output feature map passes through the activation function $f(x)$ to further Get the feature map G_j^L output by the L-layer, as shown in Eq. (1). Where M_j is the set of feature graphs under multichannel, X_i^{L-1} is the feature map of the input in layer L-1.

$$G_j^L = f\left(\sum_{i \in M_j} X_i^{L-1} * k_{ij}^1 + b_j^1\right) \tag{1}$$

The pooling layer is the downsampling layer, which is generally used in conjunction with the convolutional layer. The image after the operation of the convolutional layer has local features, and each pixel region of the image usually has a high similarity. The pooling layer represents the output in this area through a certain statistic, which can not only effectively reduce the amount of data, prevent the neural network from overfitting, but also further extract features. Common methods include mean-pooling and max-pooling, as shown in Eq. (2) and Eq. (3).

$$V_{mean\text{-}pooling} = \frac{\sum\limits_{i=1}^{N} a_i}{N} \tag{2}$$

$$V_{max\text{-}pooling} = Max(a_1, \ldots, a_i) \quad i = 1, 2 \ldots, N \tag{3}$$

The role of the fully connected layer is to reduce the dimension of the feature matrix and transform it from high-dimensional data to a one-dimensional vector. However, because the fully connected layer maps the data from one one-dimensional data to another one-dimensional data, the fully connected layer cannot operate directly for the two-dimensional data. Therefore, in the practical application process, the model usually places the Flatten layer between the fully connected layer and the convolutional layer, expanding the data, and reducing it from two-dimensional data to one-dimensional data.

Similarly, CNN play a huge role in remote sensing image processing. Based on the LeNet-5 network structure, Qu proposed a remote sensing image classification method based on the CNN model, which achieves 91% accuracy on the QuickBird remote sensing satellite image dataset. Dang trained AlexNet on surface coverage map spot data of high-resolution remote sensing images, and verified the accuracy of this model in surface coverage information classification through testing [8]. Wang has developed a high-resolution remote sensing image classification method that combines deep CNN and multi-kernel learning to improve the accuracy of image classification [9].

2.2 Fully Convolutional Neural Networks

In remote sensing image classification, the object-oriented classification method can obtain higher classification accuracy. The classification method adopts the segmentation and then classification, and extracts the segmentation objects and then uses the classifier. However, in the field of image processing, image semantic segmentation technology can extract fine-grained information and targets from image data without separate steps such as object-based feature merging, extraction and classification. This feature shows that image semantic segmentation is very suitable for remote sensing image processing [10, 11].

In 2014, Long proposed the FCN to solve the semantic segmentation problem. The FCN replaces the fully connected layer in the CNN with a convolutional layer to accept an input of any size and output pixel-level predictions of the same size.

MaggioriE et al., for the first time, the FCN network was successfully applied to the identification of remote sensing images. The network subsamples and extracts features, deconvolution samples to the original resolution, produces the final dense prediction, and can obtain more accurate classification results in a short computational time [12].

At present, the emergence of excellent semantic segmentation networks like FCN, SegNet, U-Net, DeepLabv3, MaskR-CNN, and RefineNet provides a lot of theoretical basis for the practical applications of deep learning. GuangshengChen et al. adopted DeepLabv3 architecture to increase void convolution and spatial pyramid pooling structure to improve the multi-scale expression ability of the network. The results show that this method has well improved for small target extraction and boundary accuracy of high-resolution remote sensing images [13]. ZhangK used the DenseNet structure to construct multiple circuit connections, and the network can adapt to different sizes of targets, and effectively extract road in high-resolution images [14].

3 Application and Prospect of AI Image Analysis Technology in Natural Resources Survey and Monitoring

3.1 LSTM Networks Model

At present, the means of natural resources investigation has been transformed from the traditional investigation method to a new method supported by relying on scientific and technological innovation and modern information technology construction. The rapid development of emerging technologies such as artificial intelligence is leading the innovation of natural resource survey methods.

Prof. Liangpei Zhang of Wuhan University used deep learning algorithm to identify house location and house damage, providing support for humanitarian rescue and disaster emergency response, and applied the high-resolution basic element mapping in Jiangsu Institute of Surveying and Mapping. Alibaba DAMO Academy using remote sensing AI to help the river four chaos monitoring, realize the national water, river houses, sand mining field, barrage, forest, greenhouses, photovoltaic electric field, cage farming and other typical target intelligent identification and change detection, provide technical support for law enforcement personnel; and using remote sensing image data and AI artificial intelligence technology, realize the comparison analysis of natural resources ecological environment elements, power ecological red line monitoring and supervision. Based on deep learning detection, big data analysis application and semantic segmentation algorithm, through the technical process of "positioning first before screening", Aerospace Hongtu realizes the greenhouse target detection and semantic segmentation detection, to meet the national needs of greenhouse detection and engineering detection accuracy.

3.2 Application Prospect

In the field of remote sensing, the basic software development has large investment, complex technology, high professional threshold, mainly with foreign products as the mainstream. For example, ERDAS Imagine of ERDAS, ENVI of Harris, ArcGIS of ESRI, Google Earth Engine of Google, PCIGeomatica of PCI of Canada, and eCognition of Germany are shown in Fig. 3.

With the rapid development of artificial intelligence technology in China, a number of innovative enterprises have also emerged in China to process remote sensing images through artificial intelligence technology. Taking Zhongke north latitude as an example, its independently developed remote sensing image annotation software ArcGis Pro can process a sample of a whole remote sensing image before slicing it, and export it according to the overlap degree, which can easily increase the sample data volume by multiple times. In addition, the company and Baidu have jointly developed Learth

<table>
<tr><td>(a) ERDAS Imagine</td><td>(b) Google Earth Engine</td></tr>
</table>

Fig. 3. Examples of Remote Sensing Products

Tianshu, a remote sensing intelligent vision platform based on deep learning, which integrates many services such as massive data management, data annotation, model training, model testing, and release, and remote sensing reasoning, as shown in Fig. 4.

Fig. 4. Learth Tianshu Remote Sensing Intelligent Vision Platform

4 Summary

Artificial intelligence technology, as a current social hotspot, has received a lot of attention. In the natural resources survey, various types of objects, complex environment, large data problems represented by the traditional remote sensing technology of natural resources survey, and the rise and development of artificial intelligence technology, improve the ability of remote sensing image feature extraction and change detection, to provide the natural resources survey is a more rapid, accurate and efficient method.

The production activities of the natural resources department have accumulated a large number of high-quality images and vector achievements over the years, laying a solid foundation for the construction of high-precision sample database. These high-precision sample data are also the data fuel for carry out deep learning intelligent training, which can help the more rapid development of artificial intelligence technology in the natural resource survey.

The technological revolution and industrial revolution set off by artificial intelligence will have a revolutionary impact on the development of surveying and mapping technology and the natural resource survey and monitoring business, and promote the significant improvement of the information extraction efficiency of natural resource elements, and the continuous innovation of natural resource governance mode. The future research work should focus on the difficult problems in the practical application of artificial intelligence, promote the deep integration of artificial intelligence technology and natural resource management, and improve the ability and intelligent level of natural resource governance.

References

1. Yang, Y., Newsam, S.: Bag-of-visual-words and spatial extensions for land-use classification. In: Sigspatial International Conference on Advances in Geographic Information Systems, p. 270. ACM (2010)
2. Lowe, D.G.: Distinctive image features from scale-invariant keypoints. Int. J. Comput. Vision **60**, 91–110 (2004)
3. Bei, Z., Zhong, Y., Zhang, L., et al.: The fisher kernel coding framework for high spatial resolution scene classification. Remote Sens. **8**(2), 157 (2016)
4. Perronnin, F., Dance, C.: Fisher kernels on visual vocabularies for image categorization. In: IEEE Conference on Computer Vision and Pattern Recognition, CVPR 2007. IEEE (2007)
5. Negrel, R., Picard, D., Gosselin, P.H.: Evaluation of second-order visual features for land-use classification. In: International Workshop on Content-based Multimedia Indexing. IEEE (2014)
6. Cheng, G., Han, J., Guo, L., et al.: Effective and efficient midlevel visual elements-oriented land-use classification using VHR remote sensing images. IEEE Trans. Geosci. Remote Sens. **53**(8), 4238–4249 (2015)
7. Dang, Y., Zhang, J., Deng, K., Zhao, Y., Yu, F.: Study on the evaluation of land cover classification using remote sensing images based on AlexNet. J. Geo-inf. Sci. **19**(11), 1530–1537 (2017)
8. Wang, X., Li, K., Ning, C., Huang, F.: Remote sensing image classification method based on deep convolution neural network and multi-kernel learning. J. Electron. Inf. Technol. **41**(5), 1098–1105 (2019)
9. Lin, H., Shi, Z., Zou, Z.: Fully convolutional network with task partitioning for inshore ship detection in optical remote sensing images. IEEE Geosci. Remote Sens. Lett. **14**(10), 1665–1669 (2017)
10. Chen, G., Zhang, X., Wang, Q., et al.: Symmetrical dense-shortcut deep fully convolutional networks for semantic segmentation of very-high-resolution remote sensing images. IEEE J. Sel. Top. Appl. Earth Observations Remote Sensing **11**, 1633–1644 (2018)
11. Sun, W., Wang, R.: Fully convolutional networks for semantic segmentation of very high resolution remotely sensed images combined with DSM. IEEE Geosci. Remote Sens. Lett. **15**, 1–5 (2018)
12. Maggiori, E., Tarabalka, Y., Charpiat, G., et al.: Fully convolutional neural networks for remote sensing image classification. In: Geoscience & Remote Sensing Symposium. IEEE (2016)

13. Chen, G., Li, C., Wei, W., et al.: Fully convolutional neural network with augmented atrous spatial pyramid pool and fully connected fusion path for high resolution remote sensing image segmentation. Appl. Sci. **9**(9), 1816 (2019)
14. Zhang, K., Guo, Y., Wang, X., et al.: Multiple feature reweight densenet for image classification. IEEE Access **7**, 9872–9880 (2019)

Application and Technical Analysis of Computer Vision Technology in Natural Resource Survey

Yuehong Wang and Hao Luo[✉]

Tianjin Natural Resources Investigation and Registration Center, Tianjin 300201, China
40623249@qq.com

Abstract. Natural resources are the material basis for human survival and national economic and social development. Natural resources survey is related to major national decision-making and deployment, fully supports and serves the economic and social development and ecological civilization construction in the new era, and systematically grasps the quantity, quality, spatial distribution, development and utilization, ecological status, and dynamic changes of national natural resources and landmark substrates. Computer vision technology uses computers to process, analyze and understand images to identify targets and objects in various patterns. The natural resources are photographed and imaged by industrial unmanned cameras or remote sensing satellites, and the characteristics are compared according to image processing, to realize the investigation and analysis of natural resources, improve the efficiency of natural resource management investigation, and provide a reference for planning decision-making and implementation.

Keywords: Natural Resource Survey · Image Classification · Object Detection · Semantic Segmentation

1 Introduction

The comprehensive survey of natural resources refers to the whole chain business of investigation, monitoring, evaluation, and zoning of natural resources such as land, minerals, forests, grasslands, water, wetlands, oceans, underground space, geological landscape, surface matrix, etc., to find out the quantity, quality, structure, and ecological function. The core objectives of the natural resources investigation and management plan are to meet people's living needs and protect natural resources, provide monitoring data sets for the investigation of the natural resources, provide scientific and reliable data support for the coordinated and sustainable development of resources, environment, and regions, promote green development, and promote the harmonious coexistence of human and nature.

Computer vision is one of the most popular researches in deep learning, and it is now widely used in all walks of life. Computer vision refers to simulating biological vision with cameras, computers, or other related equipment so that the computer can understand the contents of pictures or videos. In recent years, with the rapid development of computer vision technology, [1] proposed a GPU acceleration method to parallelize the

© The Author(s), under exclusive license to Springer Nature Singapore Pte Ltd. 2023
Q. Liang et al. (Eds.): AIC 2022, LNEE 871, pp. 159–166, 2023.
https://doi.org/10.1007/978-981-99-1256-8_19

eye detection algorithm based on Viola and Jones. The implemented algorithm is applied to the human-machine interface with fast feedback response to control the intelligent wheelchair specially designed for the disabled. To shorten image registration time and improve image quality, [2] proposed a vision image restoration algorithm based on a fuzzy sparse representation. The algorithm extracts the visual features of fuzzy medical computer vision images by constructing an image acquisition model. Then, the 3D vision reconstruction technology is used for the image feature registration. Next, establishing the model of the multidimensional histogram structure, and realizing the gray-level feature extraction of the image with the wavelet multidimensional scale feature detection method. Finally, the fuzzy sparse representation algorithm is used for automatic image optimization. [3] studied the application of computer vision and fuzzy logic in forensic medicine. This research not only focused on the most critical technology in forensic medicine but also deconstructed many computer vision, image processing, and fuzzy logic methods in the major subject of forensic medicine research.

The remainder of the paper is organized as follows: Sect. 2 analyzes common computer vision techniques, Sect. 3 describes the application of computer vision techniques in natural resource surveys, and Sect. 4 concludes.

2   Computer Vision Technology

Computer Vision(CV) uses cameras, computers, and other related equipment to identify, interpret and process images to simulate the human eye's recognition, tracking, and measurement of targets so that computers can understand. According to the different realization goals, computer vision can be divided into three classic tasks: Image Classification, Object Detection, and Semantic Segmentation.

2.1   Image Classification

Image classification is one of the most basic tasks in a CV. Its goal is to divide different images into several categories determined in advance with the minimum classification error. Specifically, according to the size of the categories, there are coarse-grained image classification and fine-grained image classification.

Coarse-grained image classification, also known as cross-species semantic classification, aims to identify different categories of objects at different species levels, for example, to identify whether an image belongs to buildings, bicycles, or people. There is often a large inter-class variance between classes and a small intra-class error within a class.

Fine-grained image classification is also known as sub-classification, which is more difficult than coarse-grained classification. It is often the re-classification of a large category, such as further breed identification of dogs. Fine-grained classification is to classify sub-categories more precisely based on distinguishing the basic categories. Due to the similar appearance and characteristics among images, and the presence of posture, angle of view, illumination, occlusion, background interference, and other effects during the acquisition process, the classification is also more difficult. At present, fine-grained image classification is solved by deep learning, which has achieved good results. These methods can be divided into strong-supervised fine-grained image classification and weak-supervised fine-grained image classification. The strong supervision uses additional manual annotation information such as bounding boxes and the key point to obtain the position and size of the target, which is conducive to improving the association between local and global, thus improving the classification accuracy. DeepLAC [4] model integrates the component positioning subnetwork, alignment subnetwork, and recognition subnetwork in the fine-grained image into a network, and proposes the valve connection function (VLF) to optimize the connection of the positioning subnetwork and recognition subnetwork, which can effectively reduce the error generated during recognition and alignment and ensure the accuracy of recognition. DeepLAC first unifies the resolution of the input image and then inputs it into the positioning subnetwork. The positioning subnetwork consists of five convolution layers and three full connection layers. The last full connection layer is used to regress the coordinate values of the upper left and lower right corners of the target box. The alignment sub-network receives the target position of the positioning sub-network, aligns the template, and inputs the aligned part image into the recognition sub-network for recognition. In the alignment process, basic operations such as offset, translation, scaling, and rotation are performed on the aligned subnetwork to generate the part area with attitude alignment, which is very important for recognition. In addition to attitude alignment, the identification and alignment results are used to further refine the positioning of the homogeneous sub-network in the backpropagation of the DeepLAC model. The recognition sub-network is the last module. The fine-grained category of the target is obtained by inputting the components with attitude alignment (Fig. 1).

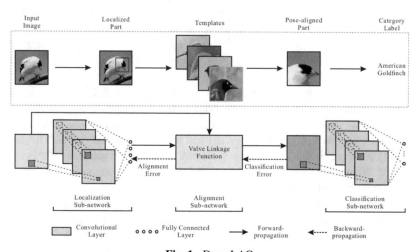

Fig. 1. Deep LAC

Weak supervision only uses the category annotation information of the image and does not use additional annotation. It can be divided into image filtering and bi-linear network. Image filtering filters the object-independent modules in the image with the help of image category information, such as Two Attention Levels (TAL) [5]. TAL uses object-level and local-level information to filter out irrelevant background information through the Selective Search algorithm, and then sends the filtered background to the CNN network for training to obtain object-level classification results. Then, it uses the clustering algorithm to refine the features of different locations and splices the features of different regions into SVM for training. The bilinear network coordinates the local feature extraction and classification by constructing two linear networks. The algorithm achieves 84.1% accuracy on the CUB-200 dataset. However, this method will generate high dimensions in the merging stage, making the whole calculation very expensive. The subsequent improvement direction of the bilinear network is to optimize or simplify bilinear convergence and reduce overhead calculation, such as Compact Bilinear CNN, Multivision Bilinear Convergence, etc.

2.2 Object Detection

Classification tasks care about the whole, describing the content of the entire image, while detection focuses on specific object targets, determining their location and size. Due to the different appearances, shapes, and postures of various objects, as well as the interference of factors such as illumination and occlusion during imaging, object detection has always been the most challenging problem in the field of machine vision. The target detection algorithm can be divided into two-stage target detection and one-stage target detection.

After feature extraction, the two-stage target detection algorithm first performs region proposal (RP) (a pre-selection box that may contain objects to be detected) and then performs sample classification through a convolutional neural network. Common two-stage algorithms are R-CNN, SPP-Net, Fast R-CNN, Faster R-CNN, R-FCN, etc., which are characterized by high detection accuracy, but the detection speed is not as high as the one-stage method. Taking R-CNN [6] as an example, after inputting the image to be detected, select the area frame for the image input in the first step, and use the edge, texture, color, color change, and other information of the image to select the area that may contain objects in the image. Then use the CNN network extracts features from the potential regions of the selected objects, and send the extracted features to the SVM classifier to obtain a classification model, where each category corresponds to an SVM classifier, and for each category of classifiers, it is only necessary to judge whether it is of this category. If multiple results are positive at the same time, the one with the highest probability is selected. This Regression is mainly for precise positioning and is used to correct the image area obtained by the second step Region proposals. Like the SVM classifier, each category corresponds to a Regression model (Fig. 2).

One-stage object detection algorithms extract feature values directly in the network to classify objects and locate them. Common one-stage target detection algorithms include OverFeat, YOLOv1, YOLOv3, SSD, RetinaNet, etc., which are characterized by fast detection speed, but the accuracy is not as high as that of the Two-stage method.

2.3 Semantic Segmentation

As one of the three classic computer vision problems, semantic segmentation is the most informative task among the three. It assigns the high-level semantic label of the image to each pixel, that is, classifies each pixel. The high-level semantic label refers to a variety

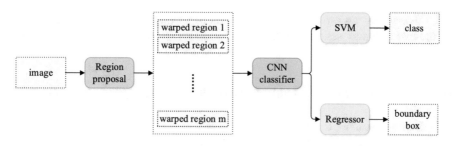

Fig. 2. R-CNN

of object categories (such as people, animals, cars, etc.) and background categories (such as the sky) in the image. Semantic segmentation tasks have high requirements for both classification accuracy and positioning accuracy: on the one hand, we need to locate the contour boundary of the object accurately, and on the other hand, we need to accurately classify the area within the contour, so that the specific objects are segmented from the background. Therefore, how maintaining the balance between localization accuracy and classification accuracy is a crucial issue in semantic segmentation. Generally speaking, to improve the classification accuracy, we need to increase the receptive field of the deep neural network, to integrate more information, but expanding the receptive field of the deep network will lead to a much loss of image details, which is not conducive to our positioning of the boundary. Therefore, an improvement goal in the direction of semantic segmentation is to fuse more global information without losing local details of the image. Before the popularity of deep learning methods, semantic segmentation methods such as TextonForest and random forest-based classifiers were common. With the popularity of deep convolutional networks, the performance of deep learning methods has greatly improved compared to traditional methods. The encoder-decoder structure is now more popular in image segmentation. The encoder gradually reduces the spatial dimension due to pooling, while the decoder restores the spatial dimension and detailed information gradually. Usually, there is a shortcut connection from the encoder to the decoder. U-net [7] is one such architecture. It is designed based on the idea of an FCN network. The entire network has only convolutional layers and no fully connected layers. In the enrichment path of FCN, the image resolution is reduced gradually, and the context information is enhanced. In the expansion path, the resolution of the feature map is increased gradually by up-sampling. At the same time, to incorporate the strong location information of low-level feature maps, the corresponding parts in the enriched path are combined into the extended path. This architecture can perform better location positioning. The modifications made by U-net are: 1) In the up-sampling part, the number of channels of the feature map is large, and the author believes it can transfer context information to higher resolution layers. A consequence of this is that it is essentially symmetrical to the enrichment path, thus looking like a U-shaped structure; 2) To predict the pixels in the image boundary area, the overlap-tile strategy is used to complete the missing context; 3) Since the training data is too small, a large amount of elastic deformation is used to enhance the data, which allows the model to better learn deformation invariance; 4) Another challenge in the cell segmentation task is how to separate objects of the same class that are in contact with each other. This paper proposes to use a weighted loss. In the loss function, segmenting the pixels of cells that touch each other gets a larger weight.

3   Application of CV Technology in Natural Resource Survey

3.1   Remote Sensing Image

Land and resources survey and monitoring are mainly completed by using optical satellite remote sensing with sub-meter spatial resolution. However, with the development of society and the economy, the demand for land and mineral resources is increasing day by day. Only using optical satellite remote sensing data can not meet the requirements of land and resource surveys, planning, supervision, and remediation requirements for the accuracy and effectiveness of information. Satellite remote sensing data combined with computer vision can be used to extract new hypothetical land use and occupied land information that is focused on monitoring, and provide a new method for the comprehensive application of multi-source remote sensing images in land use remote sensing monitoring. After being processed by GIS technology, through image fusion, geometric correction, image enhancement, image stitching, and analysis and comparison, changes in the utilization of land resources can be displayed (Fig. 3).

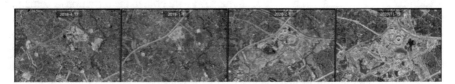

Fig. 3. Remote Sensing Image Examples

3.2   Wildlife Monitoring

Wildlife resources are also an important part of natural resources. The reduction of wild animals will lead to serious ecological problems, such as natural disasters, pests and diseases, and soil settlement and degradation. A variety of wild animals constitute a complete food chain. Once a node is broken, it will trigger a series of chain reactions, resulting in irreversible serious consequences. Therefore, the investigation of wild animal resources is also an indispensable part of natural resources investigation. However, compared with other natural resources, there are certain difficulties in the investigation of wild animal resources. First, the population of many wild animals is gradually decreasing, and it is difficult to find them in the wild; secondly, many animals have different habits from humans, like nocturnal and nocturnal, and whereabouts. It is relatively secretive and difficult to track and observe; in addition, wild animals are widely distributed and have a variety of living environments. Environmental factors make monitoring difficult and costly. Therefore, compared with traditional wildlife survey methods, such as the transect survey method and trace survey method, using cameras to track and survey them is the most suitable and timely method. Combined with artificial intelligence technology, important information such as species distribution, population size, and behavior of wild animals can be effectively analyzed, helping us to better monitor and protect wildlife resources (Fig. 4).

Fig. 4. Wildlife Monitoring

4 Conclusion

As a common hot topic of science and technology and society in today's world, artificial intelligence technology widely exists in all aspects of people's production and life and has greatly changed people's way of life. In natural resource surveys, computer vision technology is used to analyze and process images to achieve unified scheduling and management intelligently. Provide scientific and reliable data support for the coordinated and sustainable development of resources, environment, and regions, promote green development, and promote harmonious coexistence between man and nature. With the support of CV technology, the investigation of various natural resources has undergone revolutionary changes. A large number of data resources combined with a variety of artificial intelligence algorithms can efficiently and comprehensively analyze various natural resources. A large number of case studies can give a more reasonable decision-making deployment plan. Future research work should focus on "innovation" and "integration", integrating new technologies and new needs, continuously improving the integration, integration, and intelligence of facilities, and enhancing the comprehensive management and governance capabilities of natural resources.

References

1. Ghorbel, A., Ben Amor, N., Abid, M.: GPGPU-based parallel computing of viola and jones eyes detection algorithm to drive an intelligent wheelchair. J. Signal Process. Syst. 1–15 (2022)
2. Tang, Y., Qiu, J., Gao, M.: Fuzzy medical computer vision image restoration and visual application. Comput. Math. Methods Med. (2022)
3. Thakkar, P., Patel, D., Hirpara, I., et al.: A comprehensive review on computer vision and fuzzy logic in forensic science application. Ann. Data Sci. 1–25 (2022)
4. Lin, D., Shen, X., Lu, C., et al.: Deep lac: deep localization, alignment and classification for fine-grained recognition. In: Proceedings of the IEEE Conference on Computer Vision and Pattern Recognition, pp. 1666–1674 (2015)

5. Xiao, T., Xu, Y., Yang, K., et al.: The application of two-level attention models in deep convolutional neural network for fine-grained image classification. In: Proceedings of the IEEE Conference on Computer Vision and Pattern Recognition, pp. 842–850 (2015)
6. Girshick, R., Donahue, J., Darrell, T., et al.: Rich feature hierarchies for accurate object detection and semantic segmentation. In: Proceedings of the IEEE Conference on Computer Vision and Pattern Recognition, pp. 580–587 (2014)
7. Ronneberger, O., Fischer, P., Brox, T.: U-net: convolutional networks for biomedical image segmentation. In: Navab, N., Hornegger, J., Wells, W.M., Frangi, A.F. (eds.) MICCAI 2015. LNCS, vol. 9351, pp. 234–241. Springer, Cham (2015). https://doi.org/10.1007/978-3-319-24574-4_28

Review of Generative Adversarial Networks in Object Detection

Chenyang Zhou[✉], Siman Kong, and Jianzhi Sun[✉]

College of Computer Science, Beijing Technology and Business University, Beijing 100048, China
563588355@qq.com, sunjz@th.btbu.edu.cn

Abstract. Generative Adversarial Network (GAN) has become a research focus in the field of deep learning, and its research output has grown exponentially. This brand-new technology provides new ideas and methods for object detection, and has achieved remarkable success. Firstly, this paper introduces the basic GAN model and its derivative models in the field of object detection. Then analyzes the application status of GAN from object detection fields, such as industrial defect detection, medical image detection, remote sensing image detection, and face detection. Finally, summarize and prospect the technology development of generative adversarial networks.

Keywords: deep learning · generative adversarial network · adversarial learning · object detection

1 Introduction

In recent years, with the surge of data volume, the research of neural network algorithm and the improvement of computing power, the field of artificial intelligence has developed rapidly. Unsupervised learning has received more and more attention due to its unique methodology, becoming a hot direction of deep learning in recent years. Variational Auto-Encoder, Deep Belief Network, Flow-based model have emerged, however, the generalization ability of these models is insufficient. The proposal of Generative Adversarial Network (GAN) brings a new breakthrough in the field of artificial intelligence in the direction of generation. At present, generative adversarial network has become a research hotspot in deep learning, which has been applied in computer vision, language processing, anomaly detection and localization, information security, object detection and other fields.

2 GAN and Its Derivative Models in the Field of Object Detection

In 2014, Goodfellow et al. [1] proposed a new framework, Generative Adversarial Network (GAN), to estimate generative models through adversarial training.

Q. Liang et al. (Eds.): AIC 2022, LNEE 871, pp. 167–176, 2023.
https://doi.org/10.1007/978-981-99-1256-8_20

The generative adversarial network consists of a generator network and a discriminator network. The generator G captures the original data distribution and transforms the random noise into pseudo samples that are close to the real samples, while the discriminator D is used to determine whether the input data comes from the original data distribution or from the pseudo data generated by the generator. The output of the discriminator D is fed back to the generator G, which is constantly trained to make the generated data closer to the real data. The core idea of generative adversarial network is derived from the two-person zero-sum game [2]. When training GAN, the generator and the discriminator play a minimax game. After training, iteratively optimizes and finally reaches a Nash equilibrium [3]. The model structure of GAN is shown in Fig. 1.

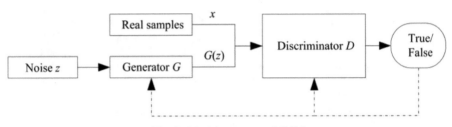

Fig. 1. Model structure of GAN

Compared with traditional generative models, GAN has many advantages: 1) It does not need to use Markov chain, only need to use back propagation to obtain gradient. 2) No inference is required in the learning process, and a large number of training tricks and loss functions that have been proposed can be used. 3) The generator of GAN is trained indirectly by discriminator, which means that the input source data is not directly copied into the parameters of the generator. 4) GAN can represent very sharp or even degenerate distributions.

GAN solves many problems of generative models, but it also has some limitations: 1) GAN training needs to achieve Nash equilibrium, but how to find the Nash equilibrium point is a difficulty and challenge [4, 5]. 2) The training process needs to ensure that the discriminator and the generator are trained synchronously, otherwise there will be a mode collapse problem. 3) Poor interpretability, unable to use mathematical formulas or parameters to represent the sample distribution generated. In addition, there are also problems such as vanishing gradients, too free models, difficulty in evaluating model training, and unsuitability for processing discrete data.

2.1 Perpetual GAN

Jianan Li et al. [6] proposed the Perceptual GAN in 2017, which is mainly used for small object detection. By mining structural correlation between objects of different scales, the feature representation of small objects is improved to make them similar to large objects. Perceptual GAN includes a generator network and a perceptual discriminator network. The generator network is a deep residual feature generation model, which converts the original poor features into high-score deformed features by introducing low-level fine-grained features. On the one hand, the discriminator network distinguishes the high-resolution features generated by small objects from the real large objects, and on the other hand, it could use the perceptual loss to improve the detection rate. Experiments on Traffic-sign Detection Datasets and Pedestrian Detection Datasets demonstrated the effectiveness of Perceptual GAN for small object detection.

2.2 MTGAN

Yancheng Bai et al. [7] proposed an end-to-end multi-task generative adversarial network (MTGAN). The model consists of a generator network and a discriminator network. In the generator, a super-resolution network (SRN) is introduced, which can up-sample smaller target images to a larger scale, and SRN can generate higher quality images. The discriminator is a multi-task network that simultaneously distinguishes real and generated super-resolution images, predicts object categories, and refines predicted bounding boxes. Furthermore, in order for the generator to recover more details for detection, the classification and regression losses in the discriminator are back-propagated into the generator during training. Extensive experiments on the COCO dataset demonstrate the effectiveness of the method in recovering sharp super-resolved images from small blurred images, and show improved detection performance over the new technique. Based on this model, reference [8] combines the reconstruction error with the discriminator output to improve the performance of anomaly detection.

2.3 CGAN

Mirza et al. [9] proposed Conditional Generative Adversarial Networks (CGAN) in 2014. The main contribution is to add extra information y to the input of the generator and discriminator of the GAN. In the generator, prior input noise $P_z(z)$ and conditional information y are combined as joint hidden layer representation. In the discriminator, the real data x and extra information y are used as input to the discriminant function. Reference [10] proposed an image-to-image conversion framework based on CGAN, using CGAN loss LcGAN and reconstruction loss LL1 to learn and observe the normal internal characteristics of crowd moving scenes. Alarge number of training G and D are performed on the normal frame of the moving scene and its corresponding optical flow images, and anomalies are detected by calculating the local difference between the generated content and the real frame.

3 GAN'S Application in the Field of Object Detection

Object detection is one of the classic tasks in the field of computer vision. With the large-scale application of deep learning in the field of object, the accuracy and speed of detection technology have been greatly improved, so it has been widely used in industrial defect detection, medical image detection, remote sensing image detection, face detection and other fields. Although the current object detection algorithm has achieved good results compared with traditional methods, it still cannot meet the needs of some special detection problems. The proposal of generative adversarial network provides a certain solution to the challenges in the field of object detection. Table 1 summarizes the models applied in different fields of object detection.

Table 1. GAN's application in the field of target detection

Field	Model
Industrial Defect Detection	AnoGAN [11], GANomaly [12], MVAE-GAN [13], Two-stage material surface defect detection based on CycleGAN [14], Defect simulation algorithm based on GLS-GAN [15]
Medical Image Detection	Sparse-GAN [16], MADGAN [17], Detection of lesion area based on GAN [18]
Remote Sensing Image Detection	Attention-GAN-Mask R-CNN [19], Ship identification based on GAN [20], FPN-GAN [21], EESRGAN [24]
Face Detection	SRGAN [29], GAN-based end-to-end convolutional neural network [30], C-GAN [31], GAN-based de-occlusion architecture [32]

3.1   Industrial Defect Detection

With the in-depth integration of the new generation of information technology and the manufacturing industry, people have higher and higher requirements for product quality. Product surface defects not only destroy the appearance quality of the product, but also may cause serious damage to the performance of the product. Surface defect detection is very important in order to detect problems in time, so as to effectively control product quality. The challenge of surface defects lies in the lack of sufficient training samples, especially defective samples. Insufficient training samples are prone to the problem of overfitting of deep learning models, so the research process of unsupervised learning methods is greatly accelerated. The manifestations of deep learning methods based on unsupervised learning are defect-free sample training methods and simulated defect sample training methods.

The defect-free sample training method obtains the defect detection result by learning the sample distribution, reconstructing the defect-free sample, and comparing the difference between the reconstructed sample and the input sample. Schlegl T et al. [11] proposed AnoGAN, the first method to introduce generative adversarial networks into defect detection. In this method, non-defective samples are used as training samples for unsupervised training. The idea is to learn the distribution of normal samples through GAN, and then map defective samples to the latent variables, and then reconstruct the samples from the latent variables. The reconstructed image will eliminate the defective part on the basis of retaining the original image characteristics, so the defect could be located by the residual between the reconstructed image and the input image. Akcay et al. [12] proposed a new anomaly detection algorithm, GANomaly, which utilizes conditional generative adversarial networks to jointly learn the generation and latent reasoning of high-dimensional image space and uses encoder-decoder-encoder in the generator network. The network, by comparing the latent variables obtained by coding and the latent variables obtained by reconstructing coding, could judge whether it is an abnormal sample. Through experiments on datasets from different fields, the validity of the model is verified. Li [13] proposed an image reconstruction model MVAE-GAN based on the generative adversarial network and the variational autoencoder. By training non-defective samples, it could learn the latent feature information of non-defective samples and make it have the reconstruction ability of normal samples. Experiments show that the model performs better in various indicators such as structural similarity and peak signal-to-noise ratio.

The simulation defect sample training method solves the problem of insufficient defect samples in practical applications by generating annotated simulation defect samples and training the defect detection model. Tsai et al. [14] proposed a two-stage Cycle-GAN to automatically synthesize and annotate local defective pixels. The first stage uses two CycleGAN models to automatically synthesize and annotate defective pixels in images. Then, the defect images synthesized in CycleGAN model and their corresponding annotation results are used as input-output pairs to train U-NET network. Experiments show that the scheme has sufficient generality for industrial detection applications. Liu [15] proposed a defect simulation algorithm based on GLS-GAN, which fused the network structure of U-shaped network and residual network characteristics. The region training strategy for local defect generation enables the generator network to create simulation defects based on the real image. Using simulation samples to train defect identification model and defect segmentation model can greatly reduce the number of necessary real defect samples.

3.2 Medical Image Detection

Medical images reflect the internal structure or internal function of anatomical area, and are one of the main bases for modern medical diagnosis. There are many types of images in the medical field, and they are greatly affected by the environment of the equipment, which will affect the doctor's diagnosis to a certain extent. Introducing deep learning into medical image detection, training the network based on imaging data and theoretical guidance, and improving the accuracy of diagnosis. Traditional segmentation and classification methods are mainly based on supervised learning and good matching of images or voxel labels, relying on large-scale unlabeled images of healthy subjects. 2D/3D single medical image reconstruction to detect outliers in the learned feature space or from high reconstruction loss.

Deep learning has been successful in retinal disease detection, but usually relies on large-scale labeled data. To break this limitation, Kang et al. [16] proposed a sparse constrained generative adversarial network (Sparse-GAN) for image anomaly detection using only health data. Sparse-GAN maps the reconstructed image into the latent space and attaches an encoder to reduce the effect of image noise, is able to predict anomalies in the latent space rather than image-level anomalies, and is also constrained by a novel sparse regularization network. The feasibility of OCT image anomaly detection and the effectiveness of the method are verified by public datasets, and the abnormal activation map of lesions is displayed, which makes the results more interpretable. Han et al. [17] proposed MADGAN, a two-step unsupervised medical anomaly detection method based on GAN-based multi-slice reconstruction. Combined with the WGAN-GP gradient penalty term and the L1 loss, train on three healthy brain MRI axial slices and reconstruct the next three slices, the L1 loss only generalizes well to unseen images with similar distribution to the training images, and the WGAN-GP loss captures recognizable structures. Since squared error is sensitive to outliers, L2 loss is used to clearly distinguish healthy samples from abnormal samples. Using 1133 healthy T1 MRI scans for training, the AUC was 0.727 when AD was detected in early MCI and 0.894 when AD was detected late. Based on a GAN model, Chen et al. [18] realized the detection of diseased regions in an unsupervised manner by learning the brain MRI data distribution of healthy subjects. The model is trained using T2-weighted health MRI images extracted from the Human Connectome Project dataset. The generator uses an adversarial autoencoder (AAE) and a variational autoencoder (VAE) to generate the health data distribution. The discriminator detects the lesion area by the pixel-wise intensity difference between the original image and the reconstructed image. The results showed that the AUC of the AAE model reached 0.923.

3.3 Remote Sensing Image Detection

Remote sensing image object detection has a wide application prospect in environmental supervision, military, transportation, civil industry and other fields. With the development of remote sensing platforms and high-performance sensors, the detailed information of ground objects obtained is more abundant. However, the traditional object detection algorithm is not ideal in the case of variable environment, complex background, object aggregation, too many small objects and so on, and could not extract valuable information.

Li et al. [19] proposed a remote sensing image object detection model Attention-GAN-Mask R-CNN based on attention mechanism and generative adversarial network. The model introduced a generative adversarial network in the Mask branch. The generators in the adversarial network are defined the same, so use a separate generative adversarial network to pre-train the Mask generation network of the Mask branch, thereby improving the accuracy of the generator in the original Mask branch. Lin et al. [20] proposed a SAR image ship object recognition method based on GAN pre-training CNN. Under the condition of limited training data, GAN was used to generate samples of corresponding categories, and then real samples with category annotations were used for fine-tuning to achieve higher feature extraction capability. The MSTAR dataset proves that the algorithm has good classification and recognition performance for multi-class objects.

To solve the problem of low detection performance of small objects in remote sensing images, Ahmad et al. [21] constructed a novel end-to-end FPN-GAN network architecture to solve the problem of small object detection. In the generator network, the feature pyramid is combined with the convolution layer, and the least squares loss is used for both global and local images in the discriminator network [22]. In order to improve the quality and efficiency of the model, Resnet-50 is used as the backbone network architecture. Through the experiments on the large-scale benchmark dataset DIOR [23] of optical remote sensing image object detection, the performance of the model in terms of accuracy, precision, recall, and validation loss is analyzed, and the superiority of the method is verified. Rabbi et al. [24], inspired by Edge Enhancement GAN (EEGAN) and ESRGAN, studied a novel Edge Enhancement Super-Resolution GAN (EESRGAN) to improve the quality of remote sensing images and train the network in an end-to-end manner. The whole architecture consists of EESRGAN network and detector network. For generator and edge enhancement network, residual dense block (RRDB) [25] is used. These blocks contain multi-level residual networks with dense connections and perform well in image enhancement. And the Charbonnier loss [26] is used in the edge-enhanced network, and finally different detectors are used to detect small objects from SR images. The method is applied to the created oil and gas storage tank (OGST) dataset [27] and COWC dataset [28], and the detection performance of different use cases is compared. The results show that the method is superior to the latest research results.

3.4 Face Detection

In recent years, face detection has been applied to people's daily life. With the rapid development of deep learning, a large number of face detection algorithms have emerged. However, due to the gradual expansion of the application scope and the complex use scenarios, the current technology have problems of misjudgment in the case of low resolution, angle, occlusion, different face image styles, and face forgery. The emergence of generative adversarial networks has played an important role in solving the above problems.

Generative adversarial network improves the face detection effect in different environments and scenes by using context information, super-resolution reconstruction, image enhancement and other methods. SRGAN [29] is the first deep learning algorithm to apply generative adversarial network to the field of super-resolution reconstruction. However, in low-resolution face images, the obtained images are blurry and lack details. Bai et al. [30] proposed based on GAN end-to-end convolutional neural network. In the generator, using a super-resolution network (SRN) to upsample small faces, and make use of the surrounding region information of the face cropped by enlarged window to train GAN, but there are still problems of ambiguity and lack of detailed information. Then further introduce an improved sub-network (RN) to restore the missing information and generate high-resolution images. In the discriminator, designed a new loss function to complete the discrimination of real/fake face and face/non-face. Zhang et al. [31] proposed a Contextual based Generative Adversarial Network(C-GAN), which added a regression branch to improve the border detection of difficult faces. Through ablation experiments, the effectiveness of C-GAN on blurring small faces is verified. Gu [32] proposed a de-occlusion architecture with generative adversarial network as the main body. By improving U-net network, convolution with padding was used, and the edge information was fully utilized to improve the network performance. The improved SU-net network is used as generator network, which effectively improves the face detection accuracy.

4 Conclusion

This paper reviews the basic theory of GAN and its research progress, focusing on a systematic review of its application in object detection fields, such as industrial defect detection, medical image detection, remote sensing image detection, and face detection. As a generative model, GAN provides a good solution to the problems of insufficient samples, low image resolution, and difficulty in feature extraction, and it could make the network more robust to occlusion and deformation problems, improve the accuracy of detection.

Object detection has very important value for the current information society. The application of GAN in object detection could be deeply explored. It could innovate the algorithm, select an appropriate loss function, optimize the network structure, and then combine with specific application scenarios to improve the real-time and accuracy of detection. GAN could be used for sample generation to expand the datasets to address the lack of training data for many object detection scenarios. Improving the interpretability and the evaluation criteria of the model also have very important research value.

References

1. Goodfellow, I., Pouget-Abadie, J., Mirza, M., et al.: Generative adversarial nets. In: Neural Information Processing Systems. MIT Press (2014)
2. Mertens, J.F., Zamir, S.: The value of two-person zero-sum repeated games with lack of information on both sides. Int. J. Game Theory 1(1), 39–64 (1971)
3. Ratliff, L.J., Burden, S.A., Sastry, S.S.: Characterization and computation of local Nash equilibria in continuous games. In: Proceedings of the 51st Communication, Control, and Computing (Allerton), Monticello, IL, USA, pp. 917–924. IEEE (2013)
4. Arjovsky, M., Bottou, L.: Towards principled methods for training generative adversarial networks. arXiv:1701.04862. https://arxiv.org/abs/1701.04862
5. Wang, B., Liu, K., Zhao, J.: Conditional generative adversarial networks for commonsense machine comprehension. In: Proceedings of the 26th International Joint Conference on Artificial Intelligence, 19–25 August 2017, Melbourne, Australia, Melbourne: IJCAI, pp. 4123–4129 (2017)
6. Li, J., Liang, X., Wei, Y., Xu, T., Feng, J., Yan, S.: Perceptual generative adversarial networks for small object detection. In: 2017 IEEE Conference on Computer Vision and Pattern Recognition (CVPR), pp. 1951–1959 (2017). https://doi.org/10.1109/CVPR.2017.211
7. Bai, Y., Zhang, Y., Ding, M., Ghanem, B.: SOD-MTGAN: small object detection via multitask generative adversarial network. In: Ferrari, V., Hebert, M., Sminchisescu, C., Weiss, Y. (eds.) Computer Vision – ECCV 2018. LNCS, vol. 11217, pp. 210–226. Springer, Cham (2018). https://doi.org/10.1007/978-3-030-01261-8_13
8. Song, K.: Research and implementation of abnormal detection in aircraft agent based on machine learning. University of Electronic Science and Technology of China (2022). https://doi.org/10.27005/d.cnki.gdzku.2022.002510
9. Mirza, M., Osindero, S.: Conditional generative adversarial nets. arXiv:1411.1784 (2014)
10. Yarlagadda, S.K., Güera, D., Bestagini, P., Maggie Zhu, F., Tubaro, S., Delp, E.J.: Satellite image forgery detection and localization using GAN and one-class classifier. Electron. Imaging 2018(7), 214-1–214-9 (2018)
11. Schlegl, T., Seeböck, P., Waldstein, S.M., Schmidt-Erfurth, U., Langs, G.: Unsupervised anomaly detection with generative adversarial networks to guide marker discovery. In: Niethammer, M., Styner, M., Aylward, S., Zhu, H., Oguz, I., Yap, P.-T., Shen, D. (eds.) IPMI 2017. LNCS, vol. 10265, pp. 146–157. Springer, Cham (2017). https://doi.org/10.1007/978-3-319-59050-9_12
12. Akcay, S., Atapour-Abarghouei, A., Breckon, T.P.: Ganomaly: semi-supervised anomaly detection via adversarial training. In: Jawahar, C.V., Li, H., Mori, G., Schindler, K. (eds.) ACCV 2018. LNCS, vol. 11363, pp. 622–637. Springer, Cham (2019). https://doi.org/10.1007/978-3-030-20893-6_39

13. Li, J.: Research on defect inspection of wooden floor based on unsupervised learning. Beijing Jiaotong University (2021). https://doi.org/10.26944/d.cnki.gbfju.2021.000622
14. Tsai, D.M., Fan, S.K.S., Chou, Y.H.: Auto-annotated deep segmentation for surface defect detection. IEEE Trans. Instrum. Meas. **70**, 1–10 (2021)
15. Liu, L.: Research on surface defect detection algorithm based on deep learning. Huazhong University of Science & Technology (2019). https://doi.org/10.27157/d.cnki.ghzku.2019. 001438
16. Zhou, K., et al.: Sparse-Gan: sparsity-constrained generative adversarial network for anomaly detection in retinal OCT image. In: 2020 IEEE 17th International Symposium on Biomedical Imaging (ISBI), pp. 1227–1231 (2020). 10.1109 /ISBI45749.2020.9098374
17. Han, C., Rundo, L., Murao, K., et al.: MADGAN: unsupervised medical anomaly detection GAN using multiple adjacent brain MRI slice reconstruction. BMC Bioinform. **22**, 31 (2021). https://doi.org/10.1186/s12859-020-03936-1
18. Chen, X., Konukoglu, E.: Unsupervised detection of lesions in brain MRI using constrained adversarial auto-encoders (2020). https://arxiv.org/abs/1806.04972
19. Li, J., Deng, Y., Wu, X., et al.: Object detection in remote sensing image based on attention mechanism and GAN. Comput. Syst. Appl. **31**(6), 182–191 (2022). http://www.c-s-a.org.cn/ 1003-3254/8490.html
20. Lin, Z.: Ship detection and recognition in remote sensing images based on deep learning. National University of Defense Technology (2018). https://doi.org/10.27052/d.cnki.gzjgu. 2018.000972
21. Ahmad, T., Chen, X., Saqlain, A.S., Ma, Y.: FPN-GAN: multi-class small object detection in remote sensing images. In: 2021 IEEE 6th International Conference on Cloud Computing and Big Data Analytics (ICCCBDA), pp. 478–482 (2021). https://doi.org/10.1109/ICCCBD A51879.2021.9442506
22. Jolicoeur-Martineau, A.: The relativistic discriminator: a key element missing from standard GAN (2018)
23. Li, K., Wan, G., Cheng, G., Meng, L., Han, J.: Object detection in optical remote sensing images: a survey and a new benchmark. ISPRS J. Photogram. Remote Sens. **159**, 296–307 (2020)
24. Rabbi, J., Ray, N., Schubert, M., Chowdhury, S., Chao, D.: Small-object detection in remote sensing images with end-to-end edge-enhanced GAN and object detector network. Remote Sens. **12**(9), 1432 (2020). https://doi.org/10.3390/rs12091432
25. Wang, X., et al.: ESRGAN: enhanced super-resolution generative adversarial networks. In: Leal-Taixé, L., Roth, S. (eds.) ECCV 2018. LNCS, vol. 11133, pp. 63–79. Springer, Cham (2019). https://doi.org/10.1007/978-3-030-11021-5_5
26. Charbonnier, P., Blanc-Féraud, L., Aubert, G., Barlaud, M.: Two deterministic half-quadratic regularization algorithms for computed imaging. In: Proceedings of International Conference on Image Processing, vol. 2, pp. 168–172 (1994)
27. Rabbi, J., Chowdhury, S., Chao, D.: Oil and Gas Tank Dataset. In Mendeley Data, V3 (2020). https://data.mendeley.com/datasets/bkxj8z84m9/3
28. Mundhenk, T.N., Konjevod, G., Sakla, W.A., Boakye, K.: A large contextual dataset for classification, detection and counting of cars with deep learning. In: Leibe, B., Matas, J., Sebe, N., Welling, M. (eds.) ECCV 2016. LNCS, vol. 9907, pp. 785–800. Springer, Cham (2016). https://doi.org/10.1007/978-3-319-46487-9_48
29. Ledig, C., Theis, L., Huszár, F., et al.: Photo-realistic single image super-resolution using a generative adversarial network. In: Proceedings of the IEEE Conference on Computer Vision and Pattern Recognition, pp. 4681–4690 (2017)
30. Bai, Y., Zhang, Y., Ding, M., et al.: Finding tiny faces in the wild with generative adversarial network. In: Proceedings of the IEEE Conference on Computer Vision and Pattern Recognition, pp. 21–30 (2018)

31. Zhang, Y., Ding, M., Bai, Y., et al.: Detecting small faces in the wild based on generative adversarial network and contextual information. Pattern Recogn. **94**, 74–86 (2019)
32. Gu, J.: Design and Implementation of Yujiao Robot Face Recognition System. Chongqing University (2019). https://doi.org/10.27670/d.cnki.gcqdu.2019.001616

A Review of the Application of Convolutional Neural Networks in Object Detection

Siman Kong, Chenyang Zhou, and Jianzhi Sun[✉]

College of Computer Science, Beijing Technology and Business University, Beijing 100048, China
ksm1075177520@163.com, sunjz@th.btbu.edu.cn

Abstract. Convolutional neural networks have become one of the important research directions in the field of computer vision based on deep learning, and they are widely used in object detection with excellent performance. Based on the research results of many scholars, the paper reviews the application research of convolutional neural networks in object detection, introduces its application research from three aspects: object detection based on candidate region, object detection based on regression and video object detection, and finally summarizes the development of convolutional neural networks in the field of object detection.

Keywords: convolutional neural networks · deep learning · computer vision · object detection

1 Introduction

In recent years, Convolutional Neural Networks (CNN) have made a series of break-throughs in the field of computer vision and received wide attention from all walks of life. In the field of computer vision, object detection is a fundamental problem, which aims at object classification and object localization of images by object detection algorithms. The traditional object detection methods mainly include three stages of region selection, feature extraction and classification, but they cannot effectively meet the application requirements at this stage. After continuous attempts, many scholars have proposed the model of CNN and it is widely used in object detection. Based on this, the article reviews the research on the application of CNN in object detection, mainly from three aspects: object detection application based on candidate region, object detection application based on regression and video object detection application, and concludes with an outlook on the development of CNN in the field of object detection.

2 Application of Object Detection Based on Candidate Region

The object detection algorithm based on the candidate region firstly searches the region, and then classifies the candidate region to obtain the final detection result. The main algorithms are shown in Table 1.

© The Author(s), under exclusive license to Springer Nature Singapore Pte Ltd. 2023
Q. Liang et al. (Eds.): AIC 2022, LNEE 871, pp. 177–186, 2023.
https://doi.org/10.1007/978-981-99-1256-8_21

Table 1. Object detection algorithms based on candidate region

Years	Scholar	Algorithm	Applicable scene
2014	Girshick et al. [1]	R-CNN	Object Detection
2015	He et al. [2]	SPPNet	Object Detection
2015	Girshick et al. [4]	Fast R-CNN	Object Detection
2015	Ren et al. [6]	Faster R-CNN	Object Detection
2016	Dai et al. [10]	R-FCN	Object Detection
2017	He et al. [13]	Mask R-CNN	Object Detection and Instance Segmentation
2018	Cai et al. [18]	Cascade R-CNN	High precision object detection
2019	Li et al. [21]	TridentNet	High-precision, multi-scale object detection

2.1 R-CNN

In 2014, Girshick et al. [1] proposed the R-CNN (Regions with CNN features) algorithm, which is the beginning of using CNN in the field of object detection. Its structure is shown in Fig. 1. R-CNN is based on CNN, linear regression and support vector machine (SVM) algorithms to implement object detection techniques. Its detection is mainly divided into three stages, firstly, candidate regions are generated by selective search, then CNN-based feature extraction is performed on the candidate regions, and finally, SVM is used for result prediction. R-CNNN is not only able to extract advanced features, but also speeds up detection. However, R-CNN is not perfect and has defects such as incomplete picture information and low detection accuracy.

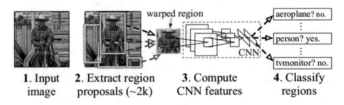

1. Input image   2. Extract region proposals (~2k)   3. Compute CNN features   4. Classify regions

Fig. 1. R-CNN network structure diagram [1]

2.2 SPPNet

In 2015, in order to overcome some difficulties of R-CNN, He et al. [2] artificially equipped the network with "Spatial Pyramid Pooling (SPP)" and proposed Spatial Pyramid Pooling Network (SPPNet). Specifically, an SPP layer is added to the last convolutional layer, and the SPP layer pools features in a single convolutional operation and produces a fixed-length output, which is then fed to the fully connected layer. Although SPPNet achieved good results in the 2014 ILSVRC competition, it has shortcomings, such as the need to store a large number of features and high space consumption. Since then, CNN has been widely used in network traffic classification.

The traditional CNN model has some drawbacks in using traffic data set in the network. Therefore, Zhou et al. [3] proposed a new CNN traffic classification model based on SPPNet. Based on the CNN model of LeNet-5, the new model replaces the maximum pool with a spatial pyramid pool in the pooling layer before the fully connected layer, which can achieve network traffic with indeterminate long datasets. It is also demonstrated that this model reduces the influence of human factors on traffic classification for traffic applications.

2.3 Fast R-CNN

In 2015, Girshick et al. [4] developed Fast R-CNN (Fig. 2), which overcomes the shortcomings of R-CNN and SPPnet, and enhances its performance and accuracy. The original image and multiple regions of interest are fed into the fully convolutional network, and each region of interest is in turn integrated into a feature mapping of constant size, which is then mapped to a feature vector through a fully connected layer. Fast R-CNN has faster training speed and testing speed than R-CNN and SPPnet, so it is widely used in object detection.

In face detection applications, to avoid the drawbacks associated with complex environments, Lin et al. [5] proposed a multiscale Fast R-CNN method based on upper and lower layers (UPL-RCNN). The network consists of a spatial affine transform component and a feature region component (ROI). This method plays a crucial role in face detection.

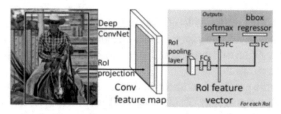

Fig. 2. Fast R-CNN network structure [4]

2.4 Faster R-CNN

In 2015, Ren et al. [6] introduced the Region Proposal Network (RPN) and proposed Faster R-CNN. Figure 3 is a diagram of the detection process of Faster R-CNN. RPN is implemented by a fully convolutional network and has the ability to simultaneously predict the object boundary and occurrence probability at each location. Compared with Fast R-CNN, Faster R-CNN has improved both accuracy and speed, and is widely used in the field of object detection. For example, Zhu et al. [7] applied Faster R-CNN to train detection network on book digital image database and achieve automatic recognition and localization of books.

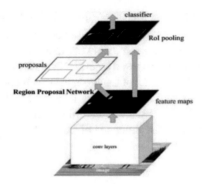

Fig. 3. Detection process diagram of Faster R-CNN [6]

In order to meet the application requirements, some researchers have also proposed improvements to Faster R-CNN. Gong Shengbin et al. [8] proposed a lightweight improvement method of Faster R-CNN algorithm, that is, to improve the feature extraction network of the algorithm Inception-V2, and improve the defects of traditional object detection methods such as difficult feature extraction and low detection accuracy. Xu Degang et al. [9] proposed an improved Faster R-CNN grainworm object detection method based on deep learning to improve the performance of grainworm detection by introducing a pyramid pooling module after the feature map and optimizing the loss function.

2.5 R-FCN

In 2016, Dai et al. [10] designed R-FCN based on regional full convolutional network and proposed a new convolutional layer, Position Sensitive Score Map, which guarantees the invariance of position information and is widely used in vehicle object detection and pedestrian object detection.

In response to the problems of traditional vehicle object recognition algorithms and pedestrian detection algorithms, Zhou et al. [11] proposed a vehicle object detection method based on R-FCN. The method is based on the idea of full convolutional network and applies the R-FCN framework to avoid the feature selection problem that occurs in traditional detection. Liu Wanjun et al. [12] introduced R-FCN algorithm in performing pedestrian detection and proposed a small-scale pedestrian detection algorithm with improved R-FCN model to improve the accuracy rate of pedestrian detection.

2.6 Mask R-CNN

In 2017, He et al. [13] proposed Mask R-CNN, which combines the ideas of Faster R-CNN and FCN, and extends the faster R-CNN by adding a branch to predict the segmentation mask on each region of interest, performing classification and bounding box regression in parallel with existing branches. Mask R-CNN can not only perform the object detection task, but can also perform the instance segmentation task. Wang Zhibo et al. [14] applied the Mask R-CNN model to road traffic sign recognition for instance segmentation of traffic signs. Safonova et al. [15] completed Mask R-CNN based multi-resolution image olive tree biovolume segmentation by using Mask R-CNN and UAV images for olive tree crown and shadow segmentation to further estimate biological volume of individual trees.

In order to meet the application, some improvement methods have also been proposed based on the existing Mask R-CNN. Shi et al. [16] improved on the original Mask R-CNN, removed the Mask branch, and adjusted the ratio of the anchor boxes in the RPN network, which greatly improved the speed of the algorithm while ensuring the detection accuracy. To address the problem of poor detection accuracy when Mask R-CNN algorithm is applied to pedestrian instance segmentation, Yin Song et al. [17] proposed an improved Mask R-CNN algorithm. Add the concatenated feature pyramid network module to the Mask branch, and then fuse the multi-layer features generated by the network to make full use of the semantic information of different feature layers.

2.7 Cascade R-CNN

In 2018, Cai et al. [18] proposed a multi-stage object detection framework Cascade R-CNN for improving the quality of object detectors. Cascade R-CNN can be viewed as a cascaded Faster R-CNN structure consisting of a series of detectors trained with increasing cross-joint thresholds. This architecture was shown to avoid the problems of training overfitting and inference quality mismatch [18].

Wu Jun et al. [19] applied the Cascade R-CNN algorithm to the detection of transmission line defects, and used a multi-level cascade detector to discriminate and classify the small objects of the transmission line. Aiming at the problem that the defects of anti-vibration hammer components of high-voltage transmission lines are difficult to accurately locate and identify, Bao et al. [20] proposed a method for detecting anti-vibration hammer defects based on the improved Cascade R-CNN algorithm, and proved that in the anti-vibration hammer defect test set It has a high detection accuracy rate, which is better than other mainstream object detection algorithms.

2.8 TridentNet

Scale variation is one of the key challenges in object detection.In 2019, Li et al. [21] proposed a novel network structure-TridentNet, a three-branch network, which aims to generate scale-specific feature mappings with uniform representation capabilities.TridentNet constructs a parallel multi-branch architecture, in which each branch shares the same TridentNet achieves a significant accuracy improvement over previous multi-scale object detection algorithms, with an accuracy of 48.4% on the COCO dataset [21].

3   Application of Object Detection Based on Regression

The regression-based object detection algorithm does not directly generate the region of interest, but regards the object detection task as a regression task for the entire image, which truly achieves end-to-end. Its main algorithms include YOLO series and SSD series.

3.1 YOLO Series

In 2015, R. Joseph et al. [22] proposed YOLO (You Only Look Once), which is a pioneering one-stage detection algorithm. Its network structure is shown in Fig. 4. The network regards the problem of object detection as a regression problem to solve. It only needs to process the input image once to obtain the category and position of the object. The advantage is that the detection speed is very fast.

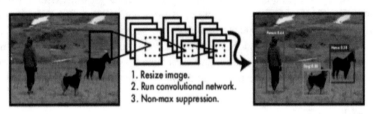

Fig. 4. YOLO network structure diagram [22]

In 2017, an improved version of YOLO, YOLOv2 [23], was proposed, mainly improving on the 2015 YOLO model by (1) using BN operations for each convolutional layer, (2) using Darknet-19 as the YOLOv2 backbone network, and (3) using anchor box and convolution to do prediction.

In 2018, YOLOv3 [24] was proposed to be used mainly for multi-scale object detection. YOLOv3 uses Darknet-53 as the backbone network. In order to enhance the accuracy of small object detection, the FPN architecture is applied to three feature maps of different scales.

In 2020, Bochkovskiy et al. proposed YOLOv4 [25], which is mainly used for high-precision real-time object detection. This model uses CSPDarknet53 as the backbone network, increases the receptive field by introducing the SPP idea, and replaces the FPN with PANet to improve the detection effect of small object detection objects.

The YOLO series is a commonly used object detection algorithm, which is widely used in various scenarios. In order to solve the problem of low accuracy of infrared object detection in complex ground scenes, Liu Dong et al. [26] proposed a lightweight infrared real-time object detection model MCA-YOLO. Mobilenet-v2 is used as the backbone network, combined with deep separable convolution, and the coordinate attention module CA (Coordinate Attention) is embedded in the neck network of the model to enhance the feature extraction ability of the model. Yu et al. [27] proposed a face recognition and standard wear detection algorithm based on improved YOLOv4 to solve the problems of low accuracy, low real-time performance and poor robustness caused by complex environment.

3.2 SSD Series

In 2016, Liu et al. [28] proposed the SSD (Single Shot Multibox Detector) model, which for the first time realized object detection with multi-scale feature maps. SSD discretizes the output space of bounding boxes into a default set of boxes with different aspect ratios and scales at each feature map location. At prediction time, the network generates scores for each object class present in each default box and adjusts the box to better match the object shape [28]. And SSD ensures the accuracy of object detection by adopting the anchor point mechanism in Faster R-CNN.

There are two deficiencies in SSD: one is that the same object is easily detected by frames of different sizes at the same time; the other is the missed detection of small-scale objects. To address these deficiencies, some optimization models are proposed. In 2017, Jeong et al. [29] proposed the RSSD algorithm to fuse the feature information of different layers by combining up and down sampling. In the same year, Fu et al. proposed DSSD [30], which used ResNet-101 as the backbone network, introduced a residual module and added some auxiliary convolutional layers, which effectively improved the accuracy of small object detection.

The feature pyramid detection method of SSD is difficult to integrate features of different scales. Li et al. proposed an enhanced SSD, FSSD [31], which mainly borrows the idea of FPN and reconstructs a set of pyramid feature maps to connect features of different scales and layers. Although the detection speed did not change much, the accuracy of object detection was improved.

4 Video Object Detection Application

Video is composed of continuous images, and its object detection can be regarded as an extension of image object detection in the field of video, which is more challenging and practical than image object detection.

With the explosive growth of video data, more and more researchers are working on the field of video object detection. In 2016, the team of Dai Jifeng of Microsoft Research Asia proposed the deep feature flow algorithm DFF [32] (Deep Feature Flow). Its network structure can be roughly divided into feature extraction network and detection network. The feature network only acts on the feature extraction of key frames, and the detection network is used to detect the feature map. Since object detection in video is affected by various environmental factors, it is difficult to successfully detect objects in a single frame image. In 2017, Dai Jifeng's team proposed a new algorithm FGFA [33] (Flow-Guided Feature Aggregation) based on DFF, which is mainly composed of two modules: feature extraction and feature fusion, and won the championship in the 2017 ILSVRC competition. In 2018, Gedas Bertasius et al. [34] proposed Spatiotemporal Sampling Networks (STSN), which adopted the Deformable Conv structure to extract adjacent frame spatial features to perform video object detection. In 2020, a learnable spatio-temporal sampling module (LSTS) was proposed in [35] to achieve the ultimate in accuracy and speed, and to achieve the purpose of spreading advanced features between frames.

In the last two years, several researchers have been improving the accuracy and speed of video object detection in their improvement innovations. Gao Hui [36] used YOLOv5 as the basic network for video object detection in video object detection, and a subset of the ImageNet VID dataset as the dataset. Using the great similarity and temporal correlation between adjacent frames of video data, the detection speed is improved by selecting key frames and frame-level propagation, and the global semantic information of video data is used to improve the detection accuracy through memory module. Shi Yuhu et al. [37] proposed a lightweight video object detection method. By designing a feature propagation model and constructing a dynamic allocation key frame module, the detection speed is accelerated and the real-time online video object detection is realized on the premise of ensuring the detection accuracy.

5 Conclusion

Object detection based on CNN has become a research hotspot, and new object detection results appear every year, with great progress in both accuracy and speed. In general, the object detection process based on candidate regions is simple and has high accuracy, but the speed is slow. The regression-based object detection is slightly faster, but the accuracy is not as good as the former. Although some achievements have been made in object detection in the video field, there are still many difficulties. How to use the temporal context information of the current image frame to improve the performance of the detection model remains to be solved. Although object detection has been widely used in various fields, there are some difficult problems to be broken through, and the future can be approached in the direction of small object detection problem, weakly supervised object detection problem, 3D and video object detection, etc.

References

1. Girshick, R., Donahue, J., Darrell, T., Malik, J.: Rich feature hierarchies for accurate object detection and semantic segmentation. In: 2014 IEEE Conference on Computer Vision and Pattern Recognition, pp. 580–587 (2014). https://doi.org/10.1109/CVPR.2014.81
2. He, K., Zhang, X., Ren, S., Sun, J.: Spatial pyramid pooling in deep convolutional networks for visual recognition. IEEE Trans. Pattern Anal. Mach. Intell. **37**(9), 1904–1916 (2015). https://doi.org/10.1109/TPAMI.2015.2389824
3. Zhou, H., Wang, Y., Ye, M.: A method of CNN traffic classification based on Sppnet. In: 2018 14th International Conference on Computational Intelligence and Security (CIS), pp. 390–394 (2018). https://doi.org/10.1109/CIS2018.2018.00093
4. Girshick, R.: Fast R-CNN. In: 2015 IEEE International Conference on Computer Vision (ICCV), pp. 1440–1448 (2015). https://doi.org/10.1109/ICCV.2015.169
5. Jiang, L., et al.: Application of a fast RCNN based on upper and lower layers in face recognition. Comput. Intell. Neurosci. **2021** (2021)
6. Ren, S.Q., He, K.M., Girshick, R., et al.: Faster R-CNN: towards real-time object detection with region proposal networks. In: Proceedings of the Annual Conference on Neural Information Processing Systems, pp. 91–99. NIPS Foundation Press, Montreal (2015)
7. Zhu, B., Wu, X., Yang, L., Shen, Y., Wu, L.: Automatic detection of books based on faster R-CNN. In: 2016 Third International Conference on Digital Information Processing, Data Mining, and Wireless Communications (DIPDMWC), pp. 8–12 (2016). https://doi.org/10.1109/DIPDMWC.2016.7529355
8. Shengbin, G., Shaojie, W., Liang, H., Ronghui, Z., Lin Xiaohan, W., Binyun.: Lightweight improvement based on faster-RCNN algorithm and its application in beach waste detection. J. Xiamen Univ. (Nat. Sci. Ed.) **61**(02), 253–261 (2022)
9. Xu, D., Wang, L., Li, F., Guo, Y., Xing, K.: Application research of improved faster RCNN in grain insect object detection. Chin. J. Cereals Oils **37**(04), 178–186 (2022)

10. Dai, J., Li, Y., He, K., et al.: R-FCN: object detection via region-based fully convolutional networks. In: Conference on Neural Information Processing Systems, pp. 379–387 (2016)
11. Zhigang, Z., Huan, L., Pengcheng, D., Guangbing, Z., Nan, W., Wei-Kun, Z.: Vehicle target detection based on R-FCN. In: 2018 Chinese Control and Decision Conference (CCDC), pp. 5739–5743 (2018). https://doi.org/10.1109/CCDC.2018.8408133
12. Liu, W., Dong, L., Qu, H.: Small-scale pedestrian detection with improved R-FCN model. Chin. J. Image Graph. **26**(10), 2400–2410 (2021)
13. He, K., Gkioxari, G., Dollár, P., Girshick, R.: Mask R-CNN. In: 2017 IEEE International Conference on Computer Vision (ICCV), pp. 2980–2988 (2017). https://doi.org/10.1109/ICCV.2017.322
14. Zhibo, W., Shuangming, Z.: Road traffic sign recognition based on mask R-CNN. Surv. Mapp. Geogr. Inf. **47**(03), 119–122 (2022). https://doi.org/10.14188/j.2095-6045,2019444
15. Safonova, A., et al.: Olive tree biovolume from UAV multi-resolution image segmentation with mask R-CNN. Sensors **21** (2021)
16. Shi, J., Zhou, Y., Zhang, W.X.Q.: Object detection based on improved mask RCNN in service robot. In: 2019 Chinese Control Conference (CCC), pp. 8519–8524 (2019). https://doi.org/10.23919/ChiCC.2019.8866278
17. Yin, S., Chen, X., Bei, X.: Improved mask RCNN algorithm and its application in pedestrian instance segmentation. Comput. Eng. **47**(06): 271–276+283 (2021). https://doi.org/10.19678/j.issn.1000-3428.0058058
18. Cai, Z., Vasconcelos, N.: Cascade R-CNN: delving into high quality object detection. In: 2018 IEEE/CVF Conference on Computer Vision and Pattern Recognition, pp. 6154–6162 (2018). https://doi.org/10.1109/CVPR.2018.00644
19. Wu, J., et al.: Defect detection method for transmission line small object based on Cascade R-CNN algorithm. Power Grid Clean Energy **38**(04), 19–27+36 (2022)
20. Wenxia, B., Yangxun, R., Dong, L., Xianjun, Y., Qiuju, X.: Defect detection algorithm of anti-vibration hammer based on improved cascade R-CNN. In: 2020 International Conference on Intelligent Computing and Human-Computer Interaction (ICHCI), pp. 294–297 (2020). https://doi.org/10.1109/ICHCI51889.2020.00070
21. Li, Y., Chen, Y., Wang, N., Zhang, Z.-X.: Scale-aware trident networks for object detection. In: 2019 IEEE/CVF International Conference on Computer Vision (ICCV), pp. 6053–6062 (2019). https://doi.org/10.1109/ICCV.2019.00615
22. Redmon, J., Divvala, S., Girshick, R., Farhadi, A.: You only look once: unified, real-time object detection. In: 2016 IEEE Conference on Computer Vision and Pattern Recognition (CVPR), pp. 779–788 (2016). https://doi.org/10.1109/CVPR.2016.91
23. Redmon, J., Farhadi, A.: YOLO9000: better, faster, stronger. In: 2017 IEEE Conference on Computer Vision and Pattern Recognition (CVPR), pp. 6517–6525 (2017). https://doi.org/10.1109/CVPR.2017.690
24. Redmon, J., Farhadi, A.: YOLOv3: an incremental improvement. arXiv:1804.02767 (2018)
25. Bochkovskiy, A., Wang, C.Y., Liao, H.: Yolov4: optimal speed and accuracy of object detection. arXiv preprint arXiv:2004.10934 (2020)
26. Liu, D., Li, T., Du, Y., Cong, M.: Lightweight infrared real-time object detection algorithm based on MCA-YOLO. J. Huazhong Univ. Sci. Technol. (Nat. Sci. Ed.) 1–7 (2022). https://doi.org/10.13245/j.hust.239405
27. Yu, J., Wei, Z.: Face mask wearing detection algorithm based on improved YOLO-v4. Sensors **21** (2021)
28. Liu, W., Anguelov, D., Erhan, D., Szegedy, C., Reed, S., Fu, C.-Y., Berg, A.C.: SSD: single shot multibox detector. In: Leibe, B., Matas, J., Sebe, N., Welling, M. (eds.) ECCV 2016. LNCS, vol. 9905, pp. 21–37. Springer, Cham (2016). https://doi.org/10.1007/978-3-319-46448-0_2

29. Jeong, J., Park, H., Kwak, N.: Enhancement of SSD by concatenating feature maps for object detection. In: Proceedings of the British Machine Vision Conference. BMVA Press, London (2017)
30. Fu, C.Y., Liu, W., Ranga, A., et al.: DSSD: deconvolutional single shot detector (2017). https://arxiv.org/pdf/1701.06659.pdf
31. Li, Z., Zhou, F.: FSSD: feature fusion single shot multibox detector. arXiv:1712.00960 (2017)
32. Zlu, X., Xiong, Y., Dai, J., et al.: Deep feature flow for video recognition. In: 2017 IEEE Conference on Computer Vision and Pattern Recognition (CVPR), pp. 511–521 (2017)
33. Zlu, X.Z., Wang, Y.J., Dai, J.F., et al.: Flow-guided feature aggregation for video object detection. In: International Conference on Computer Visior (ICCV), Venice, Italy, pp. 408–417 (2017)
34. Bertasius, G., Torresani, L., Shi, J.: Object detection in video with spatiotemporal sampling networks. In: ECCV, Munich, Germany, pp. 342–357 (2018)
35. Jiang, Z., Liu, Y., Yang, C., et al.: Learning where to focus for efficient video object detection. In: ECCV, Glasgow, United Kingdom (2020)
36. Gao, H.: Research on video object detection algorithm based on deep learning. University of Electronic Science and Technology of China (2021). https://doi.org/10.27005/d.cnki.gdzku.2021.003628
37. Yuhu, S., Qigui, Z.: A fast video object detection method based on local attention. Comput. Eng. **48**(05), 314–320 (2022). https://doi.org/10.19678/j.issn.1000-3428.0061362

Optimal Deployment of Large-Scale Wireless Sensor Network Based on Topological Potential Adaptive Graph Clustering

Hefei Gao and Wei Wang[✉]

Tianjin Key Laboratory of Wireless Mobile Communications and Power Transmission, Tianjin Normal University, Tianjin 300387, China
weiwang@tjnu.edu.cn

Abstract. This paper proposed an optimal deployment algorithm for large-scale wireless sensor networks based on graph topology. The algorithm mines the intrinsic property of sensor measurement data as node quality through auto encoder and defines topological potential to describe the relationship between sensor nodes as the distribution of the topological potential field, thereby dividing the large-scale graph into several subgraphs. Next, singular value-QR decomposition is used to find the crucial nodes in each sub-network, and achieve the optimal deployment of large-scale sensor networks. We experiment with this algorithm on the CIMIS dataset. The results show that the mean square error of this algorithm is only 0.02116.

Keywords: Large-scale Wireless Sensor Network · Topological Potential · Auto Encoder · Singular-Value-QR Decomposition

1 Introduction

A Large-scale Wireless Sensor Network (LWSN) is a wireless network constructed by massive sensor nodes through self-organization and multi-hop. A large number of researches have shown that there are many redundant nodes in the sensor network. These redundant nodes seriously affect the transmission efficiency and data accuracy of the LWSN. Therefore, it is necessary to reduce redundant nodes as much as possible and achieve optimal deployment to prolong network life and reduce resource waste without losing crucial information about the sensor network. There has been a lot of research works devoted to this in recent years. In [1], an improved swarming spider algorithm based on chaos is proposed. The algorithm searches for the optimal node deployment scheme by simulating the behaviors of cooperation and attraction between spider populations, effectively reducing energy consumption and improving network coverage performance. Research [2] builds a Hebbian neural network model to solve the data redundancy between sensor nodes. Specifically, when two sensors perceive the same data, the correlation between the data increases the weight of the Hebbian model, which also represents a strong connection between the two nodes. If the nodes are fully

Q. Liang et al. (Eds.): AIC 2022, LNEE 871, pp. 187–194, 2023.
https://doi.org/10.1007/978-981-99-1256-8_22

connected and the environment has not changed, these sensor nodes can be replaced by one. Malathy et al. proposed an electronic track topology algorithm for the seismic monitoring sensor network, which covers a vaster area with fewer sensor nodes. Firstly, the multiple sectors are divided around targets, and a "supervisor" is selected for each sector. Next, a "cluster supervisor" is selected from a group of sector supervisors that are farthest from the target and have the maximum remaining energy. Several cluster supervisors are connected to form a polygonal network, and the optimal positions of all cluster supervisors are determined by calculating the centroid of the polygon to achieve the optimal deployment of the sensor network [3].

Graphs are usually used to represent the relationship between multiple entities. With the continuous development of multi-sensor technology, the number of sensors and the complexity of the network structure continues to increase. Such irregular data relationships can be analyzed and processed efficiently and conveniently based on graphs. Therefore, this paper proposes a set of optimal deployment algorithms for large-scale wireless sensor networks that combine graph clustering and matrix factorization. The algorithm first uses an auto encoder to learn the inherent properties of sensor nodes. Next, the topological potential is defined, and the intrinsic relationship between sensor nodes is transformed into the distribution of the topological potential field, and then the sensor nodes are clustered. Then, singular value-QR decomposition is used to find the crucial nodes in each sub-network to achieve the optimal deployment of large-scale wireless sensor networks.

The rest of this paper is organized as follows: Sect. 2 introduces the theory and process of the proposed algorithm in detail; Sect. 3 verifies the reliability and effectiveness of the proposed algorithm through experiments; Sect. 4 is the conclusion and discussion.

2   Theory and Algorithm

2.1   Adaptive Graph Clustering Based on Topological Potential

This paper defines a topological potential to measure the interaction between the sensor node i and surrounding neighbor nodes

$$\varphi_i = \sum_{j \in \mathcal{N}} m_j e^{-\left(\frac{d_{ij}}{\sigma}\right)^2} \tag{1}$$

where d_{ij} is the practical distance between node i and node j, which can be calculated by the latitude and longitude coordinates of the two nodes; σ is called the influence factor, which is used to control the influence range of each node. From the properties of the Gaussian function, the influence range of each sensor node is the neighborhood space with the center of the node, and the radius of $\frac{3\sigma}{\sqrt{2}}$; m_j is the node mass of neighbor node j. As shown in Fig. 1, the node mass of each sensor node is obtained by representation learning through a multilayer nonlinear under-complete auto encoder, which represents the intrinsic relationship between feature samples.

Specifically, the encoder extracts features from the original graph signal and embeds them as a node mass

$$\boldsymbol{m} = \alpha(\boldsymbol{W}_{\text{enc}}\boldsymbol{x} + \boldsymbol{b}_{\text{enc}}) \tag{2}$$

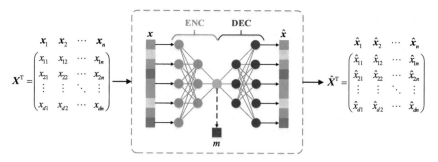

Fig. 1. Multilayer nonlinear under-complete auto encoder model

where $W_{\text{enc}} \in \mathbb{R}^{d' \times d}$ is the transformation matrix of the encoder, d is the graph signal dimension, and d' is the embedded dimension; $x \in \mathbb{R}^d$ is the row of the graph signal matrix $X = [x_1, x_2, \cdots, x_N]^{\text{T}} \in \mathbb{R}^{N \times d}$, which represents the measurement values of the sensor node, and N is the number of sensor nodes; $b_{\text{enc}} \in \mathbb{R}^{d'}$ is the bias; $\alpha(\cdot)$ is the nonlinear activation function.

The decoder reconstructs the graph signal based on the extracted features

$$\hat{x} = \alpha(W_{\text{dec}}m + b_{\text{dec}}) \tag{3}$$

where $W_{\text{dec}} \in \mathbb{R}^{d \times d'}$ is the transformation matrix of the decoder; $b_{\text{dec}} \in \mathbb{R}^d$ is the bias; $\alpha(\cdot)$ is the nonlinear activation function.

An auto encoder is an unsupervised neural network model that does not require labels for supervised learning. Instead, iterative training is performed by minimizing the reconstruction error between input and output. This paper uses the mean squared error as the reconstruction loss function

$$\mathcal{L} = \frac{1}{N} \sum_{i=1}^{N} (\hat{x} - x)^2 \tag{4}$$

Then the parameters W_{enc}, W_{dec}, b_{enc}, and b_{dec} are continuously optimized using the Adam optimization algorithm. Finally, the embedded representation of sensor nodes is realized, and the node mass is obtained.

In addition, in order to calculate the topological potential better, we normalize the obtained node mass

$$m_i^* = \frac{m_i}{\sum_{i=1}^{N} m_i} \tag{5}$$

After constructing a complete topological potential field, we use the attraction of the high-potential region to the low-potential in the potential field to get the cluster division. First, we find all topological potential local maximum nodes as sources in the potential field

$$s = \arg\max_i \varphi_i \quad \text{s.t} \quad i \in \mathcal{N}_i \cup \{i\} \tag{6}$$

where $\mathcal{N}_i$ is the neighborhood of node i.

In order to achieve better clustering, we merge several source nodes whose distance is less than the topological potential's influence range $\frac{3\sigma}{\sqrt{2}}$ into one Attractor k. Each attribute of an attractor (e.g. mass and position) is determined by the average value of those source nodes that generate the attractor. The remaining nodes that cannot be merged become attractors naturally and their attributes are inherited from the corresponding source nodes. The attractor has an attractive effect on all nodes in the field, which is shown as a topology attractive force

$$f_{ki} = m_k^* m_i^* e^{-\left(\frac{d_{ki}}{\sigma}\right)^2} \tag{7}$$

A node in the field will be subject to the topology attractive force exerted by all attractors simultaneously, and the cluster corresponding to the attractor that exerts the maximum topology attractive force component on it is the cluster to which the node belongs

$$C_k = \arg\max_k f_{ki} \tag{8}$$

We summarize the proposed Adaptive Graph Clustering based on Topological Potential in Algorithm 1.

Algorithm1: Adaptive Graph Clustering based on Topological Potential

Input: Adjacency matrix A, Graph signal matrix X, Node location list P
Output: Clusters $C = \{C_1, C_2, \cdots, C_K\}$

1: for $epoch = 1, 2, \cdots, 50$ do
2:       Encode the graph signal with Eq(2)
3:       Reconstruct the original graph signal with Eq(3)
4:       Calculate the reconstruction loss function value $\mathcal{L}$ with Eq(4)
5:       Back-propagate loss value and update model parameters by Adam
6: end for
7: Obtain the normalized node mass m^* with Eq(5)
8: Calculate the topological potential value φ with Eq(1)
9: Determine the field source nodes s with Eq(6)
10: Determine K attractors
11: for $i = 1, 2, \cdots, N$ then
12:     for $k = 1, 2, \cdots, K$ then
13:             Calculate the topology attractive force with Eq(7)
14:     end for
15:     Determine the cluster to which node i belongs with Eq(8)
16:     $C_k = C_k \cup \{i\}$
17: end for
18: Output clusters $C = \{C_1, C_2, \cdots, C_K\}$

2.2 Optimal Nodes Selection Based on SVD-QR

After dividing the sensor network according to the clustering algorithm proposed in Sect. 2.1, we continue to obtain the subgraphs $\{G_1, G_2, \cdots, G_K\}$ of each cluster, where $G_k = G[C_k]$ is the induced subgraph of the original sensor network.

We perform singular value decomposition on the graph signal matrix X_k of the subgraph G_k

$$X_k = U_k \Sigma_k V_k^{\mathrm{T}} \tag{9}$$

where $U_k \in \mathbb{R}^{N_k \times N_k}$ and $V_k \in \mathbb{R}^{d \times d}$ are the singular vector matrices of X_k; N_k is the number of subgraph nodes; Σ_k is the singular value matrix.

Then we partition U_k according to the optimized number of sensor nodes

$$U_k = \begin{bmatrix} U_{k1} & U_{k2} \end{bmatrix} \tag{10}$$

where $U_{k1} \in \mathbb{R}^{N_k \times M_k}$, $U_{k2} \in \mathbb{R}^{N_k \times (N_k - M_k)}$, and M_k is the number of nodes after optimization selection.

Next, perform column pivoting QR decomposition on the singular vector sub-matrix U_{k1} corresponding to the first M_k singular values, and a permutation matrix P_k can be obtained

$$U_{k1}^{\mathrm{T}} P_k = Q_k \, R_k \tag{11}$$

where Q_k is a unitary matrix and R_k is an upper triangular matrix.

The M_k columns with a value of 1 in the permutation matrix P_k correspond to the M_k most crucial sensor nodes in the least square sense of the sub-network. So we can get the graph signal matrix of the optimized sub-network

$$X'_k = P_k X_k \tag{12}$$

Finally, we combine the optimized graph signal matrices of all sub-networks to achieve the optimized deployment of the LWSN.

3 Experiment

3.1 Dataset and Parameter Settings

The algorithm proposed in this paper is tested on the California Irrigation Management Information System (CIMIS)[1]. In this paper, the hourly average air temperature data of 31 days is selected as the measurement value of sensor nodes to build a wireless sensor network for experiments, in which there are 144 active nodes and the data feature dimension is 744.

---

[1] http://www.cimis.water.ca.gov.

The K-nearest neighbor algorithm is used to construct the sensor network as a graph topology, where the value of K is 8, and the constructed graph has 576 edges. The number of neurons in the input layer of the auto encoder is 744, the hidden layer has 2 layers, the numbers of neurons are 128 and 64 respectively, the number of neurons in the output layer is 1, and the number of neurons in the decoder part is set corresponding to the encoder. Except that the activation function of the decoder output layer is the hyperbolic tangent function, the activation functions of the other layers are all Leaky Rectified Linear Unit functions. The learning rate of the Adam optimizer is 0.001, and the number of iterations is 50. The topological potential influence factor is 172.79.

3.2 Experimental Results and Analysis

This paper learns the intrinsic relationship between the features of each sensor node based on the auto encoder described in Sects. 2.1 and 3.2 to obtain the node mass. The result is shown in Fig. 2(a), where the node position in the figure is determined by latitude and longitude, and the node mass can be judged by color. Next, calculate the topological potential value of each sensor node and find 5 attractors, as shown in Fig. 2(b), the topological potential value of each node can be judged by color, the purple star in the figure is the attractor.

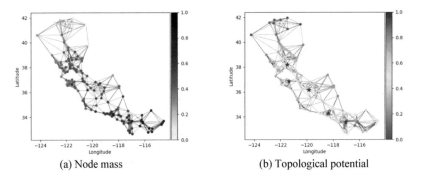

(a) Node mass   (b) Topological potential

Fig. 2. Node mass and Topological potential distribution

All sensor nodes are divided into 5 clusters under the attraction of 5 attractors, the topological attractive force components of each node under different attractors are shown in Fig. 4, and the clustering results are shown in Fig. 5.

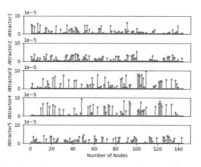

Fig. 3. Topological attractive component

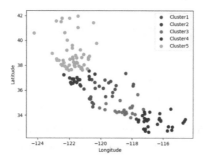

Fig. 4. Clustering results

In order to prove the applicability of the proposed algorithm in this paper, it is compared with five classical wireless sensor network optimal deployment schemes, namely MinPinv [4], MaxVol [4], MinSpec [5], Entropy [6], MI [7]. The results are shown in Fig. 6. The number of optimized nodes increases by 10 from 5 to 140. It can be seen from the figure that when the number of optimized nodes is 15, a good optimization effect can be achieved. Moreover, the optimal deployment algorithm proposed in this paper outperforms other algorithms, and its mean square error is as low as 0.02116.

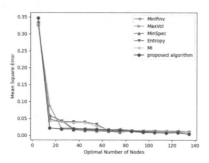

Fig. 5. Performance comparison

4   Conclusion

Aiming at the optimal deployment of large-scale wireless sensor networks, this paper proposes a solution based on graph topology. In this solution, an adaptive graph clustering algorithm based on topological potential is firstly proposed, which divides a large-scale sensor network into several sub-networks. By performing singular value-QR decomposition on the graph signals of each sub-graph divided, a set of independent node subsets that minimize the residual in the least square sense can be found. The nodes in this subset are not only significantly less correlated with each other but also retain the crucial information of the wireless sensor network. However, the optimal deployment algorithm proposed in this paper is mainly designed for vast-area sensor networks, and further research is needed for the optimal deployment of high-density sensor networks.

Acknowledgement. This work is supported by National Natural Science Foundation of China (No. 61731006, 61971310).

References

1. Cao, L., Yue, Y., Cai, Y., et al.: A novel coverage optimization strategy for heterogeneous wireless sensor networks based on connectivity and reliability. IEEE Access **9**, 18424–18442 (2021)
2. Kusuma, S.M., Veena, K.N., Aparna, N.: Effective deployment of sensors in a wireless sensor networks using hebbian machine learning technique. In: 2021 International Conference on Computing, Communication, and Intelligent Systems (ICCCIS), pp. 268–274. IEEE (2021)
3. Sathyamoorthy, M., Kuppusamy, S., Nayyar, A., Dhanaraj, R.K.: Optimal emplacement of sensors by orbit-electron theory in wireless sensor networks. Wireless Netw. **28**(4), 1605–1623 (2022). https://doi.org/10.1007/s11276-022-02919-9
4. Tsitsvero, M., Barbarossa, S., Di Lorenzo, P.: Signals on graphs: Uncertainty principle and sampling. IEEE Trans. Signal Process. **64**(18): 4845–4860 (2016)
5. West, P.W.: Sampling theory. In: West, P.W. (eds.) Tree and Forest Measurement, pp. 105–118. Springer, Cham (2015). https://doi.org/10.1007/978-3-319-14708-6_10
6. Shewry, M.C., Wynn, H.P.: Maximum entropy sampling. J. Appl. Stats **14**(2), 165–170 (1987)
7. Krause, A., Singh, A., Guestrin, C.: Near-optimal sensor placements in Gaussian processes: theory, efficient algorithms and empirical studies. J. Mach. Learn. Res. **9**(3), 235–284 (2008)

INSL: Text2SQL Generation Based on Inverse Normalized Schema Linking

Tie Jun[1,2,3], Fan Ziqi[1,2,3]([✉]), Sun Chong[1,2,3], Zheng Lu[1,2,3], and Zhu Boer[1,2,3]

[1] College of Computer Science, South-Central Minzu University,
Wuhan 430074, China
[2] Hubei Provincial Engineering Research Center for Intelligent Management of
Manufacturing Enterprises, Wuhan 430074, China
m17755528298@163.com
[3] Hubei Provincial Engineering Research Centre for Agricultural Blockchain and
Intelligent Management, Wuhan 430074, China

Abstract. Structured Query Language (SQL) is a query language widely used in databases, Text2SQL automatically parses natural language into SQL, which has great potential to facilitate non-expert users to query and mine structured data using natural language. Current research focuses on improving the matching accuracy of SQL clause tasks, but ignores the correctness of SQL syntax generation, and SQL generation involving multi-table joins still suffers from a large number of errors. Therefore, a neural network-based Text2SQL approach is proposed. To implement a practical Text2SQL workflow, the model associates natural language queries with an inverse normalized database schema, called INSL (Inverse Normalized Schema Link Generation Network). Through theoretical analysis and experimental validation on the public dataset Spider, INSL can effectively improve the quality of Text2SQL tasks.

Keywords: inverse normalization · schema linking · semantic parsing · Text2SQL

1 Introduction

Text2SQL is at the heart of intelligent question and answer systems for relational databases and is an essential task in natural language processing. To implement a practical Text2SQL workflow, the model must associate a natural language query with a given database.

Text2SQL consists of two subtasks, the first is to construct the correct framework for the SQL statement (which can be referred to as the syntax generation task), and on top of this, each subelement of the framework is matched to the corresponding value (which can be referred to as the value matching task). As shown in Fig. 1, the single-round Text2SQL task takes a single natural statement (Query)–Find the number of pets for each student who has any pet and student

Industry-University-Research Innovation Fund Project of Science and Technology Development Center of Ministry of Education (No. 2020QT08).

Q. Liang et al. (Eds.): AIC 2022, LNEE 871, pp. 195–202, 2023.
https://doi.org/10.1007/978-981-99-1256-8_23

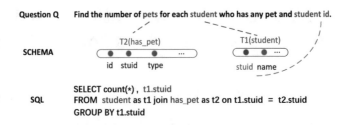

Fig. 1. Multi-table Text2SQL problem example

id, a database schema as input, and a SQL query statement as output. Among them, solutions to syntactic problems are dedicated to constructing suitable slot-filling structures bearing string matching and heuristic fusion coding to support the transformation of complex papers into SQL statement models. RYANSQL [1] converts syntactic problems into concrete nested syntactic problems, converting nested SQL statements to nonnested SQL statements by annotating statement location information to improve the accuracy of sketching slot-filling methods; however, table join selection errors still exist in abundance.

This paper proposes a schema inverse normalization method based on database schema information mining, which is used to optimize the Text2SQL technique. Through the inverse normalization method, the database schema is reconstructed to reduce the schema boundaries in the database and further generate queries by decomposing associative queries. The reconstructed database schema is encoded-decoded by a neural network. The advantages of inverse normalization are the reduction in the need for join operations, the reduction in the number of indexes and foreign keys in the database schema.

2   Related Work

The application of Text2SQL to deep neural networks relies heavily on the encode-decode framework. The encoding part of the encode-decode framework is to transform the input sequence into vectors, while the decoding part of the encode-decode framework is to transform the generated vectors into output sequences. The encode-decode framework has been used in many applications, from machine translation and text generation in the early stages of development, to document extraction and question and answer systems.

RYANSQL proposes a sketch-based slot-filling method to generate SQL statements that convert a nested SQL query into a set of non-nested SELECT statements for complex Text2SQL problems. RYANSQL presents a detailed sketch of complex SELECT statement predictions and statement location code to handle nested queries.

The introduction of schema linking is one of the new research priorities, such as entity linking in the domain of knowledge graphs [8–10] and slow filling in dialogue systems [11,12], where a large amount of annotated data and models have been proposed to address their specific properties. In the general area of semantic analysis, decoupling the underlying structure from the vocabulary has been shown to facilitate cross-domain semantic analysis [13]. However, when it comes to text-to-SQL problems, existing approaches often treat schema linking as a secondary preprocessing process using simple heuristics, such as string matching between natural language statements and column or table names [14]. Simple heuristics have difficulty accurately identifying the columns or tables involved in natural language discourse and understanding the relationship between the discourse and the corresponding database schema. Therefore, they take the first step towards schema linking as a particular research problem. However, the difficulty of this problem is further illustrated by the limited improvements they achieve on the challenging Spider dataset due to the lack of direct supervision of schema linking. In contrast to previous approaches, recent models [15–16] have made pattern linking sets learnable.

This paper proposes a Text2SQL method based on inverse normalization, which introduces inverse normalization technology to link database schema and build an encode-and-decode semantic parser, INSQL, to simplify the multi-table join query problem into a single-table query problem by inverse normalization, reducing table joins, thereby improving query accuracy and query speed, and effectively improving the quality of the solution.

3 Definition

The Text2SQL task is expressed formally as Eq. (1):

$$sql_q = SL(q, D) \tag{1}$$

The input consists of a natural language statement and a database schema set, the output represents a SQL statement, and the database schema set contains table names, column names, primary key tags, foreign key relationships, and other information, which represents a collection of table names in the database, represents a collection of column names in the database, PK represents a collection of primary key tags and FK represents a collection of foreign key relationships. The work in this paper can be expressed formally as Eq. (2):

$$f = \max \sum_{\forall q_i \in Q} G(sql_{q_i} | sql_{q_i} = SL(q, D), sql_{q_i'} | sql_{q_i'} = SL(q, D')) \tag{2}$$

D^C is the database schema after inverse normalization, sql_{q_i} is the SQL statement generated after inverse normalization, and is the inverse normalization process, which is formalized in Eq. (3).

$$D' = IN(D) \tag{3}$$

The output of the inverse normalized SQL statement is correct when $EMA(sql_q) = 1$, and incorrect when $EMA(sql_q) = 0$. The aim of this work is to find the IN model that maximizes the sum of any input function G in Eq. (2), i.e., the optimal quality improvement of the SQL query statement generation.

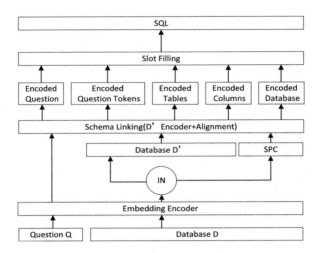

Fig. 2. Input encoder construction diagram

4   Database Schema Inverse Normalization

To investigate in depth the role of database schema on multi-table joins, this paper proposes the INSL model, which is based on an encoding-decoding framework. The encoding part of the model is shown in Fig. 2 and consists of a five-layer structural input encoder base model, with an inverse normalized schema reconstruction component (IN).

4.1   Basic Model

The basic model is based on an encoding-decoding framework, with the encoding part consisting of an embedding encoder layer,an inverse normalization partand a schema linking layer, and the decoding part consisting of ten padded slots designed according to SQL syntax.

The slot-filling based decoder predicts the values of the slots in the template, based on the basic structure of the SELECT statement, i.e. the presence of constituent clauses and the number of conditions in each clause. The statement encoding vectors are first combined to obtain the statement encoding vector which uses a heuristic matching method to splice the function. An algebraic representation of the existence of clauses and the number of each clause is obtained by applying two fully concatenated layers on top of each other to classify the presence of the FROM and SELECT clauses that must be present to form a valid selection statement, indicating that one of INTERSECT, UNION or EXCEPT. If the value is one of INTERSECT, UNION or EXCEPT, the corresponding SPC (query statement location code) is created and the SELECT statement structure is generated recursively, with the resulting SQL statement combining encoded column and table vector to predict the value of the filled slot in each of the existing subtasks. Output the generated SQL query statement.

4.2 Inverse Normalization

There are three types of multi-table join errors that occur more frequently in the base model. The first is a multi-table join error, which is reflected in reversing the order of table joins, the second is a multi-table join error, which is reflected in connecting extra tables when the query task can be completed, resulting in unnecessary table joins, or incorrect table selection due to multiple tables containing the same name resulting in incorrect multi-table join paths, and the third is a special type where the SQL query statement has multiple errors but the query results are correct.

The base model error case points to significant room for improvement in the multi-table join problem. Primary and foreign key relationships are an integral part of the multi-table join problem. In the base model, foreign key relationships are used to construct join paths, and the inverse normalized schema reconstruction component mines the foreign key information to exploit it (Fig. 3).

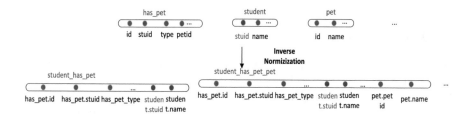

Fig. 3. Example of inverse normalization

Each set of relationships in the database schema set is inverse normalized to obtain multiple sets of schema, and the reconfiguration is named after the IN relationships, and the inverse normalized reconfiguration schema set is added to the original database schema and put into the base network.

4.3 Schema Linking

The schema linking section covers two important tasks, which are question column alignment and question table alignment. The table encoding part integrates the column vectors of each table to obtain the encoded table vectors. In the table alignment part of the problem, which has the same network architecture as the question-to-column part of the problem, contextual information from the problem statement is used to model the table vectors and output the encoded table vectors. The words in the question and column names are encoded in the column encoding layer model using DenseCNN to capture the semantic features and output the question word encoding and hidden column vector. At the problem column alignment part, the problem tokens are associated with the corresponding columns, and the column vectors are modeled with contextual information about the problem. The problem tokens are aligned with the column vectors using scaled dot product attention. For larger values of the scaling factor, the dot product exhibits a large increase in order of magnitude, pushing the softmax function into regions with very small gradients. To reduce the impact of the large growth of the dot product, an expansion of the dot product is done. Heuristic fusion is to obtain the fused column matrix, and Transformer is applied on the fused column matrix to capture the contextual column information and output the column encoding vector.

5   Experiments and Analysis of Results

5.1   Datasets

The experimental dataset uses the Spider dataset, in which the training, development and test sets are not duplicated and the test set is hidden from the public. The Spider dataset was introduced in 2018, which is in the Text2SQL dataset and reaches the highest difficulty and complexity level available.

5.2   Evaluation Indicators

The Spider dataset provides two validation metrics-Exact Match (EMA) and Execution Accuracy (EA). The exact match metric means that the predicted SQL query statement matches the correct SQL query statement provided by the dataset exactly on each keyword subtask module, while the execution accuracy metric means that the predicted SQL query statement is executed so that the database can be executed correctly and with consistent execution results.

5.3   Quality Assessment

The WikiSQL dataset is a large scale Text2SQL corpus containing 80,654 samples on 24,241 tables from Wikipedia [17], which is a substantial simplification compared to the actual SQL query generation scenario. On WikiSQL, a question is associated with only one table, whereas on the Spider dataset, a question is associated with multiple tables. The experiments on the dataset Spider based on the sub-task division, listing the number of training set samples for the five sub-tasks in single-table and multi-table queries. The Spider dataset is unevenly distributed between single-table and multi-table, and across different tasks, and the validation set is divided into one-time table connection subset, two-time table connection, and mixed subset in this paper.

In summary, the performance of the INSQL model on the validation set is shown in Table 1, with a large improvement in matching accuracy compared to RYANSQL, and a slight disadvantage in execution accuracy compared to RYANSQL, the underlying reason being the huge redundancy caused by the inverse normalization of the database schema causing significant interference in execution accuracy. An accuracy of 68.55% was achieved on the validation set.

Table 1. Model assessment

MODEL	DEV:EMA(%)
RYANSQL	43.40
RYANSQL(+BERT)	66.60
INSL	68.55
INSL(once join)	71.39
INSL(twice join)	59.22

According to Eq. (2), the combined index value of the inverse normalization process is, which proves that Text2SQL generation based on inverse normalization outperforms the base model and brings a new and effective solution to the multi-table join problem.

6   Summary

This paper proposes a Text2SQL approach based on inverse normalization for the multi-table join problem, introducing inverse normalization techniques, reconstructing the database schema and building an encode-decode semantic parser, INSQL, which reduces table joins by reducing the multi-table join query problem to a single table query problem through inverse normalization. The model was experimented on the Spider dataset and the quality of SQL statement generation was effectively improved. The selection of the inverse normalized pattern reconstruction pattern is also an important influence factor on the degree of redundant information interference.

Acknowledgments. This work is supported and assisted by the Industry-University-Research Innovation Fund Project of Science and Technology Development Center of Ministry of Education (No. 2020QT08), National People's Committee Training Program for Young and Middle-aged Talents (MZR20007), Hubei Science and Technology Major Special Project (2020AEA011), Wuhan Science and Technology Plan Applied Basic Frontier Project (2020020601012267).

References

1. Choi D.H., Shin, M.C., Kim, E.G.: RYANSQL: recursively applying sketch-based slot fillings for complex text-to-SQL in cross-domain databases, pp. 309-332. CL (2021)
2. Zhen, D., Sun, S.Z., Liu, H.Z., Lou, J.G., et al.: Data-anonymous encoding for text-to-SQL generation. In: Proceedings of the 2019 Conference on Empirical Methods in Natural Language Processing and the 9th International Joint Conference on Natural Language Processing (EMNLP-IJCNLP), pp. 5405-5414. ACL, Hong Kong (2019)

3. Fu, X.Y., Shi, W.J., Yu, X.D., Zhao, Z., et al.: Design challenges in low-resource cross-lingual entity linking. In: Proceedings of the 2020 Conference on Empirical Methods in Natural Language Processing (EMNLP), pp. 6418-6432 (2020)
4. Rijhwani, S., Xie, J., Neubig, G., et al.: Zero-shot neural transfer for cross-lingual entity linking. In: Proceedings of the AAAI Conference on Artificial Intelligence, pp. 6924-6931 (2019)
5. Logeswaran, L., Chang, M.W., Lee, K., et al.: Zero-shot entity linking by reading entity descriptions. http://arxiv.org/abs/1906.07348
6. Xu, P.Y., Hu, Q.: An end-to-end approach for handling unknown slot values in dialogue state tracking. In: Proceedings of the 56th Annual Meeting of the Association for Computational Linguistics, pp. 1448-1457. ACL, Australia (2018)
7. Nouri, E., Hosseini-Asl, E.: Toward scalable neural dialogue state tracking model. http://arxiv.org/abs/1812.00899
8. Herzig, J., Berant, J.: Decoupling structure and lexicon for zero-shot semantic parsing. In: Proceedings of the 2018 Conference on Empirical Methods in Natural Language Processing, pp. 1619-1629. ACL, Belgium (2018)
9. Guo, J., Zhan, Z., Gao, Y.: Towards complex text-to-SQL in cross-domain database with intermediate representation, pp. 4524-4535. ACL, Florence (2019)
10. Bogin, B., Berant, J., Gardner, M.: Representing schema structure with graph neural networks for text-to-SQL parsing. In: Proceedings of the 57th Annual Meeting of the Association for Computational Linguistics, Association for Computational Linguistics, pp. 4560-4565. ACL, Florence (2019)
11. Wang, B.L., Shin, R., Liu, X.D., et al.: RAT-SQL: relation-aware schema encoding and linking for text-to-SQL parsers. In: Proceedings of the 58th Annual Meeting of the Association for Computational Association for Computational Linguistics, pp. 7567-7578. ACL (2020)
12. Zhong, V., Xiong, C., Socher, R.: Seq2SQL: generating structured queries from natural language using reinforcement learning. https://arxiv.org/abs/1709.00103

The Research on Prediction for Financial Distress in Car Company Listed Combining Financial Indicators and Text Data

Yu Du[1], Fengyi Wang[2(✉)], Yongchong Wang[1], and Jingjing Jia[1]

[1] Business School, No. 15, Xueyuan Road, Haidian District,
Beijing 100083, People's Republic of China
[2] School of Information Science, Beijing Language and Culture University, No. 15, Xueyuan
Road, Haidian District, Beijing 100083, People's Republic of China
wangfengyi428@163.com

Abstract. In recent years, China's auto manufacturing industry has been developing rapidly and Chinese car companies are facing both internal and external pressures. As a result, managers of automobile enterprises are now paying more attention to the financial situation of their enterprises, hoping to avoid the risk of bankruptcy by predicting financial distress in advance. Most of the existing research findings are based on financial index data or annual financial reports to build quantitative and qualitative models to predict the risk of financial distress of automobile enterprises. This study innovatively proposes to combine financial indicators data, financial text data and non-financial text data to predict the risk of financial distress of automobile enterprises, and constructs a set of prediction models based on deep learning algorithms, which is a useful attempt of cross-disciplinary research in this field. In terms of experimental analysis, we found that the introduction of financial text data and non-financial text data into the model significantly improves the prediction performance compared with the traditional prediction model based on financial indicator data, which indicates that the combination of online user reviews and financial annual reports can be more helpful for predicting the financial distress risk of car companies.

Keywords: Deep Learning · Natural Language Processing · Business Intelligence · Financial Distress Prediction

1 Introduction

In recent years, the Chinese automobile manufacturing industry has achieved better results in competing with foreign brands, but in the face of the rapidly changing market, Chinese car companies are facing both internal and external pressure. Therefore, at this stage, managers of car companies are paying more attention to the financial situation of their companies, hoping to avoid the risk of corporate bankruptcy by predicting financial difficulties in advance [1].

With the advent of the big data era, more and more researchers and managers have started to apply machine learning algorithms to corporate financial distress prediction.

© The Author(s), under exclusive license to Springer Nature Singapore Pte Ltd. 2023
Q. Liang et al. (Eds.): AIC 2022, LNEE 871, pp. 203–210, 2023.
https://doi.org/10.1007/978-981-99-1256-8_24

Tsai and Wu compared single-layer neural network algorithms with multiple neural network algorithms, and the results showed that the performance of multiple NN algorithms was not greatly improved in the field of financial distress prediction [2].Besides, Lee and Wu confirmed the superior performance of neural network algorithms in the field of financial distress prediction compared to traditional statistical methods in their subsequent studies [3]. Hosaka used convolutional neural network models to extract features from financial indicators of insolvent firms [4] to predict the financial risk status of firms and determine whether they are in financial distress, while Ristolainen compared neural network algorithms with traditional machine learning algorithms and found that ANN models outperformed logistic regression and weak classifiers such as SVM in the area of financial distress crisis prediction in the banking industry [5]; in addition, a growing number of researchers have found that text-based data related to business operations can help in corporate financial distress prediction. Rastin used CNN and RNN models for deep prediction of corporate financial distress risk and mined the feature patterns [6]. Li's study investigated online reviews and combined with empirical analysis to conclude that online reviews have a close relationship with corporate business performance [7]. However, there is a lack of studies combining financial data with textual data such as financial statements and public opinion reviews to predict the financial distress of enterprises.

Therefore, we will combine financial indicator data, textual data from corporate annual reports and textual data from online user reviews to predict the risk of financial distress of Chinese car companies by building a deep learning model. This study finds that the introduction of text data will effectively improve the model prediction accuracy.

2 Methodology

The overall framework of the model is shown in Fig. 1. The data input mainly contains three parts, namely financial indicator data, financial text data and non-financial text data. The model first processes the three data sets and then stitches them together, then inputs them into a CNN network for feature extraction, and finally inputs them into an RNN network for prediction.

2.1 Financial Text Feature Extraction Model

As the financial text data contains long chapters with complex contextual links, a BiLSTM model was used to extract deep features and generate financial text feature vector data [8].

$$h_t = LSTM(x_t, h_{t-1}) \tag{2.1}$$

$$h'_t = LSTM\left(x_t, h'_{t-1}\right) \tag{2.2}$$

$$o_t = wh_t + vh'_t + b_t \tag{2.3}$$

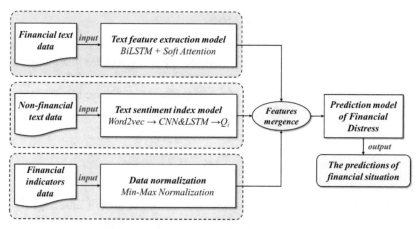

Fig. 1. Financial distress prediction model based on financial index data and text data

where x_t is the model input, h_t is the forward hidden state, h_t' is the backward hidden state, and o_t is the model output. In addition, in order to improve the training efficiency of the model, we also introduced a soft attention mechanism in modelling [8].

$$U_t = tanh(Wh_t) \tag{2.4}$$

$$\alpha_t = \frac{exp(score(U_t))}{\sum_{i=1}^{T} exp(score(U_i))} \tag{2.5}$$

$$S_t = \Sigma_{i=0}^{n}\alpha_t h_t \tag{2.6}$$

where h_t is the hidden state and α_t constitutes the probability vector for the attention distribution and S_t is the attention scoring function.

2.2 Non-financial Text Sentiment Indicator Model

We first used word2vec to vector transform the non-financial text data, and then used CNN-LSTM model to calculate the sentiment tendency of the text data [9]. After obtaining the sentiment tendency of the text data, we constructed a sentiment index model of online user reviews by borrowing the calculation of the bullish index [10] to generate non-financial text sentiment index data, which reflects the evaluation and opinion of online users on the future development trend of car companies within a certain time period.

$$Q_i = ln\left(\frac{1 + D_t^{pos}}{1 + D_t^{neg}}\right) \tag{2.7}$$

where D_t^{pos} represents the number of positive comments after sentiment analysis, while D_t^{neg} represents the number of negative comments obtained after sentiment analysis.

2.3   Financial Distress Forecasting Model

Finally, we stitch the financial text feature vector data, non-financial text sentiment indicator data and financial indicator data into a CNN network for feature extraction, and input the obtained feature vectors into an RNN network to predict the financial distress of car companies. Evaluation of model performance we used $F_1 - score$:

$$R = \frac{TP}{TP + FN}, P = \frac{TP}{TP + FP} \tag{2.8}$$

$$F_1 = \frac{2 \times P \times R}{P + R} \tag{2.9}$$

where TP is the number of true case samples, FN is the number of false negative case samples, FP is the number of false positive case samples, R is the recall of the model, and P is the precision rate, F_1 is $F_1 - score$.

3   Experiment and Analysis

3.1   Data Preparation

We selected 188 listed car companies in China for our study, with data spanning 2015–2020, of which 56 had experienced financial distress (i.e. were labelled as ST). We mark the period in which the ST sample is located as period T, thus predicting their financial distress.

The data for the financial indicators were mainly obtained from the balance sheets, income statements and cash flow statements of the vehicle companies, totalling 47 indicators. After obtaining the data, we used the min-max method to standardise the data:

$$X' = \frac{X - \min(X)}{\max(X) - \min(X)} \tag{3.1}$$

where $max(X)$ and $mim(X)$ are the maximum and minimum values corresponding to the financial indicator in the sample, and X' represents the standardised financial indicator data.

The financial text data was obtained from the annual report data on Juchao Information Website (http://www.cninfo.com.cn), and we extracted the "Business Operation Discussion" and "Important Matters" sections from the annual report for model training. The non-financial text data was obtained from online user reviews on the Oriental Wealth website (https://www.eastmoney.com/) and the Car Quality website (http://www.12365a uto.com/). For the textual data, the following steps were taken:

1) Remove punctuation, line breaks, tabs and numbers from the text
2) Remove non-important words such as stop words
3) Build a glossary and remove words that occur less than 10 times

In addition, previous research findings have found that it is more effective to predict the T-period financial status of a company using its T-3 period financial indicator data [1], so we trained the model with the T-3 period financial indicator data of the car company as the model input, and the output was set to the T-period financial status (1 being ST and 0 being normal).

3.2 Experiment I - Based on Financial Indicator Data

First, we predict the financial distress of car companies based on financial indicator data only, and select XGboost, random forest and CNN with good classification performance to construct the prediction model and tune the parameters by grid search technique. The core parameters of the model are shown in Table 1, and the experimental results are shown in Table 2.

Table 1. Experiment I key model parameters

Xgboost		*Random Forest*		*CNN*	
model parameters	value	model parameters	value	model parameters	value
eta	0.2	n_estimatiors	51	conv kernel size	4×4
max_depth	4	criterion	gini	conv kernel number	300
max_child_weight	6	max_depth	8	pooling layer	4×4
subsample	0.08	max_samples_split	4	CNN layer number	4
regularization lambda	1	max_features	16	dense layer neurons	128/64

Table 2. Experiment I model prediction performance

Xgboost		*Random Forest*		*CNN*	
Evaluation Metrics	value	Evaluation Metrics	value	Evaluation Metrics	value
accuracy rate	77.14%	accuracy rate	70.93%	accuracy rate	78.01%
precision rate	89.12%	precision rate	75.66%	precision rate	89.12%
recall rate	54.58%	recall rate	63.58%	recall rate	54.58%
F1-score	0.677	F1-score	0.691	F1-score	0.677

Comparing the results of the above experiments, we can find that the deep learning algorithm CNN is better in terms of accuracy and precision when predicting the financial distress of car companies using only financial indicator data, reaching 78.01% and 89.12%, while the traditional machine learning algorithm Random Forest is better in terms of recall and F1-score, reaching 63.58% and 0.6910. Since this experiment is to predict the financial distress risk of Chinese car companies' financial distress risk, more attention is paid to the recall and F1-score in the model's prediction performance.

Therefore, we believe that the performance of the model based on the random forest algorithm is more representative, and use it as the baseline model to compare with the results of the subsequent experiments.

3.3 Experiment II - Based on Financial Indicator Data, Financial Text Data and Non-financial Text Data

We used the model framework mentioned in Chapter 2 to construct a prediction model for financial distress of car companies based on financial indicators data, financial text data and non-financial text data,with core algorithms including deep learning algorithms CNN and RNN,etc.The main experimental parameters are shown in Table 3 and the experimental results are shown in Table 4.

Table 3. Experiment II main model parameters

CNN		RNN	
model parameters	value	model parameters	value
convolution kernel size	5 × 5	input dimension	200
convolution kernel number	256	hidden layer number	5
pooling layer	5 × 5	dropout	0.3
CNN layer number	6		
dense layer neurons	128/64		

Table 4. Experiment II model prediction performance

Random Forest		Combine Model	
Evaluation Metrics	value	Evaluation Metrics	value
accuracy rate	70.93%	accuracy rate	85.86%
precision rate	75.66%	precision rate	93.38%
recall rate	63.58%	recall rate	77.67%
F1-score	0.691	F1-score	0.848

Comparing the Above Experimental Results, It Can Be Found that the Experimental Results in Experiment II Outperformed the Baseline Model in Experiment I in All Aspects, with the Accuracy, Precision and Recall Rates Being 21.0%, 23.4% and 22.7% Higher Respectively, and the F1-Score Being 22.7% Higher. The Experimental Results Demonstrate that the Inclusion of Financial Text Data and Non-financial Text Data Effectively Improves the Performance of the Model for Predicting the Financial Distress Risk of Vehicle Companies.

4   Conclusion and Future Works

In terms of model construction, we innovatively proposed to combine financial indicator data, financial text data and non-financial text data to predict the risk of financial distress of vehicle enterprises, and built a set of prediction model framework based on deep learning algorithms, which is a useful attempt in cross-disciplinary research in this field and also lays the foundation for subsequent research.

In terms of experimental analysis, we found that the introduction of financial text data and non-financial text data into the model significantly improved the predictive performance of the model compared to the traditional predictive model based on financial indicator data, which indicates that text data such as online user reviews and financial annual reports are more helpful in predicting the risk of financial distress of car companies. Therefore, in addition to paying attention to their own financial indicators, car company managers should also pay attention to the online user reviews of the relevant companies when choosing their business strategies, so as to avoid financial risks; and industry investors should also extract and analyse the information from the texts in annual reports when conducting investment analysis, which will provide useful directional guidance for their investment behaviour.

Acknowledgments. This research project is supported by Science Foundation of Beijing Language and Culture University (supported by "the Fundamental Research Funds for the Central Universities") (Approval number: 22YJ090006).

References

1. Du, Y., Wei, K., Wang, Y., Jia, J.: New energy vehicles sales prediction model combining the online reviews sentiment analysis: a case study of Chinese new energy vehicles market. In: Liang, Q., Wang, W., Mu, J., Liu, X., Na, Z. (eds.) Artificial Intelligence in China. LNEE, vol. 854, pp. 424–431. Springer, Singapore (2022). https://doi.org/10.1007/978-981-16-9423-3_53
2. Tsai, C.F., Wu, J.W.: Using neural network ensembles for bankruptcy prediction and credit scoring. Expert Syst. Appl. **34**(4), 2639–2649 (2008)
3. Lee, S., Choi, W.S.: A multi-industry bankruptcy prediction model using back-propagation neural network and multivariate discriminant analysis. Expert Syst. Appl. **40**(8), 2941–2946 (2013)
4. Hosaka, T.: Bankruptcy prediction using imaged financial ratios and convolutional neural networks. Expert Syst. Appl. **117**, 287–299 (2019)
5. Ristolainen, K.: Predicting banking crises with artificial neural networks: the role of nonlinearity and heterogeneity. Scand. J. Econ. **120**(1), 31–62 (2018)
6. Matin, R., Hansen, C., Hansen, C., Mølgaard, P.: Predicting distresses using deep learning of text segments in annual reports. Expert Syst. Appl. **132**, 199–208 (2019)
7. Li, X., Wu, C., Mai, F.: The effect of online reviews on product sales: a joint sentiment-topic analysis. Inf. Manag. **56**(2), 172–184 (2019)
8. Trueman, T.E., Kumar, A., Narayanasamy, P., Vidya, J.: Attention-based C-BiLSTM for fake news detection. Appl. Soft Comput. **110**, 107600 (2021)

9. Gandhi, U.D., Malarvizhi Kumar, P., Chandra Babu, G., Karthick, G.: Sentiment analysis on twitter data by using convolutional neural network (CNN) and long short term memory (LSTM). Wirel. Pers. Commun., 1–10 (2021)
10. Du, Y., Wang, Y., Wei, K., Jia, J.: The sentiment analysis and sentiment orientation prediction for hotel based on BERT-BiLSTM model. In: Liang, Q., Wang, W., Mu, J., Liu, X., Na, Z. (eds.) Artificial Intelligence in China. LNEE, vol. 854, pp. 498–505. Springer, Singapore (2022). https://doi.org/10.1007/978-981-16-9423-3_62

Frame Interpolation for Weather Radar Data

Hao Ge[(⊠)], Xi Chen, Yungang Tian, Hui Ding, Ping Chen,
and Flora Kumama Wakolo

State Key Laboratory of Air Traffic Management System and Technology,
Nanjing Research Institute of Electronic Engineering, Nanjing 210007, China
gehao1@cetc.com.cn

Abstract. Rain and snow contain particles generated by different physical and chemical processes. Weather radars send directional pulses of microwave radiation, on the order of a microsecond long. Between each pulse, the radar station serves as a receiver as it listens for return signals from particles in the air. This is the measurement principle of microwave weather (meteorological) radar. Limited by the information processing capability, the weather radar data published by the meteorological website often have a large time interval, such as 6 min. Video frame interpolation technology has made great progress in recent years with the development of deep learning technology. The frame interpolation of weather radar charts will bring users more accurate weather descriptions and more intuitive decision-making references. This paper propose a novel video frame interpolation algorithm for weather radar data, which is able to generate the intermediate frames between the frames at the sampling time.

Keywords: Weather radar · Convolutional neural networks · Deep learning · Machine learning

1 Introduction

Weather radar is more sensitive to rain and snow in the air, so its reflectivity can intuitively reflect the number of water molecules in the air. Weather radar currently plays a huge role in many fields. Taking severe weather forecast as an example, for areas with a particularly high incidence of hail severe weather, hail will reduce the total output of grain and oil in the entire region to a large extent, sometimes even as high as tens of thousands of catties. In aviation, accurate weather forecasts will determine whether or not flights can take off on time, affecting the travel plans of thousands of travelers [1,2].

Figure 1 shows a weather radar data of China at 10:30 a.m. on July 4, 2022. The reflectance is marked with false color in the figure. High reflectivity indicates that the size or number of precipitation particles per unit volume is large, which in turn indicates that the region has a high possibility of precipitation.

© The Author(s), under exclusive license to Springer Nature Singapore Pte Ltd. 2023
Q. Liang et al. (Eds.): AIC 2022, LNEE 871, pp. 211–218, 2023.
https://doi.org/10.1007/978-981-99-1256-8_25

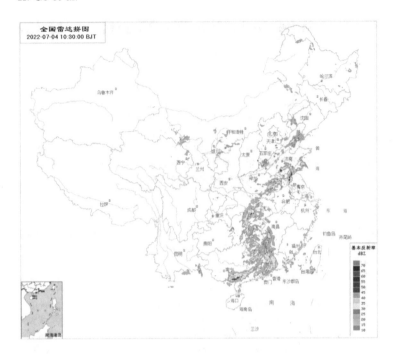

Fig. 1. Weather radar data.

The weather radar data published by meteorological websites often have a large time interval. Taking the Central Meteorological Observatory (http://www.nmc.cn/publish/radar/chinaall.html) as an example, the data published by it exists 6 min intervals. For some fields with high time precision, the time interval of 6 min is difficult to meet its requirements.

In the field of aviation, the operation of aircraft is heavily dependent on weather conditions. Whether it is flight time slot allocation, flight planning or real-time diversion, the participation of the meteorological department is required. With the development of 4D trajectory technology [3], the trajectory planning of aircraft has reached the second level. Therefore, the 6-minute update frequency of weather radar data has been difficult to support the needs of real-time aircraft trajectory planning. Therefore, we propose a frame interpolation algorithm based on optical flow and depth of field, which is used to provide meteorological radar data with shorter time intervals, and is used for aircraft trajectory planning in the aviation field and similar fields which require real-time meteorological data.

In the field of computer vision, researchers have done a lot of research on the problem of video frame interpolation. Video frame interpolation methods are divided into three categories: flow-based, kernel-based, and phase-based. These methods all estimate the optical flow field between two frames, and generally obtain the intermediate frame by warping the previous frame. Video frame interpolation is particularly difficult in some scenarios, such as scenes and objects that are constantly moving and changing, or when there is occlusion, there will be multiple solutions to the frame interpolation problem.

The contribution of our method can be summarised as:

(1) We propose a optical flow based frame interpolation framework to predict the motion trajectories of weather phenomena between images with 6-minute intervals.
(2) We propose to use the depth-of-field to optimize the effect of frame insertion algorithm in weather radar data.

The rest of the paper is organized as follows. In Sect. 2, we first review related work. The details of the proposed method are illustrated in Sect. 3. In Sect. 4, we would presents and discuss the experimental results. Section 5 provides conclusions.

2   Related Work

In this section, we briefly introduce data processing methods for weather radar data and different video frame interpolation algorithms.

2.1   Data Processing Methods for Weather Radar Data

YL Qin [4] proposes a method based on motion estimation and compensation, which analyzes the evolution of the predicted radar date by calculating the motion vector and chromaticity changes on the texture of the first two or more images, so as to predict the texture content of the radar data in the next one or more images.

Cică R [5] use three-dimensional single polarization weather radar data to detect the hail clouds, using data including composite reflectivity, vertically integrated liquid density and so on. The results show that radar variables have a good detection effect on the areas where hail clouds were reported.

It can be seen that the weather radar data contains a lot of information, which can provide help for many applications in many fields, so the data processing of weather radar data is of great significance.

2.2   Video Frame Interpolation

Video frame interpolation is a fundamental computer vision task. The goal of Video Frame Interpolation is to composite several frames between two adjacent frames of the original video. It has become a compelling strategy for numerous applications, such as frame rateup-conversion [6], video compression [7], and slow motion generation [8–12].

In recent years, with the rapid development of deep neural networks, a large number of researches have been conducted in this area. The mainstream approach is to estimate the optical flow field between two frames, and generally obtain the intermediate frame by warping the previous frame [8,9,11,12]. In this work, we use the model of Bao W [13], which is a video frame interpolation framework, which utilized flow estimation, depth estimation, context extraction and kernel estimation to generate the output frame.

2.3 Depth Estimation

Divided from the number of signal sources, depth estimation can be divided into two types: depth estimation from stereo images [14] and depth estimation from single image [15]. According to the type of input signal, depth estimation can be divided into two types: depth estimation for image [14] and depth estimation for video [16].

In this work, we use the model of Casser V [16], which is an depth estimation network trained on the KITTI dataset [17]. This work addresses the monocular depth problem by modeling individual objects' motion in 3D.

3   The Proposed Approach

3.1 Algorithm Overview

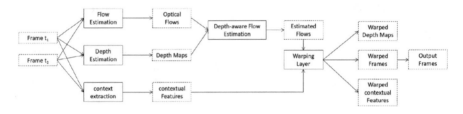

Fig. 2. Architecture of the proposed video frame interpolation model. Solid boxes represent models and dashed boxes represent images.

Given two frames $F_{t(0)}$ and $F_{t(1)}$, our goal is to estimate an intermediate frame F_t at time $t \in [0, 1]$. In this work, we adopt the flow estimation layer in Bao et al. [6] to estimate the flow vector. Similar to Bao W et al. [13], after estimate the intermediate flows, we adapt a warping layer to warp the input frames, contextual features and depth maps. The warped frames is the output of our framework (Fig. 2).

3.2 Flow Estimation

We adopt the state-of-the-art model ARFLOW [18] as our flow estimation netrowk. We initialize our flow estimation network from the pre-trained ARFLOW. It is an unsupervised learning of optical flow, which leverages the supervision from view synthesis, has emerged as a promising alternative to supervised methods.

3.3 Depth Estimation

We adopt the depth prediction model proposed by Casser et al. [16] as our depth estimation network. The model is trained on KITTI dataset, and fine-tuned on our weather radar data to obtain better performance on radar data.

3.4 Context Extraction

Unlike many traditional video frame interpolation models that use Resnet for context ectraction, we adopt a context extraction network [13] with one 7×7 convolutional layer and two residual blocks, as shown in Fig. 3.

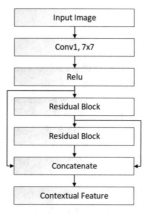

Fig. 3. Structure of the context extraction network.

3.5 Implimentation Details

Training Dataset. We constructed a dataset using the weather radar data published by Central Meteorological Observatory (http://www.nmc.cn/publish/radar/chinaall.html). This database contains weather radar data for 2020 with a sampling interval of 6 min. The resolution of a single radar data is 1024*880. We train our network to predict the frame that is not at the sampling time. We augment the training data by horizontal flipping.

Training Strategy. We use a batch size of 10 to train the proposed network. The initial learing rates of the context extraction is set to 1e−4. The flow estimation and depth estimation networks are initialized from pre-trained models, so we use learning rates of 1e-6 to train them. We train the network for 20 epochs on an NVIDIA 3080Ti GPU card, which takes 2 days to converge.

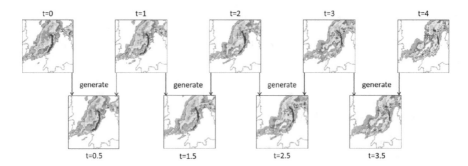

Fig. 4. The experimental results on our weather radar data. The proposed method is able to generate the intermediate frame between the original frames.

4   The Experiments

We tested the proposed method on our weather radar dataset. In Fig. 4, we show the experimental results of our method. As can be seen, we successfully generate the intermediate frame between the frames at the sampling time points. In order to be able to see clearly, the pictures in Fig. 4 are all cropped. The original picture contains the entire territory of China. The size of the cropped picture is about 1% of the original picture.

Also, we are able to interpolate more frames between frames, like 3 frames or 5 frames or even more. The effect of frame interpolation is difficult to fully display in the form of a small number of pictures. More detailed experimental results can be seen in my github(https://github.com/ghghgh0001/frame-interpolation/tree/main/examples) in the form of more pictures or gifs (Fig. 5).

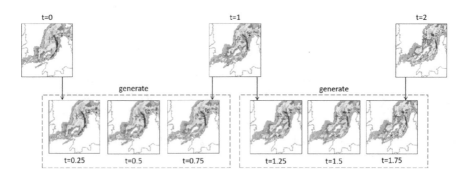

Fig. 5. The proposed method is able to generate 3/5/7 or even more intermediate frame between the original frames.

In Table 1, we provide quantitative performances on our weather radar dataset. Our approach performs favorably against existing methods. "Avg" means fill all the pixels with the mean of the corresponding pixels of adjacent frames. The experiments are tested on 1024×800 images for each method in Table 1. It can be seen that our method performs better than the current mainstream video frame interpolation algorithms on our weather radar dataset.

Table 1. Quantitative performances on our weather radar dataset.

Method	PSNR	SSIM
Avg	25.79	0.9441
DVF [9]	34.27	0.9645
SepConv-l_f [11]	34.76	0.9661
SepConv-l_1 [11]	34.81	0.9668
Our method	35.02	0.9684

5 Conclusion

In this work, we propose a novel video frame interpolation algorithm for weather radar data, which is able to generate the intermediate frames between the frames at the sampling time. Quantitative evaluations shows that the proposed method outperforms existing frame interpolation algorithms on our weather radar dataset. The effectiveness of the proposed method provides ideas for future research on frame interpolation algorithms in the field of weather radar data.

Acknowledgement. This work is financially supported by the National Key R&D Program of China, Project Number 2018YFE0208700.

References

1. Mao, Y.Q., Ding, Y.B., Cheng, X.F.: Thunderstorm service and decision support technology based on composite reflectivity information. Command Inf. Syst. Technol. **10**(6), 14–19 (2019)
2. Liu, Y.: Application of aeronautical meteorological technology in air traffic management. Command Inf. Syst. Technol. **1**(2), 43–47 (2010)
3. Zeng, W., Chu, X., Xu, Z., et al.: Aircraft 4D trajectory prediction in civil aviation: a review. Aerospace **9**(2), 91 (2022)
4. Qin, Y., Zhang, B., Yang, D., et al.: Prediction and analysis of radar mosaic based on motion estimation and compensation. Comput. Digit. Eng. **48**(3), 678–681 (2020)
5. Cică, R., Burcea, S., Bojariu, R.: Assessment of severe hailstorms and hail risk using weather radar data. Meteorol. Appl. **22**(4), 746–753 (2015)
6. Bao, W., Zhang, X., Chen, L., et al.: High-order model and dynamic filtering for frame rate up-conversion. IEEE Trans. Image Process. **27**(8), 3813–3826 (2018)

7. Wu, C.-Y., Singhal, N., Krähenbühl, P.: Video compression through image interpolation. In: Ferrari, V., Hebert, M., Sminchisescu, C., Weiss, Y. (eds.) ECCV 2018. LNCS, vol. 11212, pp. 425–440. Springer, Cham (2018). https://doi.org/10.1007/978-3-030-01237-3_26
8. Jiang, H., Sun, D., Jampani, V., et al.: Super SloMo: high quality estimation of multiple intermediate frames for video interpolation. In: Proceedings of the IEEE Conference on Computer Vision and Pattern Recognition, pp. 9000–9008 (2018)
9. Liu, Z., Yeh, R.A., Tang, X., et al.: Video frame synthesis using deep voxel flow. In: Proceedings of the IEEE International Conference on Computer Vision, pp. 4463–4471 (2017)
10. Niklaus, S., Liu, F.: Context-aware synthesis for video frame interpolation. In: Proceedings of the IEEE Conference on Computer Vision and Pattern Recognition, pp. 1701–1710 (2018)
11. Niklaus, S., Mai, L., Liu, F.: Video frame interpolation via adaptive separable convolution. In: Proceedings of the IEEE International Conference on Computer Vision, pp. 261–270 (2017)
12. Peleg, T., Szekely, P., Sabo, D., et al.: IM-NET for high resolution video frame interpolation. In: Proceedings of the IEEE/CVF Conference on Computer Vision and Pattern Recognition, pp. 2398-2407 (2019)
13. Bao, W., Lai, W.S., Ma, C., et al.: Depth-aware video frame interpolation. In: Proceedings of the IEEE/CVF Conference on Computer Vision and Pattern Recognition, pp. 3703-3712 (2019)
14. Wang, Y., Lai, Z., Huang, G., et al.: Anytime stereo image depth estimation on mobile devices. In: 2019 International Conference on Robotics and Automation (ICRA), pp. 5893-5900. IEEE (2019)
15. Mertan, A., Duff, D.J., Unal, G.: Single image depth estimation: an overview. Digit. Sig. Process. 103441 (2022)
16. Casser, V., Pirk, S., Mahjourian, R., et al.: Depth prediction without the sensors: leveraging structure for unsupervised learning from monocular videos. In: Proceedings of the AAAI Conference on Artificial Intelligence, vol. 33, no. 01, pp. 8001-8008 (2019)
17. Geiger, A., Lenz, P., Stiller, C., et al.: Vision meets robotics: the KITTI dataset. Int. J. Rob. Res. 32(11), 1231–1237 (2013)
18. Liu, L., Zhang, J., He, R., et al.: Learning by analogy: reliable supervision from transformations for unsupervised optical flow estimation. In: Proceedings of the IEEE/CVF Conference on Computer Vision and Pattern Recognition, pp. 6489-6498 (2020)

A Hybrid Neural Network for Music Generation Using Frequency Domain Data

Huijie Wang[1], Shuang Han[3], Guangwei Li[2(✉)], and Bin Zhao[2(✉)]

[1] Unit 95169 of the People's Liberation Army, Beijing, China
[2] School of Artificial intelligence, Guilin University of Electronic Technology, Guilin, Guangxi 541004, China
liguangwei0304@126.com, zhaobinnku@gmail.com
[3] School of Information Engineering, Huanghuai University, Zhumadian 463000, Henan, China

Abstract. Currently, many deep learning based methods for music generation have been proposed. However, these methods use time-domain audio data to train their networks; consequently, these methods generate low-quality music. This paper proposes using frequency-domain data that are transformed from time-domain data to train a hybrid neural network that combines a recurrent neural network with a generative adversarial network to generate music in an end-to-end way. The automatic music generation method proposed in this paper explores a new representation of music data, and the results prove the effectiveness of our method.

Keywords: Deep neural network · long short-term memory(LSTM) · generative adversarial network(GAN) · music generation

1 Introduction

In recent years, deep learning based methods have been widely used in automatic music generation [1]. A deep learning-based generator can automatically learn models and styles from any music corpus and then generate music by making predictions or classifications based on the learned distributions and correlations [2].

Although convolutional neural network(CNN)-based methods have performed well on music generation, they cannot express emotive music [3]. Because recurrent neural networks (RNNs) can better learn long-term correlation, they have been used in music generation and prediction [4]. To overcome the shortcomings of RNNs, variants known as LSTM networks [5] have been developed. Eck et al. [6] proposed an LSTM method for automatic music generation by learning the long-term correlation between data and the corresponding music style. Furthermore, Goodfellow et al. [7] proposed a new framework network, namely, the GAN, to learn data information through the mutual confrontation between the generator and the discriminator. GANs have achieved good results in both image generation and data augmentation by extracting salient features

Q. Liang et al. (Eds.): AIC 2022, LNEE 871, pp. 219–228, 2023.
https://doi.org/10.1007/978-981-99-1256-8_26

from complex data and understanding the true underlying data generation distribution [8]. However, these methods use time-domain audio data to train their networks; consequently, these methods generate low-quality music. This paper proposes using frequency-domain data to train a hybrid neural network that combines a RNN with a GAN to generate music in an end-to-end way. The main contributions of this work can be concluded as follows:

1) A new data processing algorithm that can convert time-domain waveform data into frequency-domain spectrogram data is proposed.
2) To better learn the long-term correlation between the spectrogram data, a hybrid neural network that combines LSTM with a generative adversarial network is proposed to generate music in an end-to-end way.

2   Related Works and Preliminaries

Some traditional algorithms, such as hidden Markov models [9], random sampling generation and statistical models based on composition rules [10], are used for music generation. However, they depend heavily on handcrafted image features, as their modelling capabilities are still significantly limited. Recently, deep learning has been used for music generation. However, deep learning methods need time-domain audio data to train their networks; consequently, these methods generate low-quality music.

2.1   Music Data Representation

With the continuous deepening of the research on automatic music generation, the data used for training are constantly changing in various forms. For instance, the earliest form of data used for deep learning music generation tasks is the one-hot vector [11]. By converting each note in a piano piece into a one-hot vector, the piano pieces are represented in matrix form for subsequent processing and network training. Other ways of processing music data into symbolic form include the MIDI form [12], which represents music as note-on and note-off information, corresponding to the start and end of notes. Each note-on and note-off message is also accompanied by a MIDI note number, key velocity value, channel specification and timestamp. MusicXML [13] is an extensible markup language that converts music data into textual data. A rule is defined to encode documents in a human- and machine-readable manner. In addition, CSV [14] and Rendering [15] are both symbolic data forms. Later, time-domain waveform data appeared [16]; these data are sorted by time and follow a certain standard or standard coding.

This work proposes transforming time-domain data into frequency-domain data for training. The difference between our method and the existing method is that the audio data are converted into spectrograms; thus, our method explores a new way of processing music data.

2.2 Discrete Short-Time Fourier Transform

Speech signal is a typical non-stationary signal, and the characteristics of non-stationary signal will change with time where the characteristic is frequency in signal processing [17,18]. To learn this time-varying characteristic, the time-frequency analysis of the signal is performed using a short-time Fourier transform (STFT). The main idea of the STFT is to consider not the whole signal but only a small part of the signal. An STFT window function, which is nonzero for a short period, is used. A window function in STFT is used, which is is non-zero for a short period of time. Then multiply the original signal by the window function to get the window signal. Then, the original signal is multiplied by the window function to obtain the window signal. To obtain frequency information at different instants, the window function is moved in time, and the Fourier transform is calculated for each resulting window signal. The definitions are given as follows:

$$\mathcal{X}(m, k) := \sum_{n=0}^{N-1} x(n + mH)w(n) \exp(-2\pi ikn/N) \tag{1}$$

let $x : [0 : L - 1] := \{0, 1, \ldots, L - 1\} \to \mathbb{R}$ be a real-valued discrete-time (DT) signal of length L obtained by equidistant sampling with respect to a fixed sampling rate F_s given in Hertz. Furthermore, let $w : [0 : N - 1] \to \mathbb{R}$ be a sampled window function of length $N \in \mathbb{N}$, with $m \in [0 : M]$ and $k \in [0 : K]$. The number $M := \lfloor \frac{L-N}{H} \rfloor$ is the maximal frame index such that the window's time range is fully contained in the signal's time range. (Later, some variants will use padding strategies.) Furthermore, $K = N/2$ (assuming that N is even) is the frequency index corresponding to the Nyquist frequency. The complex number $\mathcal{X}(m, k)$ denotes the kth Fourier coefficient for the mth time frame.

The spectrogram is a two-dimensional representation of the squared magnitude of the STFT:

$$\mathcal{Y}(m, k) := |\mathcal{X}(m, k)|^2 \tag{2}$$

the spectrogram can be visualized by a two-dimensional image, where the horizontal axis represents time and the vertical axis represents frequency. In this image, the spectrogram value $\mathcal{Y}(m, k)$ is represented by the intensity or colour in the image at the coordinate (m, k).

3   Model and Formulation

The overall architecture of this study is shown in Fig. 1. First, the data are preprocessed into the frequency-domain spectrogram. Then, we use the combined LSTM-GAN network we used for training. Through the system proposed in this paper, we can better learn the long-term and spatial correlation of audio data, and the combination of the two results in smoother and more harmonious music. The method and model adjustment strategy are described in detail.

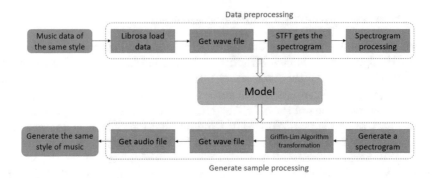

Fig. 1. Overall research framework flow chart.

3.1   Data Preprocessing

In this study, songs of the same genre were collected from the same singer for training; each of these songs was processed into segments of the same length. Eighty percent of these segments are used for training, and twenty percent are used for testing to learn the frequency-domain features of music and to generate the same music style based on the frequency-domain features.

Because of the nature of the human perception of sound, we know that each fixed note corresponds to a fixed frequency value. We can associate to each pitch $p \in [0 : 127]$ a centre frequency $F_{pitch}(p)$ (measured in Hz) defined as follows:

$$F_{pitch}(p) = 2^{(p-69)/12} \cdot 440 \tag{3}$$

With the intrinsic link between music and frequency, we propose to process the data in the frequency domain. The overall process of the data preprocessing is shown in Fig. 2.

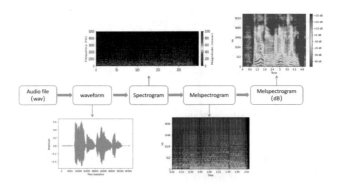

Fig. 2. Overall data processing flow chart.

The form of the time-domain waveform of each segment is obtained by using the Python standard library; a specific example is shown in Fig. 3(a).

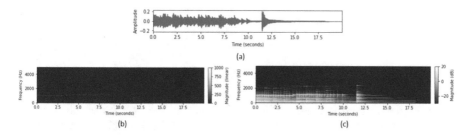

(a)

(b)   (c)

Fig. 3. (a) Sample example time-domain waveform. (b) Sample instance spectrogram(linear). (c) Sample instance spectrogram(dB).

Then, the waveform of each segment is transformed into the frequency domain by using a discrete Fourier transform; a specific example is shown in Fig. 3(b).

Finally, we convert the linear spectrum into a logarithmic spectrum; the purpose of this transformation is to pull up those components with lower amplitudes relative to those with higher amplitudes. A specific example is shown in Fig. 3(c).

3.2   Model

The neural network model proposed in this study consists of an LSTM network and a GAN. By combining LSTM with a GAN, the temporal information and spatial structure information of the spectrogram can be better learned. An LSTM network can learn the long-term correlation between the time series data better than a CNN. GANs are very effective generative model. GANs can be used in the constant game between the generator and the discriminator to compare the generated samples with the real data, and GANs obtain samples with characteristics similar to those of the real data. The overall architecture of the proposed model is shown in Fig. 4.

Generator Network. The structure of the generator network is shown in Fig. 4 Generator. By training the generator, the distribution characteristics of real data are learned. The LeakyReLU function is used as the activation function. We input random data into the network in the form of a matrix, and the corresponding sample matrix ID is generated through the network.

Discriminator Network. The structure of the discriminator network is shown in Fig. 4 Discriminator. The discriminator distinguishes the generated samples from the real data, by estimating the probability that the samples come from

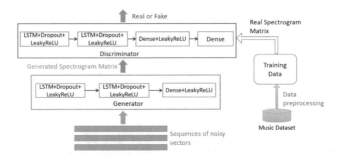

Fig. 4. Model system framework. The LSTM and GAN are combined into a hybrid model. The noise vector z is the input and the output $G(z)$ is obtained through the generator. The discriminator then discriminates between $G(z)$ and real data.

the real training dataset rather than the generator. The resulting sample matrix is fed into the discriminator network along with the original dataset for training.

3.3   Griffin-Lim Algorithm (GLA)

The GLA [19] (named after their authors) was presented in 1984 in [20]. This algorithm aims to estimate a signal from its modified short-time fourier transform. The algorithm is a phase reconstruction method based on short-time fourier transform redundancy. Using the known amplitude spectrum and the unknown phase spectrum, the phase spectrum is generated by iteration, the phase spectrum is obtained by summing the known amplitude spectrum and calculation, and the speech waveform is reconstructed. In this paper, the relationship between frames is used to estimate the phase information to reconstruct the speech waveform. The algorithm steps are as follows:

1) Randomly initialize a phase spectrum;
2) Use this phase spectrum and the known amplitude spectrum to synthesize a new speech waveform by use an inverse short-time fourier transform (ISTFT);
3) Use the synthesized speech to perform an STFT to obtain a new amplitude spectrum and a new phase spectrum;
4) Discard the new amplitude spectrum, and synthesize new speech with the known amplitude spectrum and the new phase spectrum;
5) Repeat steps 2, 3, and 4 several times until the synthesized speech achieves satisfactory results.

4 Experiment and Evaluation

This section presents the feasibility and experimental results used to verify the proposed method. The processed dataset is put into the network for training, and the optimal parameters are obtained by 500 iterations of the gradient descent method. The learning rate is gradually decreased from 0.1.

The trained generator is used to generate music, and the obtained spectrogram is shown in Fig. 5(a). The figure shows that the spectrogram has some frequencies similar to the original spectrogram.

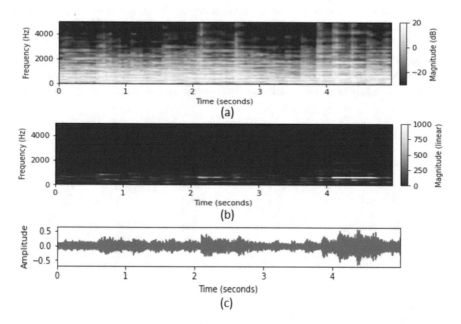

Fig. 5. (a) Generate sample spectrogram(dB). (b) Generate sample spectrogram(linear). (c) Generate sample time-domain waveforms.

Then, we use the GLA method to inversely transform the generated spectrogram to obtain its linear spectrogram and time-domain waveform, as shown in Fig. 5(b). The time-domain waveform diagram shows that the generated waveform diagram has some obvious noise, which is also a defect of this research. Finally, the time-domain waveform is converted into an audio file, as shown in Fig. 5(c). The music style is very similar to that of the original audio, but the distortion caused by the transformation is also serious.

4.1 Low-Level Feature Comparison

We use the network to extract low-level features from the generated and original data. The network structure diagram is shown in Fig. 6. The original data and the generated data are put into the network to obtain their corresponding low-level feature vectors.

Fig. 6. Low-level feature network flowchart.

Figure 7(a) shows the low-level feature map of the original real data, and we can obtain the results shown in Fig. 7(c) by numerically discretizing the data. Figure 7(b) represents the low-level features of the generated sample data, and by numerically discretizing them, we can obtain the results shown in Fig. 7(d). By representing the low-level features of the original real data and the generated data, we can analyse the similarity and error values between both types of data.

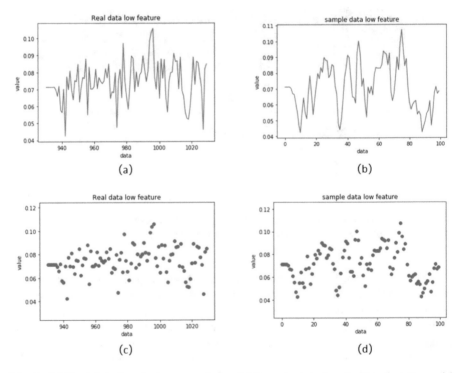

Fig. 7. (a) Real data low feature waveform. (b) Sample data low feature waveform. (c) Real data value of low feature. (d) Sample data of low feature.

Then, we compare the error value between the low-level features of the original data and the generated samples to obtain the overall error probability. The formula for the error probability is expressed as:

$$V_{error} = |Y_{sample} - Y_{real}| \qquad (4)$$

where V_{error} denotes the error value and Y_{sample} and Y_{real} represent the low-level features of the sample and the original data, respectively.

Using the above formula, we can obtain the size of the error difference, and the size of the obtained error value is shown in Fig. 8.

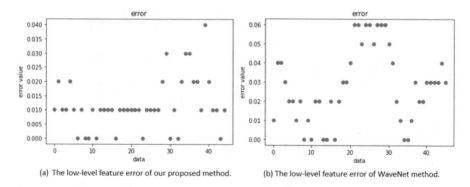

(a) The low-level feature error of our proposed method.   (b) The low-level feature error of WaveNet method.

Fig. 8. (a) The low-level feature error of our proposed method($Our - err_m = 0.0118$). (b) The low-level feature error of WaveNet method($WaveNet - err_m = 0.0286$).

As Fig. 8(a) shows, most of the low-level feature error values are below 0.02, while the feature similarity is high; this finding proves that our network has learned the corresponding feature information.

In addition, we used the same data to train the WaveNet model and our model to compare the low-level feature error rates. The representation of the low-level eigenvector error rate obtained by WaveNet is shown in Fig. 8(b). The low-level feature error values of the two methods are averaged separately. It can be obtained that the mean error value of our proposed model is 0.0118, and the mean error value of WaveNet is 0.0286. This figure shows that our proposed method has a lower error rate than WaveNet.

5   Conclusion

In this paper, we transform audio data into the frequency domain to train a hybrid neural network, and the results prove the effectiveness of our method. In addition, learning in the frequency domain provides new ideas for processing audio data multimodally.

References

1. Dannenberg, R.B.: Music representation issues, techniques, and systems. Comput. Music. J. **17**(3), 20–30 (1993)
2. Briot, J.P., Pachet, F.: Music generation by deep learning-challenges and directions. arXiv preprint: arXiv:1712.04371 (2017)
3. Li, H., Xue, S., Zhang, J.: Combining CNN and classical algorithms for music genre classification (2018)
4. Chu, H., Urtasun, R., Fidler, S.: Song from PI: a musically plausible network for pop music generation. arXiv preprint: arXiv:1611.03477 (2016)
5. Hochreiter, S., Schmidhuber, J.: Long short-term memory. Neural Comput. **9**(8), 1735–1780 (1997)
6. Eck, D., Schmidhuber, J.: A first look at music composition using LSTM recurrent neural networks. Istituto Dalle Molle Di Studi Sull Intelligenza Artificiale **103**, 48 (2002)
7. Goodfellow, I.J., Pouget-Abadie, J., Mirza, M., Xu, B., Warde-Farley, D., Ozair, S., et al.: Generative adversarial networks (2014)
8. Brunner, G., Wang, Y., Wattenhofer, R., Zhao, S.: Symbolic music genre transfer with cycleGAN. In: IEEE 30th International Conference on Tools with Artificial Intelligence (ICTAI) (2018)
9. Pachet, F., Roy, P., Barbieri, G.: Finite-length Markov processes with constraints. In: International Joint Conference on IJCAI. DBLP (2011)
10. Mark, J.S.: A generative grammar for Jazz chord sequences. Music. Percept. **2**(1), 52–77 (1984)
11. Li, G., Ding, S., Li, Y., Zhang, K.: Music generation and human voice conversion based on LSTM. In MATEC Web of Conferences, vol. 336, p. 06015. EDP Sciences (2021)
12. Huang, Y.S., Yang, Y.H.: Pop music transformer: beat-based modeling and generation of expressive pop piano compositions. In: Proceedings of the 28th ACM International Conference on Multimedia, pp. 1180-1188 (2020)
13. Good, M.: MusicXML: an internet-friendly format for sheet music. In: Xml conference and expo, pp. 03-04 (2001)
14. Shafranovich, Y.: Common format and MIME type for comma-separated values (CSV) files (No. rfc4180) (2005)
15. Kajiya, J.T.: The rendering equation. In: Proceedings of the 13th Annual Conference on Computer Graphics and Interactive Techniques, pp. 143-150 (1986)
16. Oord, A.V.D., et al.: WaveNet: a generative model for raw audio (2016)
17. Portnoff, M.: Time-frequency representation of digital signals and systems based on short-time Fourier analysis. IEEE Trans. Acoust. Speech Signal Process. **28**(1), 55–69 (1980)
18. Strawn, J.: Analysis and synthesis of musical transitions using the discrete short-time Fourier transform. J. Audio Eng. Soc. **35**(1/2), 3–14 (1987)
19. Perraudin, N., Balazs, P., Søndergaard, P.L.: A fast Griffin-Lim algorithm. In: 2013 IEEE Workshop on Applications of Signal Processing to Audio and Acoustics, pp. 1-4. IEEE (2003)
20. Griffin, D., Lim, J.: Signal estimation from modified short-time Fourier transform. IEEE Trans. Acoust. Speech Signal Process. **32**(2), 236–243 (1984)

Prediction Model for Inclusive Finance Development Considering the Impact of COVID-19: The Case of China

Yu Du, Bing Wang[(✉)], Kaiyue Wei, and Xiaoling Song

Business School, Beijing Language and Culture University, No. 15, Xueyuan Road, Haidian District, Beijing 100083, People's Republic of China
1768569791@qq.com

Abstract. Models in previous studies about inclusive finance often include economic data while excludes public online statements. In this paper Random Forest Regression (RFR) model is trained on the annual influencing factors and annual financial inclusion index to predict quarterly financial inclusion index by the quarterly influencing factors to expand the size of data. Then, BOW model tf-idf algorithm is used to convert COVID-19 – loan related online statements into word vectors. Lastly, these influencing factors of different lag periods are passed into the RFR model to compare their performance. Result of models shows that there is impact the epidemic has on the development of inclusive finance, and the lag period of the impact opinion texts on financial inclusion is 2 quarters.

Keywords: COVID-19 · Inclusive Finance · Machine Learning · Data Augmentation · Natural Language Processing

1 Introduction

Inclusive finance means that everyone has financial needs to access high-quality financial services at the right price in a timely and convenient manner [1]. Policy makers in Asian countries have taken different steps to develop inclusive finance [2] as it is important to reduce the financial crisis, improve the financial equity, reduce financial exclusion, and build a sustainable inclusive finance to develop the economy of a country [3].

Many areas have changed dramatically in China during COVID-19 since 2020. Revenues and cash flows of SMEs generally have declining. In order to stimulate economic development. China has adopted a series of policies, such as reducing taxes and fees and increasing financing support for SMEs, which likely to have an impact on financial inclusion. Whether the epidemic has had an impact on the development of financial inclusion, and if so, what kind of impact it is, are the main research questions.

Scholars have studied the influencing factors of financial inclusion before. Xiaocui Deng [4] found that the inclusive finance index is positively correlated with the degree of economic development level, deposit resources utilization level, urbanization level, and is not related to the education degree. Huang Jie, Zhang Wenshuang, Ruan Weihua [5] analyzes the network characteristic and impact factors of internet finance development

Q. Liang et al. (Eds.): AIC 2022, LNEE 871, pp. 229–235, 2023.
https://doi.org/10.1007/978-981-99-1256-8_27

in China, and found that The geographical location of different regions affects the level of inclusive financial development. Liu X, Guo S [6] consider the relationship between inclusive finance, environmental regulation and population health under epidemic conditions, but does not consider the impact of the pandemic and pandemic-loan related online statements. Thus, a quantitative approach is taken to examine the impact of the epidemic on financial inclusion in this research.

Considering the financial inclusion index is annual data and the data volume is small, Random Forest Regression [7] model will be trained for data augmentation to expand the annual data into quarterly data. We will compare the performance of RFR models includes and excludes COVID-19 related variable to confirm whether there is a relationship between the epidemic and the development of inclusive finance in this process. Then, we will transforme the COVID-19 – loan related online statements into quarterly words vectors by Bag of Words Model [8] TF-IDF Algorithm [9]. All these influencing factors will be added to the RFR model to explore the lag period of the impact COVID-19 – loan related online statements on inclusive finance.

2 Methodology and Related Theories

2.1 Bag of Words Model TF-IDF Algorithm

The bag-of-words model (BOW) [8] converts text into word vectors based on the frequency of words in the text, which ignores the information of word order. TF-IDF values [9] which represent the importance of different keywords in the text, can be used to replace the frequency of the word vector in the bag-of-words model to represent the meaning of the document better. The formula for calculating the TF-IDF value is as follows:

$$TF - IDF_{i,j} = TermFrenquency_{i,j} * log_2(\frac{D}{DocumentFrequency_i}) \qquad (1)$$

i and j represent word and document. $TermFrenquency_{i,j}$ represents the frequency of the the word i in the the document j. D represent the number of documents. $DocumentFrequency_i$ represents the number of documents in the document base that contain the word i. This paper we adopt the Bag of Words Model TF-IDF Algorithm in Sklearn module in Python.

2.2 Principles of Random Forest Regression

The Random Forest (RF) algorithm has been widely used for classification and regression. Multiple binary decision trees composed The Random Forest Regression (RFR). The selection of segmentation variables and points of every binary decision tree in this model is exhausted, and weighted impurity of each sub node $G(x_i, v_{ii})$ is generally used to measure the quality of segmentation. This paper we adopt the RandomForestRegressor function in Sklearn module in Python, in which the MSE function ($H(X_m)$) is taken as the impurity function.

$$G(x_i, v_{ij}) = \frac{n_{left}}{N_s}H(X_{left}) + \frac{n_{right}}{N_s}H(X_{right}) \qquad (2)$$

$$H(X_m) = \frac{1}{N_m} \sum_{i \in N_m} (y - \overline{y_m})^2 \tag{3}$$

x_i represents a segmentation variable, v_{ij} represent a segmentation value of the segmentation variable, n_{left}, represents the number of training samples of the left sub node after segmentation, n_{right} represents the number of training samples of the the right sub node, N_s represents the number of all training samples of after segmentation. X_{left} and X_{right} are the training sample set of the left and right sub nodes, $\overline{y_m}$ is the average value of the target variable of the current node sample. The segmentation variable and the segmentation point to minimize G will be find in the training process. The prediction result of RFR is obtained by averaging the prediction results of all internal binary decision trees [7].

2.3 Methodology

In this paper, Random Forest Regression model is trained on the annual influencing factors (X_i) and annual financial inclusion index (y). Quarterly financial inclusion index ($\hat{y}$) is predicted by the model trained on the quarterly influencing factors (X_i) to expanding the size of data. Then, BOW model tf-idf algorithm is used to convert texts into word vectors ($X_{i+1} \sim X_m$). All these influencing factors of different lag periods are passed into the RFR model to compare their performance. The framework of this paper is shown as Fig. 1.

Firstly, we performed data augmentation considering the small amount of data. The RFR model is employed to illustrate the relationship between annual financial Inclusion index and its influencing factors which include year, the code of province, annual gross domestic product (GDP), annual per capita disposable income of urban residents, fluctuation value of annual consumer price index (CPI), and number of new diagnoses. To confirm that there is impact of the epidemic on financial inclusion, the performance of model regressions that included and excluded the epidemic indicator of the number of new confirmed diagnoses are compared. Then, the RFR model is utilized to predict quarterly financial Inclusion index based on the quarterly factors from the first quarter of 2010 to the first quarter of 2022.

Secondly, variables reflecting public opinion are introduced into the RFR model because they may also have an effect on the development of inclusive finance. The BOW model tf-idf algorithm is adopted to transform the text into 15-dimensional word vectors and 15-dimensional quarterly word vectors are obtained based on the quarterly mean of the tf-idf values. Due to the possible time lag effect in public opinion on the development level of inclusive finance, texts of quarter t-1, quarter t-2, quarter t-3 are each added into the RFR model obtained to investigate the relationship between financial inclusion index and all these influencing factors of last quarter.

For all models, Root Mean Squared Error (RMSE) (in the unit of thousands) and Mean Absolute Percentage Error (MAPE) are used to measure their performance. Where m represents the total number of samples, $\hat{y_j}$ represents predicted value of samples and $\overline{y_j}$ represents the mean of true samples [7].

$$RMSE = \sqrt{\frac{1}{m} \sum_{j=1}^{m} (y_j - \hat{y_j})^2} \tag{4}$$

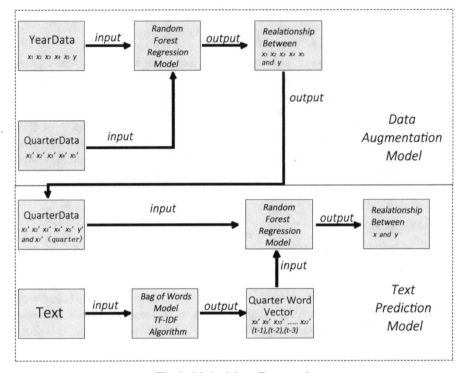

Fig. 1. Methodology Framework

$$MAPE = \frac{1}{m} \sum_{j=1}^{m} \left| \frac{y_j - \hat{y}_j}{y_j} \right| \qquad (5)$$

3   Experiment and Analysis

3.1   Data Preparation

We collect data of quarterly gross domestic product (GDP), quarterly per capita disposable income of urban residents, monthly fluctuation value of annual consumer price index (CPI) of 31 provinces from 2010 to 2022, the number of new confirmed diagnoses, financial inclusion index and 5263 sample data of epidemic-related texts from 1 quarter of 2020 to 1 quarter of 2022 from CEinet Statistics Database, Provincial Health Commission Websites, Weibo Website, and Peking University Digital Inclusive Finance Index Report.

Considering the availability of data, 31 provinces we selected includes Anhui, Beijing, Fujian, Gansu, Guangdong, Guangxi, Guizhou, Hainan, Hebei, Henan, Heilongjiang, Hubei, Hunan, Jilin, Jiangsu, Jiangxi, Liaoning, Inner Mongolia, Ningxia, Qinghai, Shandong, Shanxi, Shanxi, Shanghai, Sichuan, Tianjin, Xizang, Xinjiang, Yunnan, Zhejiang, Chongqing.

3.2 Experiment I – Data Augmentation Model

Feature Selection. On the base of previous studies, we take year (X_1), the code of province (X_2), annual gross domestic product (GDP) (X_3), annual per capita disposable income of urban residents (X_4), and fluctuation value of annual consumer price index (CPI) (X_5) as the financial inclusion index influencing factors. The correlation coefficient between financial inclusion index and the number of new confirmed diagnoses (X_6) is 0.107, with a p-value of 0.060. The performance of model regressions that included and excluded the epidemic indicator of the number of new confirmed diagnoses are compared. The comparison results are shown as the Table 1.

Table 1. Performance of Models

	Include X_6	Exclude X_6
RMSE	10.697	11.106
MAPE	4.715	4.784

Model Construction. The data are randomly divided into training set and testing set at a ratio of 9:1, the training set is used to train the Random Forest Regressor model, and the model trained is used to test the testing set. Results show that the lower RMSE of testing set in model include X_6 is 10.697, and lower MAPE of testing set in model include X_6 is 4.715.

Acquisition of Variables for Prediction. X_1, X_2, the quarterly data of X_3, X_4, X_5, X_6 and X_7 (quarter) from 31 provinces are collected from CEinet Statistics Database. The quarterly data of X_6 equals to the average of ring consumer price index during that quarter.

Prediction Results. The 1519 predicted values of quarterly financial inclusion index of 31 provinces are obtained. Prediction results are shown as Table 2. These data will be applied to the following model.

Table 2. Prediction Results of Quarterly Financial Inclusion Index

Year	Quarter	Province	F I I
2010	1	Anhui	27.444
2010	2	Anhui	27.459
2010	3	Anhui	26.911

(*continued*)

Table 2. (*continued*)

Year	Quarter	Province	F I I
2010	4	Anhui	26.911
2011	1	Anhui	26.911
......	......	......	......
2021	3	Chongqing	286.791
2021	4	Chongqing	285.565
2022	1	Chongqing	285.909

3.3 Experiment II – Text Prediction Model

Feature Selection. The quarterly words vectors on 15 dimensions (X_8 to X_{22}) are introduced into the RFR model, with X_1 to X_7 remain constant. 5263 COVID-19 – loan related online statements from 1 quarter 2020 to 1 quarter 2022 are obtained from the Weibo Website. The quarterly words vector ranging from zero to one equals to the average of all tf – idf values of the 15 most important keywords in texts grouped by quarters during that quarter. The quarterly words vector on 15 dimensions is returned from BOW Model TF-IDF Algorithm.

Model Construction. The first model (Model_t-1) takes X_3–X_6, X_8–X_{22} of the last quarter, X_1, X_2, X_7 of the current quarter as variables. The second model (Model_t-2) takes the X_8 and X_{22} of the quarter before last month, X_3–X_6 of the last quarter, and X_1, X_2, X_7 of the current quarter as variables. The third model (Model_t-3) takes takes the X_8 and X_{22} of the quarter before last 2 quarters, X_3–X_6 of the last quarter, and X_1, X_2, X_7 of the current quarter as variables. The dataset also divided into training and testing set with ratio 9:1, then the RFR model is trained on train set and used to test the test set. Performances of models with different lag periods are shown as Table 3.

Table 3. Performances of models

	Model_t-1	Model_t-2	Model_t-3
RMSE	1.630	0.163	1.454
MAPE	0.754	0.050	0.758

Comparison Results. Model_t-2 with lowest RMSE and MAPE shows the best performance, which means that the lag period of the impact COVID-19 – loan related online statements on financial inclusion index is 2 quarters.

4 Conclusion and Future Works

In this paper, we predict quarterly financial inclusion index based on the quarterly influencing factors by using the Random Forest Regression model. And the model which introduces the number of new confirmed diagnoses works better, which indicate a relationship between the epidemic and financial inclusion.

The RFR prediction model which introduces COVID-19 – loan related online statements with two-quarters lagging works better than the model with one-quarter and three-quarter lagging, as it achieves better performance with lowest RMSE and MAPE.

This paper provides a method to expand the amount of data. The result of models shows that there is impact the epidemic has on the development of financial inclusion, and the lag period of the impact COVID-19 – loan related online statements on financial inclusion index is 2 quarters. But there is still a big margin for improvement in the future work. More quarterly data and more influencing factors will be introduced to improve the performance of the prediction model.

Acknowledgments. This research project is supported by the National Social Science Foundation Project "Multi-dimensional research on the comparative evaluation and enhancement mechanism of financial inclusion under digital empowerment" (Project No: 21BJL087).

References

1. Yang, Z., Li, X.: Evaluation of the development of rural inclusive finance: a case study of Baoding, Hebei Province (2018)
2. Ratnawati, K.: The impact of financial inclusion on economic growth, poverty, income inequality, and financial stability in Asia. J. Asian Financ. Econ. Bus. **7**(10), 73–85 (2020)
3. Arif, M., Lu, Y., Mahmud, A.: Regional development of China's inclusive finance through financial technology. SAGE Open **10**(1), 1–16 (2020)
4. Deng, X.: Analysis on evaluation of inclusive finance development and influencing factors in Hubei Province. In: International Conference on Economics, Management, Law and Education (2018)
5. Huang, J., Zhang, W., Ruan, W.: Spatial spillover and impact factors of the internet finance development in China. Physica A: Stat. Mech. Appl. **527**, 121390 (2019)
6. Liu, X., Guo, S.: Inclusive finance, environmental regulation, and public health in China: lessons for the COVID-19 pandemic. Front. Public Health **9**, 662166 (2021)
7. Du, Y., Wei, K., Wang, Y., Jia, J.: New energy vehicles sales prediction model combining the online reviews sentiment analysis: a case study of Chinese new energy vehicles market. In: Liang, Q., Wang, W., Mu, J., Liu, X., Na, Z. (eds.) Artificial Intelligence in China. LNEE, vol. 854, pp. 424–431. Springer, Singapore (2022). https://doi.org/10.1007/978-981-16-9423-3_53
8. Li, J.R., Mao, Y.F., Yang, K.: Improvement and application of TF * IDF algorithm. Springer Berlin, Heidelberg (2011). https://doi.org/10.1007/978-3-642-25255-6_16

Certifiable Optimal Visual Pose Estimation for Space Applications

Yue Wang[1,2]([✉]) and Xin Zhang[2]

[1] Key Laboratory of Electronics and Information Technology for Complex Space
Systems, National Space Science Center, Chinese Academy of Sciences,
Beijing 100190, China
wangyue@nssc.ac.cn
[2] Tianjin Key Laboratory of Wireless Mobile Communications and Power
Transmission, Tianjin Normal University, Tianjin 300387, China

Abstract. Visual pose estimation is essential for many space applications, such as autonomous robots for planetary exploration, asteroid
exploration, and spacecraft pose estimation. However, the high reliability
requirement of space applications cannot be satisfied by general visual
pose estimation algorithms based on heuristic local optimization. This
paper briefly introduce certifiable optimal algorithms for the visual pose
estimation framework to the aerospace research community, including
both visual relative pose estimation problem and pose-graph optimization problem. The original optimization problem is reformulated as a
quadratically constrained quadratic program (QCQP) problem, whose
Lagrangian dual function is a semidefinite programming (SDP) problem
which is convex. Simulation results demonstrate the feasibility of certifiable optimal visual pose estimation algorithms based on the zero duality
gap for typical application scenarios.

Keywords: 3-D vision · global optimization · Lagrangian duality ·
pose estimation · quadratically constrained quadratic program
(QCQP) · semidefinite programming (SDP)

1 Introduction

Three-dimensional (3-D) vision and visual pose estimation is an enabling technology for autonomous driving [1], augmented reality (AR) devices as well as
future metaverse applications [2]. Besides that, visual pose estimation is also
essential for many space applications, such as autonomous robots for planetary
exploration [3], asteroid exploration [4], and uncooperative spacecraft pose estimation [5–7].

This work was supported by the Pre-Research Project of Space Science (No.
XDA15014700), the National Natural Science Foundation of China (No. 61601328),
and the Scientific Research Plan Project of the Committee of Education in Tianjin
(No. JW1708).

Space applications commonly require very high reliability, but visual pose estimation is normally formulated as a high-dimensional, non-convex, non-linear optimization problem and heuristic local iterative searching algorithms are usually adopted to tackle the optimization problem [8,9], which cannot provide guarantee on obtaining a global optimal solution [10]. Thus, it is urgent to propose an algorithm to get a certifiable global optimal solution.

In 2006, Banderia proposed the concept about probably certifiably correct (PCC) algorithms and probabilistic certifier [11]. Motivated by the paper, several visual pose estimation algorithms with probabilistic certifiers has been proposed [12,13]. The de-facto visual pose estimation framework is depicted in Fig. 1, where relative pose (RP) is estimated between two images containing overlapped scene, then a pose graph is constructed using these relative pose results and a pose-graph optimization (PGO) is conducted to estimate all poses [14–17].

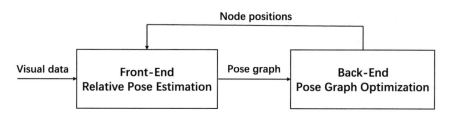

Fig. 1. The de-facto visual pose estimation framework.

In this paper, certifiable optimal visual pose estimation algorithms are introduced, containing both certifiable optimal relative pose estimation algorithms (in Sect. 2) and certifiable optimal PGO algorithms (in Sect. 3).

2 Certifiable Optimal Relative Pose Estimation

There are generally two paradigms about relative pose estimation methods, one is called the direct method based on photometric residual of pixel intensities, the other is called keypoints-based method based on salient and repeatable points (called keypoints), which can be obtained by hand-engineered methods or deep-learned methods [18]. We discuss the keypoints-based method in this paper, since it usually achieve higher accuracy in visual-feature-rich environments.

As depicted in Fig. 2, relative pose estimation problem is that given a set of N pair-wise correspondences $(\boldsymbol{f}_i, \boldsymbol{f}_i')$ between two central and calibrated cameras, we aim to estimate the relative rotation $\boldsymbol{R}$ and translation $\boldsymbol{t}$ between these two camera frames.

There are roughly two categories to estimate the relative pose, one is called the minimal solvers (e.g., the 5 point algorithm [19]), the other is called the non-minimal solvers based on direct linear transformation (DLT) algorithm [20] or eigenvalue-based algorithm [21]. Since DLT-based algorithm can leverage all

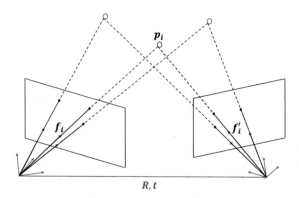

Fig. 2. Relative pose estimation problem.

the keypoints, it usually has high accuracy. To achieve higher accuracy, the local optimization algorithm on essential matrix manifold is proposed using DLT-based solution as a initial guess [22,23], where the original optimization problem is formulated as

$$f_{\text{RP}}^\star = \min_{E \in \mathcal{M}_E} = \text{vec}\,(E)^T\, C\text{vec}\,(E),\qquad\text{(RP-Original)}$$

where E is the essential matrix and defined as

$$E = \begin{bmatrix} e_1^T \\ e_2^T \\ e_3^T \end{bmatrix} = \begin{bmatrix} e_{11} & e_{12} & e_{13} \\ e_{21} & e_{22} & e_{23} \\ e_{31} & e_{32} & e_{33} \end{bmatrix},$$

$\mathcal{M}_E \triangleq \{E|E = [t]_\times R, \exists R \in \text{SO}(3), b \in \mathcal{S}^2\}$ is an essential matrix set, the special orthogonal group is defined as $\text{SO}(3) \triangleq \{R \in \mathbb{R}^{3\times3}|R^TR = I, \det(R) = 1\}$, the 2-sphere set is defined as $\mathcal{S}^2 \triangleq \{t \in \mathbb{R}^3|t^Tt = 1\}$, $\text{vec}\,(E) \triangleq [e_1; e_2; e_3]$, $C = \sum_{i=1}^N (f_i \otimes f_i')(f_i \otimes f_i')^T$ is a Gram matrix, which is positive semidefinite and symmetric and $\otimes$ means Kronecker product.

By reformulating the Equation (RP-Original) into a quadratically constrained quadratic program (QCQP) form [24,25], we have

$$f_{\text{RP}}^\star = \min_{x \in \mathbb{R}^{12}} x^T Q x,\qquad\text{(RP-Primal)}$$

$$\text{s.t.}\quad x^T A_1 x = 1,$$

$$x^T A_i x = 0,\quad i = 2, 3, ..., 6,$$

where $x \triangleq [\text{vec}\,(E)\,; t]$, $\{A_i\}_{i=1}^6$ are the 12×12 symmetric corresponding matrix forms of the quadratic constraints, and

$$Q = \begin{bmatrix} C & 0_{9\times3} \\ 0_{3\times9} & 0_{3\times3} \end{bmatrix} \in \mathbb{S}_+^{12}.$$

By using the Lagrangian duality theory, we get the dual problem of the Equation (RP-Primal) as [24,25]

$$d_{\mathrm{RP}}^{\star} = \max_{\boldsymbol{\lambda}} \lambda_1, \qquad \text{(RP-Dual)}$$

$$\text{s.t.} \quad \boldsymbol{M}(\boldsymbol{\lambda}) \succcurlyeq 0,$$

where $\boldsymbol{M}(\boldsymbol{\lambda}) \triangleq \boldsymbol{Q} - \sum_{i=1}^{6} \lambda_i \boldsymbol{A}_i$ is the so-called Hessian of the Lagrangian and $\boldsymbol{\lambda} = \{\lambda_i\}_{i=1}^{6}$ are the Lagrange multipliers.

Since the Lagrangian dual problem in Equation (RP-Dual) is a semidefinite programming (SDP) problem, which is convex and its global optimal solution can be obtained. According to the Lagrangian duality theory, if the duality gap $\Delta_{\mathrm{RP}} = f_{\mathrm{RP}}^{\star} - d_{\mathrm{RP}}^{\star}$ is numerically equal to zero, then the candidate solution can be certified as the global optimal solution.

To demonstrate the certifiable optimal relative pose estimation method, we conduct a series simulation experiments. Random scenes are generated by setting $\boldsymbol{p}_1 = \boldsymbol{0}$, $\boldsymbol{R}_1 = \boldsymbol{I}_{3 \times 3}$, random translation is set to less than 2, random Euler angles is set to less than 0.5 radians, random depth is set between 4 and 8, focal length is set to 800 pixels, number of correspondences is set to 10, and field of view (FOV) is set to 100° [33]. The initial guess is obtained using OpenGV library [26], the manifold optimization is obtained using Manopt toolbox [27], and the SDP problem is solved using CVX package [28].

Some simulation results are summarized in Table 1, where σ is the standard deviation of Gaussian noise, f_{DLT} is the object function value using DLT-based initial guess, $\varepsilon_{\mathrm{rot}}$ and $\varepsilon_{\mathrm{tran}}$ are the rotation error and the translation direction error, which are defined as [25]

$$\varepsilon_{\mathrm{rot}} = \arccos\left(\frac{\mathrm{trace}(\boldsymbol{R}_{\mathrm{true}}^T \boldsymbol{R}^{\star}) - 1}{2}\right) \cdot \frac{180}{\pi},$$

$$\varepsilon_{\mathrm{tran}} = \arccos\left(\frac{\boldsymbol{t}_{\mathrm{true}}^T \boldsymbol{t}^{\star}}{\|\boldsymbol{t}_{\mathrm{true}}\| \cdot \|\boldsymbol{t}^{\star}\|}\right) \cdot \frac{180}{\pi},$$

t_{SDP} is the runtime of the SDP solver using MATLAB 2021a installed on a laptop computer with Intel i5-12500 H processor and 16 GB RAM.

Table 1. Simulation results for relative pose estimation problem.

σ[pixel]	f_{DLT}	$f_{\mathrm{RP}}^{\star}$	$d_{\mathrm{RP}}^{\star}$	$\varepsilon_{\mathrm{rot}}$ [°]	$\varepsilon_{\mathrm{tran}}$ [°]	t_{SDP}[s]
0.5	4.4339×10^{-6}	5.2777×10^{-8}	3.9860×10^{-8}	0.0755	0.0701	0.6832
1	1.7964×10^{-5}	2.5494×10^{-7}	2.2338×10^{-7}	0.1619	0.2480	0.6845
2	6.8592×10^{-6}	2.6474×10^{-6}	1.3980×10^{-6}	0.2894	0.8684	0.6751
10	1.2947×10^{-4}	1.6305×10^{-5}	5.8175×10^{-7}	1.3185	1.8726	0.6445

From Table 1, we can observe that (i) the increasing standard deviation of Gaussian noise leads to larger object function value as well as larger rotation and translation error; (ii) optimization on essential matrix manifold leads to higher accuracy compared with the DLT-based algorithm, since $f_{\mathrm{DLT}} > f_{\mathrm{RP}}^{\star}$; (iii) for small noise scenarios (i.e., $\sigma = 0.5$ or 1 pixel), which are also the typical practical application scenarios, the duality gap $\Delta_{\mathrm{RP}} = f_{\mathrm{RP}}^{\star} - d_{\mathrm{RP}}^{\star}$ is numerically equal to zero, thus the candidate solution can be certified as the global optimal solution, and this certifiable optimal method cannot be used for large noise cases; (iv) since the time for estimating the RP is almost determined by the runtime of the SDP solver, it is possible to conduct realtime RP estimation since $t_{\mathrm{SDP}} < 0.7$ s.

3   Certifiable Optimal Pose-Graph Optimization

After obtaining the relative pose estimation, a directed graph $\mathcal{G}(\mathcal{V}, \mathcal{E})$ called the pose graph can be constructed, where each vertex represents a pose $\boldsymbol{x}_i$, $i \in \mathcal{V} = \{1, ..., n\}$ and each edge represents a relative pose estimation $\bar{\boldsymbol{x}}_{ij}$ between vertex i and j, $(i, j) \in \mathcal{E}$.

The measurement model is given as follows:

$$\bar{\boldsymbol{t}}_{ij} = \boldsymbol{R}_i^T (\boldsymbol{t}_j - \boldsymbol{t}_i) + \boldsymbol{t}_\epsilon, \quad \boldsymbol{t}_\epsilon \sim \text{Gaussian}\left(\boldsymbol{0}_3, \omega_t^2 \boldsymbol{I}_3\right),$$

$$\bar{\boldsymbol{R}}_{ij} = \boldsymbol{R}_i^T \boldsymbol{R}_j \boldsymbol{R}_\epsilon, \quad \boldsymbol{R}_\epsilon \sim \text{vonMises}\left(\boldsymbol{I}_3, \omega_R^2\right),$$

where vonMises$(\boldsymbol{S}, \kappa)$ denotes the isotropic von Mises-Fisher (or Langevin) distribution with mean $\boldsymbol{S} \in \mathrm{SO}(3)$ and concentration parameter κ [29].

Based on the assumed noise distribution, the original pose graph optimization problem can be formulated as

$$f_{\mathrm{PGO}}^{\star} = \min_{\boldsymbol{t}_i \in \mathbb{R}^3, \boldsymbol{R}_i \in \mathrm{SO}(3)} \sum_{(i,j) \in \mathcal{E}} \omega_t^2 \parallel \boldsymbol{t}_j - \boldsymbol{t}_i - \boldsymbol{R}_i \bar{\boldsymbol{t}}_{ij} \parallel^2 + \frac{\omega_R^2}{2} \parallel \boldsymbol{R}_j - \boldsymbol{R}_i \bar{\boldsymbol{R}}_{ij} \parallel_{\mathrm{F}}^2,$$

(PGO-Original)

where $\parallel \cdot \parallel_{\mathrm{F}}^2$ is the Frobenius matrix norm and $\parallel \boldsymbol{S} - \boldsymbol{R} \parallel_{\mathrm{F}}^2$ is usually referred to as the chordal distance between two rotations $\boldsymbol{S}$ and $\boldsymbol{R}$ [30].

By reformulating the Equation (PGO-Original) into a QCQP form [10], we have

$$f_{\mathrm{PGO}}^{\star} = \min_{\boldsymbol{x}, y} \parallel \boldsymbol{A}\boldsymbol{x} - \boldsymbol{b}y \parallel^2, \qquad \text{(PGO-Primal)}$$

$$\text{s.t.} \quad \boldsymbol{x}^T \boldsymbol{E}_{iuv} \boldsymbol{x} = 1, \quad u = v, \quad u, v = 1, 2, 3$$

$$\boldsymbol{x}^T \boldsymbol{E}_{iuv} \boldsymbol{x} = 0, \quad u \neq v, \quad i = 1, ..., n-1$$

$$y^2 = 1,$$

where $\boldsymbol{E}_{iuv}$ is the stacked coefficient matrices with suitable zero blocks for padding without rows and columns corresponding to the first pose.

By using the Lagrangian duality theory, we get the dual problem of the Equation (PGO-Primal) as [10]

$$d_{\mathrm{PGO}}^{\star} = \max_{\boldsymbol{\lambda}} \sum_{i=1,...,n, u=1,2,3} \lambda_{iuu} + \lambda_y, \qquad \text{(PGO-Dual)}$$

$$\text{s.t.} \quad \boldsymbol{M}(\boldsymbol{\lambda}) \succcurlyeq 0,$$

where $\boldsymbol{\lambda}$ is the vector of Lagrange multipliers, λ_{iuu} and λ_y are the associated Lagrange multipliers for the orthonormality and homogeneity constraints in Equation (PGO-Primal), respectively.

Since the Lagrangian dual problem in Equation (PGO-Dual) is a SDP problem, which is convex and its global optimal solution can be obtained. According to the Lagrangian duality theory, if the duality gap $\Delta_{\mathrm{PGO}} = f^\star_{\mathrm{PGO}} - d^\star_{\mathrm{PGO}}$ is numerically equal to zero, then the candidate solution can be certified as the global optimal solution.

To demonstrate the certifiable optimal PGO method, we conduct a series simulation experiments. A random 3D grid graph with $3^3 = 27$ vertices are generated and random loop closures with probability 0.1 are added between nearly nodes [10], as depicted in Fig. 3. We varying the noise levels of translation σ_T in meter and rotation σ_R in radian to check feasibility of the certifiable optimal PGO algorithm. The initial guess is obtained using chordal initialization [31], the PGO result is obtained via the Gauss-Newton (GN) algorithm [32], and the SDP problem is solved using CVX package [28].

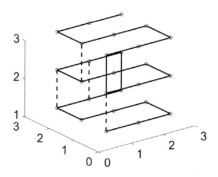

Fig. 3. A random generated 3D grid graph.

Some simulation results are depicted in Fig. 4 and summarized in Table 2, where σ represents the same noise level of both translation and rotation for simplicity, f_{chordal} is the object function value using chordal initial guess.

Table 2. Simulation results for PGO problem.

σ	f_{chordal}	$f^\star_{\mathrm{PGO}}$	$d^\star_{\mathrm{PGO}}$	$t_{\mathrm{SDP}}[\mathrm{s}]$
0.01	8.9049	6.8786	6.8786	2.4849
0.05	1.6688×10^1	1.1126×10^1	1.1126×10^1	2.4895
0.1	2.7159×10^1	2.1879×10^1	2.1870×10^1	2.5293
0.5	4.7391×10^1	2.8264×10^1	1.3927×10^1	2.5245

From Table 2, we can observe that (i) the increasing noise level leads to larger object function value; (ii) iterative GN-based local optimization leads to higher accuracy compared with the chordal initialization, since $f_{\text{chordal}} > f^{\star}_{\text{PGO}}$; (iii) for small noise scenarios (i.e., $\sigma = 0.01$ or 0.05), which are also the typical practical application scenarios, the duality gap $\Delta_{\text{PGO}} = f^{\star}_{\text{PGO}} - d^{\star}_{\text{PGO}}$ is numerically equal to zero, thus the candidate solution can be certified as the global optimal solution, and this certifiable optimal method cannot be used for large noise cases; (iv) the time consumed by the SDP solver is about 2.5 s, which may be sufficient to small-scale applications which do not require realtime PGO calculation, but the calculation time of general-purpose SDP solver exponentially increase as the number of nodes in each axis of the random 3D grid graph.

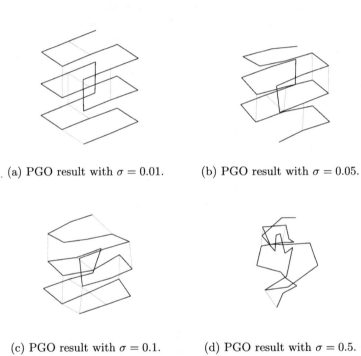

(a) PGO result with $\sigma = 0.01$.   (b) PGO result with $\sigma = 0.05$.

(c) PGO result with $\sigma = 0.1$.   (d) PGO result with $\sigma = 0.5$.

Fig. 4. PGO results with different noise level.

4   Conclusions

This paper introduces the certifiable optimal visual pose estimation algorithms for both relative pose estimation problem and pose-graph optimization problem. The core idea behind the algorithms is that the original optimization problem can be reformulated as a QCQP problem and we can derive its Lagrangian dual problem, the duality gap can be leveraged to certify the optimality of the candidate solution. Simulation results demonstrate the feasibility of the certifiable optimal visual pose estimation algorithms for typical application scenarios. Therefore, it is possible to design and implement a certifiable optimal visual pose estimation system for future space applications to enhance the reliability.

References

1. Bresson, G., Alsayed, Z., Yu, L., Glaser, S.: Simultaneous localization and mapping: a survey of current trends in autonomous driving. IEEE Trans. Intell. Veh. **2**(3), 194–220 (2017)
2. Microsoft HoloLens 2. http://www.microsoft.com/en-us/hololens
3. Cheng, Y., Maimone, M.W., Matthines, L.: Visual odometry on the Mars exploration rovers - a tool to ensure accurate driving and science imaging. IEEE Robot. Autom. Mag. **13**(2), 54–62 (2006)
4. Dor, M., Skinner, K.A., Driver, T., Tsiotras, P.: Visual SLAM for asteroid relative navigation. In: 2021 IEEE/CVF Conference on Computer Vision and Pattern Recognition Workshops (CVPRW), pp. 1–10, 19–25 June, Nashville, TN, USA (2021)
5. Kisantal, M., Sharma, S., Park, T.H., Izzo, D., Märtens, M., D'Amico, S.: Satellite pose estimation challenge: dataset, competition design, and results. IEEE Trans. Aerosp. Electron. Syst. **56**(5), 4083–4098 (2020)
6. Park, T. H., Märtens, M., Lecuyer, G., Izzo, D., D'Amico, S.: SPEED+: next-generation dataset for spacecraft pose estimation across domain Gap. In: 2022 IEEE Aerospace Conference (AERO), pp. 1–15, 5–12 March, Big Sky, MT, USA (2022)
7. Cassinis, L.P., Fonod, R., Gill, E.: Review of the robustness and applicability of monocular pose estimation systems for relative navigation with an uncooperative spacecraft. Prog. Aerosp. Sci. **110**, 1–14 (2019)
8. Grisetti, G., Kümmerle, R., Stachniss, C., Burgard, W.: A tutorial on graph-based SLAM. IEEE Intell. Transp. Syst. Mag. **2**(4), 31–43 (2010)
9. GTSAM (Georgia Tech Smoothing and Mapping). https://gtsam.org
10. Carlone L., Rosen D.M., Calafiore G., Leonard J.J., Dellaert F.: Lagrangian duality in 3D SLAM: verification techniques and optimal solutions. In: IEEE/RSJ International Conference on Intelligent Robots and Systems (IROS), pp. 1–8, 28 September, 2 October, Hamburg, Germany (2015)
11. Banderia, A.S.: A note on probably certifiably correct algorithms. C. R. Acad. Sci. Paris, Ser. I **354**, 329–333 (2016)
12. Wang, Y., Peng, X.: New trend in back-end techniques of visual SLAM: from local iterative solvers to robust global optimization. In: Liang, Q., Wang, W., Mu, J., Liu, X., Na, Z. (eds.) Artificial Intelligence in China. LNEE, vol. 854, pp. 308–317. Springer, Singapore (2022). https://doi.org/10.1007/978-981-16-9423-3_39
13. Rosen, D.M., Doherty, K.J., Espinoza, A.T., Leonard, J.J.: Advances in inference and representation for simultaneous localization and mapping. Annu. Rev. Control Robot. Auton. Syst. **4**, 215–242 (2021)
14. Ma, Y., Soatto, S., Kosěcká, J., Sastry, S.S.: An Invitation to 3-D Vision: From Images to Geometric Models. Springer Science+Business Media, NewYork (2004)
15. Hartley, R., Zisserman, A.: Multiple View Geometry in Computer Vision, 2nd edn. Cambridge University Press, Cambridge (2003)

16. Schönberger, J.L., Frahm, J.-M.: Structure-from-Motion Revisited. In: IEEE/CVF Conference on Computer Vision and Pattern Recognition (CVPR), pp. 1–10, 26 June, 1 July, Las Vegas, Nevada, USA (2016)
17. Cadena, C., et al.: Past, present, and future of simultaneous localization and mapping: toward the robust-perception age. IEEE Trans. Robotics **32**(6), 1309–1332 (2016)
18. Wang, Y., Fu, Yu., Zheng, R., Wang, L., Qi, J.: New trend in front-end techniques of visual SLAM: from hand-engineered features to deep-learned features. In: Liang, Q., Wang, W., Mu, J., Liu, X., Na, Z. (eds.) Artificial Intelligence in China. LNEE, vol. 854, pp. 298–307. Springer, Singapore (2022). https://doi.org/10.1007/978-981-16-9423-3_38
19. Nistér, D.: An efficient solution to the five-point relative pose problem. IEEE Trans. Pattern Anal. Mach. Intell. **26**(6), 756–770 (2004)
20. Hartley, R.I.: In defense of the eight-point algorithm. IEEE Trans. Pattern Anal. Mach. Intell. **19**(6), 580–593 (1997)
21. Kneip, L., Lynen, S.: Direct optimization of frame-to-frame rotation. In: IEEE International Conference on Computer Vision (ICCV), pp. 1–8, 1–8 December, Sydney, Australia (2013)
22. Tron, R., Daniilidis, K.: On the quotient representation for the essential manifold. In: IEEE/CVF Conference on Computer Vision and Pattern Recognition (CVPR), pp. 1574–1581 , 24–27 June, Columbus, Ohio, USA (2014)
23. Tron, R., Daniilidis, K.: The space of essential matrices as a Riemannian quotient manifold. SIAM J. Imaging Sci. **10**(3), 1416–1445 (2017)
24. Garcia-Salguero, M., Briales, J., Gonzalez-Jimenez, J.: Certifiable relative pose estimation. Image Vis. Comput. **109**, 1–13 (2021)
25. Zhao, J.: An efficient solution to non-minimal case essential matrix estimation. IEEE Trans. Pattern Anal. Mach. Intell. **44**(4), 1777–1792 (2022)
26. OpenGV: a library for solving calibrated central and non-central geometric vision problems. https://laurentkneip.github.io/opengv/
27. Manopt: Tool-boxes for optimization on manifolds and matrices. https://www.manopt.org/
28. CVX: Matlab Software for Disciplined Convex Programming. http://cvxr.com/cvx/
29. Boumal, N., Singer, A., Absil, P.: Cramer-Rao bounds for synchronization of rotations. Inf. Infer. **3**(1), 1–39 (2014)
30. Hartley, R., Trumph, J., Dai, Y., Li, H.: Rotation averaging. Int. J. Comput. Vision **103**(3), 267–305 (2013)
31. Carlone L., Tron R., Daniilidis K., Dellaert F.: Initialization techniques for 3D SLAM: a survey on rotation estimation and its use in pose graph optimization. In: IEEE International Conference on Robotics and Automation (ICRA), pp. 1–8, 26–30 May, Seattle, USA (2015)
32. Nocedal, J., Wright, S.J.: Numerical Optimization, 2nd edn. Springer Science+Business Media, NewYork (2006)
33. Garcia-Salguero, M., Gonzalez-Jimenez, J.: A sufficient condition of optimality for the relative pose problem between cameras. SIAM J. Imaging Sci. **14**(4), 1617–1634 (2021)

Automotive Manufacturing Revenue Prediction Using Financial and Comment Sentiment Data Based on CNN Model

Yu Du, Kaiyue Wei[✉], Bing Wang, and Meijie Du

Business School, Beijing Language and Culture University, No.15 Xueyuan Road, Haidian Distict, Beijing 100083, China
karywwy@126.com

Abstract. This research proposes a CNN-based model for automotive manufacturing companies to predict quarterly revenue that not only combines the sentiment variables of online reviews with financial indicators, but also covers four-quarter indicators of the firm itself, the industry average, and the average of the companies before and after their ranking. According to the experimental findings, our suggested model outperforms the conventional machine learning model Random Forest in terms of prediction accuracy by 75.2%. Business revenue predictions that use a sample of data from three dimensions, including the company and the industry average, perform 48.7% better than those that use only the company's own data. The model with comment variables outperforms the model without comment sentiment variables in terms of prediction performance.

Keyword: Revenue prediction · CNN · Automotive manufacturing companies · Sentiment variables

1 Introduction

The automobile industry is one of the nation's pillar industries. The revenue prediction serves as a critical foundation for businesses to understand the future market development situation and formulate business plans.

There are numerous well-known models in the area of forecasting time series data, such as SARIMA [2], VAR [1], etc. However, these models might not have the same prediction performance as some models that use multivariate forecasting. Machine learning models, including Random Forests [3, 5], SVM, etc., can implement prediction with multiple variables, but cannot use multidimensional variables. Therefore, some researchers adopt deep learning models such as LSTM [6] or use CNN for feature extraction and LSTM for time series prediction [7], but they fail to take advantage of CNN's ability to handle multi-channel data. Some researchers proposed 3D convolutional models using information in various stock markets [4], but only commercial data is included. Some combined textual data with financial information [8], but the texts are just from financial report.

© The Author(s), under exclusive license to Springer Nature Singapore Pte Ltd. 2023
Q. Liang et al. (Eds.): AIC 2022, LNEE 871, pp. 245–251, 2023.
https://doi.org/10.1007/978-981-99-1256-8_29

We therefore propose a CNN-based model that combines the sentiment variables of online comments with financial indicators. And this model covers not only the indicators that combine the three dimensions of the firm itself, the industry average, and the average data of the companies before and after their ranking, but also all indicators for the four quarters prior to the current quarter.

2 Methodology

In this research, we apply the CNN model to the task of predicting business revenue for automotive manufacturing companies.

The experiment is broken down into four steps. The first step is to convert the collected comment text into sentiment variables and calculate the corresponding industry average data as well as the corresponding average data of the companies before and after their ranking based on their own financial data and the comment sentiment variable data. The second step is to prepare samples as required by the CNN model's input layer, with all of the feature variables from the four quarters preceding each quarter and the current quarter's business revenue as a set of samples. The third step is to construct the CNN model and feed it the prepared sample training set for training and tuning. The complicated network topology may cause issues like overfitting due to the small sample size, making the model's prediction less accurate. We finally suggest a model structure with a convolutional layer and two full-connection layers after multiple multiple attempts. The framework of the proposed model is outlined in Fig. 1.

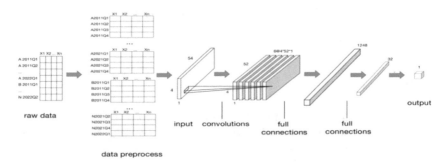

Fig. 1. Methodology Framework

The output matrix of the convolution layer is $z_{u,v}^{(l-1)}$. $x_{i+u,j+v}^{(l-1)}$ is the input feature sample, k_{rot} is the result of 180 degrees of rotation of the convolution kernel k, and b is the paranoid unit.

$$z_{u,v}^{(l-1)} = \sum_{-\infty}^{\infty} \sum_{-\infty}^{\infty} x_{i+u,j+v}^{(l-1)} \cdot k_{rot_{i,j}}^{(l)} \cdot \chi(i,j) + b^{(l)} \tag{1}$$

$$\chi(i,j) = \begin{cases} 1, 0 \le i,j \le n \\ 0, others \end{cases} \tag{2}$$

The activation function of the convolutional and fully connected layers are Relu functions (a(x)).

$$a(x) = max(0, x) \tag{3}$$

The fourth step is to predict the test set using the obtained model and to evaluate the model prediction effect. The Root Mean Squared Error (RMSE)is adopt to measure the performance. Where m is the total number of samples, $\widehat{y_j}$ represents predicted value of samples and $\overline{y_j}$ represents the mean of true samples.

$$RMSE = \sqrt{\frac{1}{m}\sum_{j=1}^{m}(y_j - \widehat{y_j})^2} \tag{4}$$

3 Experiment and Analysis

3.1 Data Preparation

The raw data which collected from CSMAR Database and Guba Website contains a total of 11,342 samples of revenue and other 16 financial indicators from 2011 to 2022 for 175 automotive manufacturing companies and 180,191 comments on these companies from July 2011 to July 2022.

The text of each comment is transformed into two variables. One is the probability that the emotion expressed by the comment is positive. The other is the emotion tag formed after the judgment of the comment, with a value of 1 or 0. After that, take the mean value of all comment emotion variables of a company in a quarter, so that the sentiment variables of the current quarter can be obtained.

All financial indicators are in quarterly units and the unit of revenue is changed to be in one hundred million. After matching each company's quarterly financial indicators to sentiment variables, 2592 usable samples remained.

For each of the 18 variables including financial indicators and sentiment variables, for each quarter the average of that variable for all companies in that quarter is taken as a measure of the industry condition of that quarter. After ranking all companies by revenue size in each quarter, for each indicator of each company in a quarter the average of the indicators of the five companies before and five companies after the company's ranking is taken as a variable to measure the business status and public opinion of the company's industry position.

After the above steps, a total of 54 features containing three dimensions of the company's own situation, industry situation, and the situation of the industry in which the company is located are available.

3.2 Experiment I - Based on Random Forest

To compare the predictive effect of our proposed model with traditional machine learning models, we build a random forest regression model.

Feature Selection. Since the input values of this model can only accept one-dimensional data, we use 54 indicators of the current quarter as the feature variables to predict the revenue of the current quarter.

Model Construction. The training set is fed into the Random Forest Regression model for training after the processed samples are randomly separated into training and test sets in a ratio of 9:1.

Prediction Results. The test set's RMSE is 172.25. In the following we will explore the CNN-based model.

3.3 Experiment II - with Enterprise's Own Indicators Only

In order to explore whether our forecast using indicators from three dimensions outperform forecast using only the company's own indicators, we first enter only the company's own financial and comment sentiment indicators into the model and forecast operating revenues below.

Feature Selection. Since only four quarters of the company's financial indicators and comment sentiment variables are passed into the model, the shape of each sample becomes (4×18). To make it easier for the samples to enter the CNN model, their shape is altered to $(4 \times 18 \times 1)$.

Model Construction. We suggest a model structure with a convolutional layer and two dense layers. Samples are randomly separated into training and test sets in a ratio of 9:1 and the training set is passed into the model.

Prediction Results. The test set's RMSE is 83.40, which is much smaller than the Random Forest model. The following experiment will contain information about the industry average and businesses close to the company's rating.

3.4 Experiment III - Without Comment Variables

In order to explore whether adding comment sentiment variables aids the model in improving the prediction performance, we will first put samples without comment variables into the proposed CNN model.

Feature Selection. We only consider financial indicators, so the shape of each sample submitted for this experiment is $(4 \times 48 \times 1)$.

Model Construction. The model structure maintains the previous convolutional layers and the dense part.

Prediction Results. The test set's RMSE is 47.90. The model with three-dimension data prediction performance is significantly improved over the results of the random forest model and the model with enterprise's own indicators. Next let's see if the inclusion of sentiment variables leads to further improvement of the model.

3.5 Experiment IV – Proposed Model

Feature Selection. We expect to forecast the company's revenue for the current quarter considering all variables from the past four quarters, so now each sample will take the shape of $(4 \times 54 \times 1)$.

Model Construction. The CNN model structure still maintains the previous convolutional layer and the dense part. The processed samples are randomly separated into training and test sets in a ratio of 9:1.

Prediction Results. The test set's RMSE is 42.79. The RMSE is much smaller than each RMSE of the previous models. The model's prediction results are displayed in the Fig. 2, where the red line corresponds to predicted data and the blue to actual data. And the losses for three CNN experiments is shown in Fig. 3. Our proposed model has the fastest loss reduction.

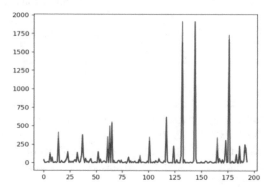

Fig. 2. Prediction results of the proposed model

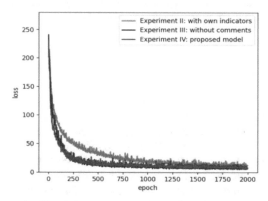

Fig. 3. Loss results of three experiments

4 Conclusion and Future Works

In this research, we not only suggest using the CNN model to predict financial data, but also expand the feature variables from the firm's own indicators to include the firm's own indicators, the industry average indicators, and the indicators of several firms before and after the firm's ranking, thus covering the status of the firm's own operation, the operation of the industry, and the status of the firm's position in the industry.

Furthermore, we expand the model input sample to a two-dimensional panel data set that includes all indicators for four quarters. We introduce comment data into the model to improve model performance even further, and we eventually obtain good business revenue prediction results. We also conduct three sets of experiments to ensure that all of our proposed innovations improve the model's prediction performance. The outcomes of all experiments are displayed below (Table 1).

Table 1. Performance comparison of different prediction models

	Proposed model	Random forest	Without comments	Own indicators
RMSE	42.79	172.25	47.90	83.40

Overall, this research offers a more accurate model for predicting quarterly revenue for automotive manufacturing companies, but there is still much room for development in the future. Comment texts will be obtained from additional platforms in the future. The method of sentiment conversion of review texts will also be further explored to obtain more accurate sentiment scores. Future developments will also see a wider range of financial metrics available.

Acknowledgments. This research project is supported by Science Foundation of Beijing Language and Culture University (supported by "the Fundamental Research Funds for the Central Universities") (Approval number: 22YJ090006 and 18PT02).

References

1. Ampountolas, A.: Forecasting hotel demand uncertainty using time series Bayesian VAR models. Tour. Econ. **25**(5), 734–756 (2019)
2. Winata, A., Kumara, S., Suhartono, D.: Predicting stock market prices using time series SARIMA. In: 2021 1st International Conference on Computer Science and Artificial Intelligence (ICCSAI), pp. 92–99 (2021). https://doi.org/10.1109/ICCSAI53272.2021.9609720
3. Yu, Du., Wei, K., Wang, Y., Jia, J.: New energy vehicles sales prediction model combining the online reviews sentiment analysis: a case study of Chinese new energy vehicles market. In: Liang, Q., Wang, W., Jiasong, Mu., Liu, X., Na, Z. (eds.) Artificial Intelligence in China: Proceedings of the 3rd International Conference on Artificial Intelligence in China, pp. 424–431. Springer Singapore, Singapore (2022). https://doi.org/10.1007/978-981-16-9423-3_53
4. Hoseinzade, E., Haratizadeh, S.: CNNpred: CNN-based stock market prediction using a diverse set of variables. Expert Syst. Appl. **129**, 273–285 (2019)

5. Qiu, X., Zhang, L., Suganthan, P.N., Amaratunga, G.A.: Oblique random forest ensemble via least square estimation for time series forecasting. Inf. Sci. **420**, 249–262 (2017)
6. Siami-Namini, S., Tavakoli, N., Namin, A.S.: A comparison of ARIMA and LSTM in forecasting time series. In: 2018 17th IEEE International Conference on Machine Learning and Applications (ICMLA), pp. 1394–1401. IEEE (2018)
7. Xue, N., Triguero, I., Figueredo, G.P., Landa-Silva, D.: Evolving deep CNN-LSTMs for inventory time series prediction. In: 2019 IEEE Congress on Evolutionary Computation (CEC), pp. 1517–1524. IEEE (2019)
8. Liu, Z.: Financial risk prediction of listed companies by combining text and financial data (Master's thesis, Jiangxi University of Finance and Economics) (2020)

Deep Learning-Enhanced ICA Algorithm for Sub-Gaussian Blind Source Separation

Zhijin Xie, Zhuo Sun$^{(\boxtimes)}$, Gang Yue, and Jinpo Fan

Beijing University of Posts and Telecommunications, Beijing 100876, China
{zhijinxie,zhuosun,yuegang,fjp}@bupt.edu.cn

Abstract. In the conventional iterative blind source separation procedure, the nonlinear function is selected as a fixed mathematical expression that cannot change with different signals. In this paper, we propose an enhanced independent component analysis (ICA) network for sub-Gaussian sources. The basic principle of the network is to replace the nonlinear function with an explainable neural network, and then to incorporate deep learning into the iterative computation for optimization. To give the explainable neural network a certain function expression ability, we design its structure and pre-train it. The gradient descent approach then be used to update the weights of the enhanced ICA network. To illustrate the effectiveness of the algorithm, we merge four different signals and perform separation experiments. Experimental results indicate that the enhanced ICA network can be consistent with the performance of the iterative algorithm, and in such conditions it can obtain even more precise waveforms.

Keywords: Independent Component Analysis · Blind Source Separation · Deep Learning

1 Introduction

Blind source separation works on the statistical independence nature of the sources by either minimizing the mutual information or maximizing the information [1]. The ICA algorithm constrained by the minimum bit error rate criterion can better achieve the separation of the mixed signal when the mixed source signal is composed of sub-Gaussian signals. Nevertheless, the iterative processes of the ICA method rely on mathematical formulas.

Theoretical analysis shows that the optimal the nonlinear excitation function of the separation matrix is related to the probability density function of the source signal [2]. If we can adjust the expression of the nonlinear function automatically according to the probability density function of different signals, we will improve the separation accuracy. In order to apply an adaptive nonlinear function to the gaussian signal problem in the ICA algorithm, Bell and

---

Z. Sun—This paper is supported by National Natural Science Foundation of China with Grant Number: 62171063 and in part by Manufacturing High Quality Development Fund Project in China with Grant Number: TC210H03D.

Sejnowski [3] proposed the Information Maximization (Infomax) algorithm in 1995, which can effectively separate the super-gaussian signal. When the source signal conforms to a gaussian distribution, the nonlinear excitation function is a tanh function. However, it performs not well in the separation of sub-Gaussian signals. Scholars such as Lee, Girolami and Sejnowski extended the Infomax algorithm [4,5] and proposed the Extended Infomax (ExtICA) algorithm, which used a dual-probability switching model to achieve the separation of the mixed signals of super-gaussian and sub-Gaussian distribution sources. The nonlinear excitation function became the $K \times tanh$ function. Shi [6] combined the respective advantages of FastICA algorithm and ExtICA. The combined algorithm used a mixed exponential model to participate in iterative operations and it improved the speed of algorithm separation.

All the blind source separation algorithms proposed above use a fixed mathematical expression as a nonlinear function to participate in the operation. This results in the nonlinear function cannot be adaptively adjusted with the difference of the signal. On the contrary, neural network has great nonlinear ability. In order to solve the issue of signal separation affected by nonlinear functions. Deep unfolding technology, which has gained popularity recently, served as inspiration. An enhanced ICA network architecture is suggested in this research. Deep unfolding takes advantages of traditional mathematical models and deep learning [7]. The unfolding network can use the neural network to replace iterative operations, get more precise results, and easy deployment. As a result, the nonlinear function can be substituted by an explicable neural network in the formula's iterative operation, leading to a more accurate representation of the nonlinear function and a better approximation of the source signal's properties.

2   Enhanced ICA Network

In this section, we build the enhanced ICA network, an alternative network model of the ICA method to address the aforementioned issues with the ICA algorithm. We introduce the steps of the algorithm after suitably constructing and pre-training the explainable neural network based on the additive exponential model.

2.1   From Math Iterative Operations to Augmented Networks

The mathematical model of ICA need to minimize or maximize an appropriate cost function [8]. The ICA algorithm usually constructs the cost function based on the maximum likelihood principle [8],

$$\widehat{\boldsymbol{W}} = \arg \min_{\mathbf{w}} \left[-\log |\det \boldsymbol{W}| - \sum_{i=1}^{M} \log f_{\boldsymbol{y}_i}(\boldsymbol{y}_i) \right]. \tag{1}$$

where log means natural logarithm, the parameter $\boldsymbol{W}$ is a matrix to be optimized within $R^{M \times M}$, M represents the number of source signals, $\boldsymbol{y}_i$ represents a single source signal, and $f_{\boldsymbol{y}_i}(\boldsymbol{y}_i)$ represents the probability density function of each component signal in the source signal.

This algorithm obtain the new loss function by constraining the cost function with the minimum error rate criterion [8],

$$
\widehat{\boldsymbol{W}} = \arg\min_{\boldsymbol{W}} \left[-\log|\det \boldsymbol{W}| - \sum_{i=1}^{M} \log f_{\boldsymbol{y}_i}(\boldsymbol{y}_i) \right.
$$
$$
\left. + \lambda \operatorname{tr}\left(\boldsymbol{W}\boldsymbol{W}^{\mathrm{T}}\right) \right].
$$

(2)

The following formula illustrates the mathematical formula applied to the iterative operations of the ICA algorithm [8],

$$
\boldsymbol{W}^{n+1} = \boldsymbol{W}^n + \mu\left(\boldsymbol{I} - \varphi(\boldsymbol{y})\boldsymbol{y} - 2\lambda \boldsymbol{W}^n {\boldsymbol{W}^n}^T\right).
$$

(3)

where μ is the learning factor, λ is a normalization factor, the function $\phi(\boldsymbol{y})$ is a nonlinear function related to the signal source, $\boldsymbol{y} = \boldsymbol{W}^{n-1}\boldsymbol{X}$ represents a matrix operation between the result of the previous matrix operation and the observed signal. The function $\phi(\boldsymbol{y})$ uses a predetermined mathematical expression that is chosen based on experience as the basis for the nonlinear function. Different signals require different adaptive signal adjustments.

Thus, we propose the enhanced ICA network. The formula's nonlinear function can be replaced by a network structure, and deep learning can be added for improved training, considerably enhancing the flexibility of nonlinear function expression. The advantage of this method is that it can automatically learn the characteristics of the probability density function based on the data provided and change adaptively.

Compared with mathematical formulas, neural networks are more flexible and easy to form nonlinear characteristics. Additionally, xNN [9] notes that explainable neural networks can fit straightforward functions. At the same time, the subnets in the explainable network can provide specific visualization properties.

2.2 Structural Design and the Pre-training of Explainable Neural Networks

The explainable neural network concept put out by Joel Vaughan claims that explainable neural networks are capable of visualizing and fitting basic functions [9]. Explainable neural networks can learn the characteristics of the data, approximating the learning goal with arbitrary precision [10]. A more important feature is that the network displays the visual appearance of the functions it learns. We can illustrate the functions learned by each sub-network in order to explore the various characteristics of the nonlinear functions of different signals.

The model consists of three parts: hidden layer, subnet, and combination layer. Each subnet has an input node and an output node that are separate from one another and each subnet is configured with a 5-layer structure to assure that

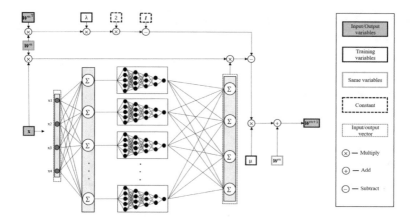

Fig. 1. Enhanced ICA network

it can effectively convey the data characteristics. Subnetworks are split into two categories in order to conserve computer resources: one category contains more nodes for learning complicated characteristics, while the other category contains fewer nodes for learning simple features. Four predicted values of the four source signals are assigned as the four output nodes at the output layer. At this layer, we combined the output values of the subnet output nodes. The design network is shown as Fig. 1.

We all know that if the node values of the neural network are arbitrarily initialized, the training time of the network will be as long as several hours or even days. Calculating a lot of parameters surely lengthens the calculation time when using the network to optimize the calculation process. Therefore, we pre-train the explainable neural network to reduce the time consumed by the training process.

2.3   Enhanced ICA Network Embedded in Deep Learning

After pre-training, the explainable neural network has a certain function expression ability. Then, for the forward propagation training of the neural network, we substituted an explainable neural network for the nonlinear function $\phi(\boldsymbol{y})$. The network weights are corrected with the gradient descent back propagation algorithm. Algorithm 1 provides a summary of the algorithm.

In the algorithm, $\boldsymbol{W}^n$, μ, *lambda* are the parameters same as ICA algorithm. Parameter n represents the data range for iterative calculation, which is divided into certain intervals; $\hat{\boldsymbol{s}}_i^n$ is the predicted signal, which is obtained by multiplying the nth separation matrix by the observed signal. And i is the ith source signal. $\boldsymbol{s}_i^n$ is the label signal, and it is used to calculate the error with the prediction signal. In the network, the parameters that can be used as training parameters

---

Algorithm 1 Enhanced ICA network algorithm

Input: Initialize $\boldsymbol{W}_0 = \boldsymbol{I}$, $\mu = 0.0001$, $\lambda = 0.005$; The observation signal $\boldsymbol{X}_0$; Pre-train explainable neural network f($\boldsymbol{y}$).

Output: Seperation matrix $\boldsymbol{W}_n$

1: $y^0 = \boldsymbol{W}^0 \boldsymbol{X}^0$

2: **for** $n = 3k - 3; k = 1,...,3200.$ **do**

3:     $\boldsymbol{W}^{n+1} = \boldsymbol{W}^n + \mu \left(I - f(\boldsymbol{y})\boldsymbol{y} - 2\lambda \boldsymbol{W}^n \boldsymbol{W}^{n^T} \right)$

4:     $\hat{\boldsymbol{s}}_i^n = \boldsymbol{W}^{n+1} \boldsymbol{X}^{n+1}$

5:     $\boldsymbol{s}_i^n = \boldsymbol{s}_i / \boldsymbol{s}_{i\,\max}$

6:     $\hat{\boldsymbol{s}}_i^n = \hat{\boldsymbol{s}}_i / \hat{\boldsymbol{s}}_{i\,\max}$

7: **end for**

---

as follow

$$\theta = \{\mu, \lambda, f(\boldsymbol{y}).trainable\_variables\}_{n=1}. \tag{4}$$

Due to the addition of explainable neural networks, we cannot extend the depth of neural networks as DetNet [11]. Because the explainable neural networks that exist at each layer have no meaning to each other. Since the sampled values of the signal are continuous, the calculation method of cross entropy cannot be used. We consider selecting the mean squared error as the loss function of the network in a single-layer neural network. Calculate the error between the label signal and the predicted signal $\hat{\boldsymbol{s}}$ which calculated by the separation matrix $\boldsymbol{W}$

$$l\left(\boldsymbol{s}_i^n; \hat{\boldsymbol{s}}_i^n \left(\boldsymbol{X}^n; \mu, \theta\right)\right) = \sum_{n=1}^{N} \left\| \boldsymbol{s}_i^n - \hat{\boldsymbol{s}}_i^n \right\|^2. \tag{5}$$

We trained the network using a variant of the stochastic gradient descent method for optimizing deep networks, named Adam Optimizer [11–13]. We implemented iterative training according to the calculation method of the ICA algorithm constrained by the minimum bit error rate criterion. In each iteration calculation, the explainable neural network is used as a function in the iterative formula to perform forward operation. The error between the predicted signal and the label signal is taken as loss, and the training parameter θ is optimized by Adam Optimizer.

3   Simulations

Firstly, we use uniformly distributed random numbers between 0 and 1 to form a random number matrix to mix four different signals. Then, we used two different methods to separate them successively. Finally, we use the similarity coefficient for mathematical analysis to numerically demonstrate the effectiveness of the proposed algorithm.

3.1   Training Process and the Result

In order to convert the data of the iterative operation into the form of one iteration operation and several groups of data comparison, the calculation mode of the data needs to be designed. The design scheme is shown in Fig. 2.

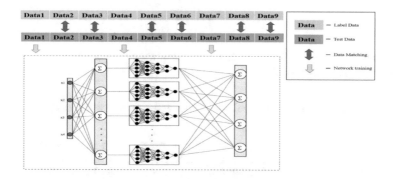

Fig. 2. Data calculation mode

In essence, fitting training of the function y^3 serves as the pre-training of the explainable neural network. Then, using enhanced ICA network and ICA bound by the least bit error rate criterion, we did a separation experiment of the mixed signals. The mixed signals consisted of sine wave, cosine wave, square wave, and triangle wave signals of different frequencies. The waveforms after the separation of the two methods are shown in the Fig. 3.

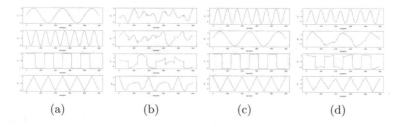

(a)	(b)	(c)	(d)

Fig. 3. Separation performance comparison of the second scenario: (a)Source signals, (b)Mixture signals, (c)Enhanced ICA network separation, (d)Infomax separation.

Comparing the output results of iterative training and network training, it can be seen that the waveform separated by enhanced ICA network is more in line with the characteristics of the source signal, and the loss of information is less. It shows that the separation matrix $\boldsymbol{W}$ obtained after the enhanced ICA net training is more accurate, and the separation matrix $\boldsymbol{W}$ can be used to separate subsequent signals.

3.2 The Effect of the Enhanced ICA Network

For demonstrating the effectiveness of the designed algorithm, simulation experiments are conducted to evaluate the performance of the enhanced ICA network. The results are verified through the similarity coefficient defined as follows

$$\xi_{ij} = \xi\left(s_i, \hat{s}_i\right) = \frac{\left|\sum_{t=1}^{T} s_i(t)\hat{s}_j(t)\right|}{\sqrt{\sum_{t=1}^{T} s_i^2(t) \sum_{t=1}^{T} \hat{s}_j^2(t)}}. \tag{6}$$

The similarity calculation of the two methods to separate the waveforms is shown in Table 1 and Table 2.

Table 1. Similarity coefficient of enhanced ICA network(sine wave, cosine wave, square wave, triangular wave signals).

	s1	s2	s3	s4
x1	0.007025182	0.9999495	0.0066374466	0.0033976801
x2	0.99982595	0.0056461245	0.0021300649	0.016851645
x3	0.00026985112	0.007652181	0.9999026	0.0060464884
x4	0.018917622	0.00082509377	0.00394196	0.9998204

Table 2. Similarity coefficient of ICA constrained by the minimum bit error rate criterion(sine wave, cosine wave, square wave, triangular wave signals).

	s1	s2	s3	s4
x1	0.07968595	0.9903527	0.051673625	0.100712225
x2	0.9772345	0.14983222	0.13201351	0.06532389
x3	0.002949858	0.012996394	0.99356425	0.1086201
x4	0.12443759	0.023711462	0.12806745	0.9832823

For each row in the table, the maximum value means that the separation signal of the current row best matches the label signal of the current column. Enhanced ICA network has higher accuracy on all four signals. Furthermore, we can see that the maximum values in the table do not form a diagonal line, which is due to the disorder of the ICA algorithm when separating the signals.

4 Conclusion

In this paper, we use an explainable deep learning-based network for ICA enhancement in blind signal separation. It can achieve better separation performance than iterative operations. The minimal bit error rate criterion-constrained enhanced ICA network is no longer strictly speaking a blind source separation network. Because the neural network must be trained with labeled data which take from the sources. In other words, training necessitated the use of known signals as data. By utilizing the enhanced ICA network bound by the minimal bit error rate requirement, we optimize the computation results after each iteration to get a more precise separation matrix W, allowing us to do further signal separation.

References

1. Fouda, M.E., Shaboyan, S., Elezabi, A., Eltawil, A.: Application of ICA on self-interference cancellation of in-band full duplex systems. IEEE Wirel. Commun. Lett. **9**(7), 924–927 (2020)
2. Amari, S.-I., Cardoso, J.-F.: Blind source separation-semiparametric statistical approach. IEEE Trans. Signal Process. **45**(11), 2692–2700 (1997)
3. Bell, A.J., Sejnowski, T.J.: An information-maximization approach to blind separation and blind deconvolution. Neural Comput. **7**(6), 1129–1159 (1995)
4. Girolami, M.: An alternative perspective on adaptive independent component analysis algorithms. Neural Comput. **10**(8), 2103–2114 (1998)
5. Lee, T.W., Girolami, M., Sejnowski, T.J.: Independent component analysis using an extended infomax algorithm for mixed subgaussian and supergaussian sources. Neural Comput. **11**(2), 417–441 (1999)
6. Shi, Z., Tang, H., Liu, W., Tang, Y.: Blind source separation of more sources than mixtures using generalized exponential mixture models. Neurocomputing **61**, 461–469 (2004)
7. Jagannath, A., Jagannath, J., Melodia, T.: Redefining wireless communication for 6G: Signal processing meets deep learning with deep unfolding. IEEE Trans. Artif. Intell. **2**(6), 528–536 (2021)
8. Luo, Z., Zhang, W., Zhu, L., Li, C.: Minimum BER criterion based robust blind separation for MIMO systems. Infocommun. J. **11**(1), 38–44 (2019)
9. Vaughan, J., Sudjianto, A., Brahimi, E., Chen, J., Nair, V.N.: Explainable neural networks based on additive index models. arXiv preprint: arXiv:1806.01933 (2018)
10. Liu, C., Shi, B., Li, C., Zou, J., Chen, Y., Xiong, H.: Deep neural network-based algorithm approximation via multivariate polynomial regression. In: 2019 IEEE Global Communications Conference (GLOBECOM), pp. 1–6. IEEE (2019)
11. Samuel, N., Diskin, T., Wiesel, A.: Deep MIMO detection. In: 2017 IEEE 18th International Workshop on Signal Processing Advances in Wireless Communications (SPAWC), pp. 1–5. IEEE (2017)
12. He, H., Wen, C.K., Jin, S., Li, G.Y.: A model-driven deep learning network for MIMO detection. In: 2018 IEEE Global Conference on Signal and Information Processing (GlobalSIP), pp. 584–588. IEEE (2018)
13. Samuel, N., Diskin, T., Wiesel, A.: Learning to detect. IEEE Trans. Signal Process. **67**(10), 2554–2564 (2019)

Research on Semantic Verification Method of AIXM Data Based on SBVR

Xiaowen Wang[1]($\boxtimes$), Yungang Tian[1], Shenghao Fu[1], and Charity Muthoni Musila[2]

[1] State Key Laboratory of Air Traffic Management System and Technology, Nanjing, Jiangsu, China
{wangxiaowen1,tianyungang,fushenghao}@cetc.com.cn
[2] East African School of Aviation, Nairobi, Kenya

Abstract. Correct and complete aviation data is a prerequisite for SWIM cross-domain information sharing, which will affect the correct operation of the application program of the air traffic control system. The current syntax-based data verification method can verify the good structure of AIXM exchange documents. And rule constraints are usually in raw textual form and cannot be automatically enforced in computerized systems. This paper proposes an automated data semantic verification method for business rules based on SBVR, constructs the semantic rules meta-model of AIXM data, forms a set of AIXM business rules writing methods, and realizes the coding conversion path for implementing this method into system development, which supports automatic verification of more complex business logic and rule constraints.

Keywords: SWIM · AIXM · SBVR · Semantic validation

1 Introduction

System wide information management (SWIM) is an important way for ICAO to realize the integration of global data governance of the next generation air traffic control system. Data consistency and data quality management are important components of system wide information management. Aiming at the problem of data consistency in the exchange and sharing of aviation information, Europe and the United States have led the construction of the aviation information exchange mode (AIXM), adopting the standard based data exchange mode instead of the special interface to standardize the data exchange, and defining the information exchanged between systems in detail at the conceptual level and the data level. Correct and complete high-quality aviation data will affect the correct operation of many applications of air traffic control system, and may further endanger the life and property safety of passengers and air traffic control support units. It is a prerequisite for cross domain information sharing. When it is required to provide or obtain AIXM data sets through SWIM, it must be verified whether the provided data meets the agreed specification requirements.

The documents shared by AIXM information exchange include the conceptual model and logical model based on Unified Modeling Language (UML) and the physical model based on extensible markup language (XML). Among them, the XML document has a high degree of flexibility in definition and structure, and supports the free exchange of structured data between organizations and applications. To ensure that the document strictly follows the agreed structure and business logic rules, the XML schema definition (XSD) can usually be used to validate the data represented by the XML document [1, 2, 3, 4], Ensure that the XML document structure of data exchange is good and the document is valid. However, the XSD model does not contain complex business logic and rule constraints, such as dependencies between attribute values in different classes, values beyond the value range, and mandatory attributes of classes.

Combined with the UML concept/logic model of AIXM, this paper proposes a complex business logic verification method based on SBVR (semantics of business vocabulary and business rules) business rules, modeling these complex constraints as business rules, using SBVR standard to write AIXM business rules Together with their definitions and data types as the basis for business vocabulary and definition of business rules, and adjusted according to the requirements of AIXM, a set of writing methods for AIXM business rules is formed, and the method for implementing AIXM business rules based on SBVR in software coding is described.

2 Data Verification Method Summary

2.1 Overview of AIXM Data Verification Method

AIXM data sets are shared among different stakeholders in the form of XML, which comes from the conceptual model and physical model of AIXM. XML data validation methods mainly include data document structure validation and business rule validation, as shown in Fig. 1.

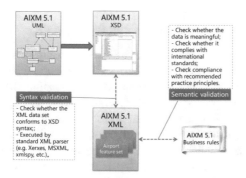

Fig. 1. AIXM data verification method1

Document structure verification is a syntax verification method [5, 6]. It uses XSD to verify the validity of the data represented by XML documents, mainly to check whether the XML data set conforms to XSD syntax. It can generally be completed by standard XML parsers, such as Xerxes, MSXML, XMLspy, etc. At present, the conceptual level model in the aeronautical information exchange model uses UML model to define the information items of the aeronautical information domain with class diagram, including classes, attributes of classes and relationships between classes. For the attributes of a class, some of its constraints can be included in the UML model, which is represented as a list of values, a range of values, or a pattern of values. The data level model in the actual information exchange adopts the XML schema (XSD) model, which is mapped from the UML model. The elements in the XSD model include classes, attributes, and class names associated with the current class. The values and data types of the elements can also be derived from the UML model.

Business rule [7] validation mainly uses semantic method [8], describes business definitions and constraints, and verifies the behaviors that maintain business structure or control and affect business [9–11]. UML class diagram can be used to describe the information structure and some semantics, such as the value range of digital attributes. However, if the description depends on the values of other attributes, it needs to be described using business rule constraints, which can be specified on a single model element. For example, in the process of cross domain sharing of aviation data, can there be runways outside the airport? If the airport has the International Air Transport Association (IATA) code, is it necessary to have the International Civil Aviation Organization (ICAO) code? What is the reasonable value of runway length? These relationships of aviation data are generally implied in ICAO annex 10, annex 11, annex 14, annex 15, FAA FNS-ND technical manual, EAD and other international standards or industry practices. These complex semantic business logic and rule constraints are not included in the UML model. In practical applications, these business rules are generally encoded in the program code, This requires that business experts who write rules need to understand the specific implementation, and when the specific implementation technology is replaced by another technology at some time in the future, they need to rewrite the rules. Rule based verification extracts the business logic processed by the system from the program code.

2.2 Semantics of Business Vocabulary and Rules (SBVR)

"Semantics" refers to the meaning or meaning relationship of a match or a group of matches. SBVR [12] is a semantic specification for describing business in natural language defined by OMG (Object Management Group). It mainly defines how to store the semantics of business vocabulary, business facts and business rules, so as to communicate among various organizations and software tools and apply to the business vocabulary and business rules of business activities of all organizations. SBVR defines a meta model for rules and vocabulary, which can describe all business activities of any organization. The main users are business personnel. SBVR is an inseparable part of OMG Model Driven Architecture (MDA). The location of SBVR in MDA [13] is shown in Fig. 2:

The main purpose of the SBVR model is to describe the business rather than the technical implementation of the information system supporting the business. SBVR includes business rules and business vocabulary, as well as related conjunctions used in the rules. Business processes and organizational structures are not included in the description scope of SBVR. SBVR is a fact based ontology description method [14]. Its basic element is fact, and fact is based on fact type, which is based on the concept of expression [15–17]. Figure 3 shows an example of a concept, fact type and fact.

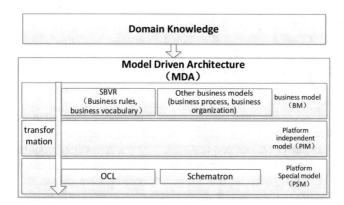

Fig. 2. SBVR structure and its position in MDA2

This method has the advantages of better readability, semantic stability, expressiveness, extensibility and flexibility.

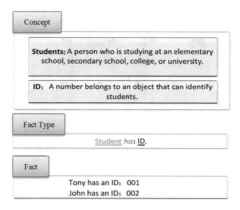

Fig. 3. Concept, fact type and fact3

2.3 Business Rule Language Analysis

In order to verify whether the domain knowledge representation is appropriate, the business rule constraint set of the conceptual model is generally explained and communicated through natural language expression. According to the maturity of business rules, the mainstream business rule language includes free text, SBVR and Schematron.

The rules of free text are usually the original English text of the rules. It describes the requirements of the rules without reference to technical details such as data model. Domain experts can understand this rule, but to some extent, it may be ambiguous or ambiguous, or at least not detailed enough to be implemented in the computerized system.

The main purpose of SBVR constraint specification language is to regularize business description rather than technical realization. SBVR rule constraint is the formal description of the pure English text rules. It is written according to a set of formal specifications to limit the possible fuzziness and fuzziness of natural language. SBVR compliant rules are also more detailed. It usually refers to formal or abstract data structures, rather than real-life objects or concepts.

Schematron constraint specification language is one of the technical implementation means of business rules. It is a simple and powerful structural pattern language. Its specification has been standardized (ISO/IEC 19757). It uses path expressions instead of syntax to define the content allowed by XML documents, and can express many structures that are difficult to express for syntax based pattern languages, And the data set can be automatically verified by the computer.

The characteristics of the three types of business rule writing languages are shown in Table 1.

Table 1. Comparison of business rule writing languages1

	Natural Language	*Normative Language*	*Programing Language*
form	Free text	SBVR specification text	Schematron code
User	Domain expert	Analysts and business personnel	Developer
Readability	High	Middle	Poor
accuracy	Low	High	Very high
Automatable	Not support	Low	High

3   AIXM Semantic Rule Meta Model Based on SBVR

3.1   AIXM Business Rule Classification

The goal of business rules is to provide a complete set of operation constraints with standard format. AIXM business rules define what is reasonable or allowable in the AIXM data set. According to the application scope of rules, business rules can be divided into structural rules and operational rules. Among them, the structural rules of AIXM mainly include the syntax rules of data type enumeration values. Most of these rules have been encoded in AIXM schema and will not be described in detail in this article. Operational rules are semantic rules mainly extracted from official documents (ICAO attachments), such as the requirements of minimum information set and frequency matching rules of VHF navigation facilities. We specifically divide them into minimum data rules, data consistency rules, recommender rules, mandatory standard rules, coding rules, etc. the classification of AIXM business semantic rules is shown in Table 2:

Table 2. Classification of AIXM Business Semantic Rules 2

Rule classification	Rule description	Rule example
Data consistency rules	Rules for correctness and integrity verification of logical relationships between related data	For the "runway. Type" attribute, it is necessary to check that the runway type 'Rwy' should not be associated with the Heliport (HP)
Minimum data rule	Rules for attribute referential integrity verification	For example, for the "runway. Associatedacportheliport" attribute, you need to verify whether the associated actual element instance exists through the instance name, and check whether it matches the element type defined by the model
Coding rules	Mandatory verification rules for measuring units	For example, for runway The length accuracy attribute needs to check whether the measurement unit is specified if the value is specified for the attribute with measurement unit
Mandatory standard rules	Relevant verification rules in ICAO and industrial mandatory standards	For example, for runway centrelinepointlocation. LevatedPoint Horizontalaccuracy attribute, it is required to check the accuracy of runway center point position, which should be better than 1m
Recommended procedure rules	Procedure verification rules in ICAO and industrial recommended standards	For example, for runway directionTruebearing attribute, it is required to check the true orientation of the runway and the final approach and take-off area. It shall be released according to the requirements of 1 / 100 degree resolution

3.2 AIXM Semantic Rule Meta Model

The vocabulary elements of SBVR include many elements such as nouns, verb concepts (fact types), name words and keywords. It is a general model developed for different fields such as transportation, finance and power. In order to meet the needs of semantic rules in the aviation field, the rule elements are expressed according to the characteristics of AIXM and meet the needs of AIXM rule description, as shown in Table 3. Among them, nouns represent objects that can be understood by business and are called classes in UML; The fact type specifies the relationship between a noun and one or more nouns. In UML, the fact type will be the relationship type between classes. The name vocabulary is an instance of a noun. For example, "CTA" is an instance of the noun "airspace. Type". The name vocabulary can also be an instance of a fact type. The keywords include many concepts such as logical operation, quant, and module.

Table 3. Concept definition of SBVR semantic rule meta model in AIXM

Essential factor	SBVR definition	SBVR summary representation in AIXM
Concept	A unit of knowledge consisting of a unique set of characteristics	No difference
Noun concept	A concept that represents a noun or noun phrase	Expressed as classes and features in the AIXM UML model, specifically, the name, role name, or attribute name of the AIXM class can be used as nouncept. Such as: Airport heliport, airspacetype, Runway. AssociatedAirportHeliport
Verb concept / Fact type	• It refers to the concept of verb phrase, which contains one or more noun concepts, and it becomes a fact after instantiation. It is also called fact type, and the expression form is: $< Fact\text{-}type >::= < concept1 > < verb > < concept2 >$ • verb concepts can be business facts or relationships between concepts	• Represents the name of the association relationship between classes in the AIXM UML model. For example: Airport heliport has name, runway hassurfacedescribedby surfacephysics
Name vocabulary	A name is a concept that refers to a single concept (object). A name is often a proper noun	• CodeList value, expressed as UML instance, enumeration literal value, and assigned feature, etc. For example, 'CHINA', 'CTA', 'CTA'_P', 'YES', 'NIL'
keyword	Used to build structured declarations, keywords can be combined with other identifiers to form declarations and definitions, including keywords such as logical operators, quant, and modality • Quant: including full name quantifiers and existential quantifiers, such as each x, at least one X, etc • Logical operation: including "and", "or" and "not"; • Modality: use different sentence patterns, such as "each… Should/shall", "it is impossible that / it is prohibited that", etc	• No difference

When expressing a rule, there will be different statements due to the use of different word orders and keywords. Therefore, the practicability of AIXM model and verification needs to be considered, and the rule writing syntax supported in the meta model should be explained from the whole to the part. The general structure of rule statements can be divided into two types according to different programs, as shown in Fig. 4.

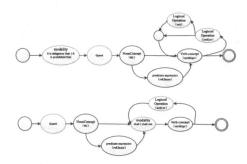

Fig. 4. Overall structure of SBVR rule statements4

It can be seen in the figure that the noun concept (NC) is preceded by a quantifier (Quant). The first noun concept lays the background for the whole sentence, which represents a class of objects. Then, a relational clause (relclause) can be added to define more constraints for the object, that is, Before verifying whether an object meets the main rules (represented by logical operations and verb concept combinations), you can first determine whether it meets the predefined preconditions.

4   AIXM Rule Writing Method and Coding

4.1   AIXM Rule Writing Method

On the basis of specifying that AIXM business rules are compiled as the meta model rules of SBVR text, the following takes the compilation of runway element class and airport element class rules as an example, and applies the SBVR meta model to compile the facts established in AIXM and ICAO standards and recommended measures and other reference documents as rules. The business rule set of AIXM meta model can be compiled using some standard office tools, Such as Microsoft Excel.

The starting point for writing AIXM rules is the UML model, which describes many different types of facts. For example, in Fig. 5, we can extract "runway isituatedat airport heliport" and "name is property of airport heliport". Some fact types mainly describe the relationship between classes. This relationship has been expressed by explicit verb concepts in the figure, such as "issituatedat"; Some represent simple properties of a class, that is, the properties encapsulated in the class. Although the name of this relationship is not written, it contains the fact type "is property of".

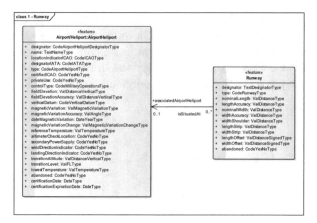

Fig. 5. Runway element class and airport element class

On the basis of these fact types, first capture some constraints used to define the scope or limit; Next, consider some operation rules that are not directly related to the fact type. For example, the accuracy requirement of the boundary point of a certain restricted area needs to depend on whether the current area is inside or outside the TMA. However, the AIXM model does not specify the fact type of the relationship between R and TMA. Both are instances of the airspace class. However, such rules can be clearly expressed by including the business concept "overlapping geometry" in the business vocabulary of the aviation information domain. In general, the method steps for writing AIXM rules are as follows:

(1) Determine which aviation element class in the AIXM UML model is targeted by the rules to be written, such as airport heliport, airspace, navaid, etc.;

(2) Determine the core fact types that the rules need to refer to in the model, such as class attributes, direct relationships between classes, and indirect relationships between classes. The aviation element class and the core fact types in the model are "runway isituatedat airport heliport", "runway has type", "airport heliport has type"; It is specified in UML that the value range of the type attribute of the airport heliport class is a predefined list (AD ',' AH ',' HP ',' LS, 'OTHER'), where 'HP' represents the heliport. At this time, a quantifier is added to the third core fact type, which becomes "airport heliport has type equal to 'HP'"; Similarly, the second fact type becomes "runway has type equal to 'FATO'";

(3) Around the core fact type, add quantifiers and logical operations for the noun concepts involved, express the rules with appropriate programs, and establish the rule text:

• Add logic operation and program to the three fact types to form "it is probable that a runway has type equal to 'FATO' and runway is situatedat airport heliport and airport heliport has type equal to 'HP'";

• Adjust the sentence to make it simple and easy to understand, "it is prohibited that a runway with type equal to 'FATO' is located at airport heliport with type equal to 'HP'". The keyword with is mainly used here.

Table 4. Example of airport runway constraint rules based on SBVR3

Rule requirements	Rule type	Free text rules	SBVR rules (text)	SBVR rules (label)
A runway with runway type 'Rwy' cannot be associated with an airport with runway type 'HP'	Data consistency rules	A Runway of type 'RWY' cannot be associated with an AirportHeliport with type 'HP'	It is prohibited that Runway with assigned type equal-to 'RWY' isSituatedAt AirportHeliport with assigned type equal-to 'HP'	\<keyword\> It is prohibited that \</keyword\> \<keyword\>a \</keyword\> \<noun_concept\> Runway\</noun_concept\> \<verb_concept\> has \</verb_concept\> \<noun_concept\> RunwayTimeSlice. type \</noun_concept\> \<keyword\> equal-to \</keyword\>\<name\> 'RWY' \</name\>\<keyword\> and \</keyword\> \<noun_concept\> AirportHeliport\</noun_concept\>\<verb_concept\> has \</verb_concept\>\<noun_concept\> AirportHeliportTimeSlice. type\</noun_concept\> \<keyword\> not \</keyword\> \<keyword\> equal-to\</keyword\> \<name\> 'HP' \</name\>

Although the above rule will be better understood semantically if it is written as "each airport helicopter with type equal to 'HP' shall not have associated runway with type equal to 'FATO'", it is noted that this rule is based on the fact type "runway issituatedat airport helicopter", It directly reflects the directionality of the relationship between the runway and the airport/heliport in the established UML model, but only writing such a statement as "a certain class is related to a certain class" cannot express this. In the XML message, the runway will point to the airport heliport, so when checking this rule, their directions are consistent. An example of airport runway rules is shown in Table 4.

4.2 Application of Business Rule Coding

Schematron is a language based on semantic rules. It uses path expressions instead of syntax to determine the validity of XML documents. By using XPath and various extension implementations provided by XSLT, it can express many structures that are difficult to verify for syntax based schema languages, such as DTD, relax ng schema and XML schema (XSD).

Shapechange [18] is a Java based open source tool for processing geographic information application patterns. It can realize automatic conversion from UML models to XML target patterns and supports GML32. ISO 19139 and inspire coding rules. We use shapechange V2.11 convert the SBVR rules of AIXM into Schematron codes, and the generated codes can be directly injected into the application code in the form of scripts, which can realize the separation of business rules and codes while realizing the automation of business rules. The coding example of AIXM business rules based on Schematron is shown in Table 5.

Table 5. Coding Example of AIXM Business Rules Based on Schematron4

Rule requirements	Free text rules	SBVR rules (text)	Schematron rules
A runway with runway type 'Rwy' cannot be associated with an airport with runway type 'HP'	A Runway of type 'RWY' cannot be associated with an AirportHeliport with type 'HP'	It is prohibited that Runway with assigned type equal-to 'RWY' isSituatedAt AirportHeliport with assigned type equal-to 'HP'	not(./aixm:type='RWY') or not(saxon:evaluate(arcext:getXPath((./aixm:associatedAirportHeliport)/@xlink:href)/aixm:timeSlice/aixm:AirportHeliportTimeSlice/aixm:type = 'HP')

At the same time, in order to make up for the problem of service, service carrying data, and the lack of relationship description, the model uses srcm:operateOn and its reverse attribute srcm:usedBy to describe the connection relationship between service and data, as shown in Fig. 3, and uses RESTful service interface Ways to provide an index in the semantic registry to improve the level of search and discovery automation.

5 Conclusion

The syntax validity of the data set in AIXM XML format can be verified by the XSD model conforming to AIXM. For some special applications, it is necessary to check whether it is correct in semantics through business rules. This article explains the concept of AIXM business rules, the method of business rule modeling, and how business rules can be implemented in system development. It should be noted that although the AIXM business rules contain a complete set of semantic verification of AIXM format messages, for different applications and users, only a subset of this complete set may be required each time. For example, as for a certain frequency attribute in the elements of the navigation equipment, we require it to meet a certain value. Such a rule is a necessary constraint in drawing aeronautical charts or other air navigation support applications, but it is not required in flight planning applications.

In addition, Not all business rules are universally applicable, and some business rules are only applicable to specific use cases. Therefore, AIXM business rules are maintained outside the AIXM UML model. In actual model application, AIXM business rules need to be loaded item by item. In addition, this method can also be applied to business rule verification of flight information exchange model (FIXM) and weather information exchange model (WXXM) of SWIM.

Acknowledgment. This work was sponsored by National Key Research and Development program (NO. 2018YFE0208700).

References

1. Rui, Z., Ling, G.: Tian MI Research on client data validation method based on JS and regular expression. J. Yan'an Univ. Nat. Sci. Edn. **27**(1), 21–24 (2008)
2. Wei, D., Wei, L.: Application of Java and XML schema in data validity verification. Comput. Eng. Des. **24**(3), 65–68 (2003)
3. Yuxiang, L., Dafang, Z.: Design of general template for data validation. Comput. Appl. Res. **22**(12), 73–75 (2005)
4. Youneng, H.: Research on Safety Processing and Verification Methods of Urban rail CBTC System Data. Beijing Jiaotong University, Beijing (2014)
5. Jian, W., Anna, Y., Chun, Y.: Data quality evaluation index system of command information system. Command Inf. Syst. Technol. **11**(2), 85–88 (2020)
6. Guangyu, Z., Qiuhui, Y., Cong, Z., Xinyu, G., Shu, Q.: Quality analysis method of open XML data. Comput. Appl. Res. **30**(7), 2082–2086 (2013)
7. Hua, C., Hui, R.: Design and implementation of simple data cleaning rule base. Command Inf. Syst. Technol. **3**(5), 79–84 (2012)

8. Erfeng, Z., Gailing, T.: Semantic based sensor observation service system. J. Xi'an Univ. Posts Telecommun. **19**(4), 63–69 (2014)
9. Dedong, M., Rui, F., Bin, Z.: Research on rule expression and verification of IEC 61850 model information. Power Syst. Prot. Control **43**, (3), 131–136 (2015)
10. Yan, C.: Application of rule engine in configuration data verification of train control products. Railway Commun. Sig. Eng. Technol. (RSCE) **17**(7), 29–34 (2020)
11. Huifen, D., Huihua, W., Dehua, Z., et al. Application of rule engine in command display system. Ordnance Autom. **36**(8), 80–83 (2017)
12. Object Management Group: Semantics of Business Vocabulary and Business Rules (SBVR), v1. 0. OMG Document dtc/08-01-02 (2008) Google Scholar
13. Miller, J., Mukerji, J., et al.: Mda guide version 1.0. OMG Document: omg/200305-01 (2003). http://www.omg.org/mda/mda_les/MDAGuideVersion1-0.pdf
14. Linehan, M.: Ontologies and rules in business models. In: Eleventh International IEEE EDOC Conference Workshop, EDOC 2007, pp. 149–156. IEEE, Los Alamitos (2008).
15. Kleiner, M., Albert, P., Bézivin, J.: Parsing SBVR-based controlled languages. In: Schürr, A., Selic, B. (eds.) MODELS 2009. LNCS, vol. 5795, pp. 122–136. Springer, Heidelberg (2009). https://doi.org/10.1007/978-3-642-04425-0_10
16. Kamada, A., Governatori, G., Sadiq, S.: Transformation of SBVR compliant business rules to executable FCL rules. In: Dean, M., Hall, J., Rotolo, A., Tabet, S. (eds.) Semantic Web Rules. RuleML 2010. Lecture Notes in Computer Science, vol. 6403, pp. 153–161. Springer, Berlin (2010). https://doi.org/10.1007/978-3-642-16289-3_14
17. Goedertier, S., Mues, C., Vanthienen, J.: Specifying process-aware access control rules in SBVR. In: Paschke, A., Biletskiy, Y. (eds.) RuleML 2007. LNCS, vol. 4824, pp. 39–52. Springer, Heidelberg (2007). https://doi.org/10.1007/978-3-540-75975-1_4
18. ShapeChange Technical Documentation (2022). https://shapechange.net

A Clustering-Based Anomaly Detection
for Unstable Approach in Terminal Airspace

Zhongrui Xu[1,2], Xiaoguang Lu[1,2(✉)], Zhe Zhang[1,2], and Zhijie Wang[1,2]

[1] College of Electronic Information and Automation, Civil Aviation University of China,
Tianjin, China
xglu@cauc.edu.cn
[2] Tianjin Key Laboratory, Civil Aviation University of China, Tianjin 300300, China

Abstract. Data-driven methods are broadly used in analyzing big data in many fields. ADS-B trajectory data provided the possibility of anomaly detection in terminal airspace. Approaching and landing accidents are usually originated from unstable approaches, which are the consequence of go-around failure. Go-around is an aborted landing process initiated in the event of an unsafe landing. In this paper, DBSCAN and HDBSCAN clustering algorithms were employed to detect spatial and energy anomalies using ADS-B trajectory data. Air traffic flow patterns are identified, thus energy safety boundaries are established. Finally, atypical scores are quantified based on the total energy, which serves as the foundation for studying, comprehending, and addressing the factors that cause unstable approach events and go-arounds. This method detects the unstable approach from the spatial and energy perspectives and quantifies the degree of anomaly. It is essential for detecting and predicting anomalies in terminal airspace, and also explores the application of data-driven methods to aviation data.

Keywords: Unstable Approach · Go-around · DBSCAN · HDBSCAN · Anomaly Detection

1 Introduction

Aviation safety is the eternal theme of air transport. The rapid growth in air traffic flow and the expected growth on a global scale pose enormous challenges to aviation safety. Among the components of the airspace system, the terminal airspace is considered to be the most critical airspace with respect to safety and efficiency at both the individual flight-level and the system-level as it contains the most critical phases of flight, the take-off and landing [1]. In the past few decades, 56% of the fatalities and 62% of the accidents occurred during the final approach and landing phase, even though that phase accounted for only 16% of the flight time. Generally, unstable approach and failed go-around are the number one risk factors for approach and landing accidents [2]. Therefore, the study of unstable approach and go-around in terminal airspace is of great significance to aviation safety.

In 2018, Haodong Zhang analyzed the trajectory points from coarse to fine according to the direction distribution information for flight anomaly detection [3]. In 2020, Simon

Q. Liang et al. (Eds.): AIC 2022, LNEE 871, pp. 272–280, 2023.
https://doi.org/10.1007/978-981-99-1256-8_32

Richard Proud proposed a new method for automatic detection of go-arounds based on flight phase characteristics [4]. In 2021, Lu et al. modeled aircraft mass variation, wind speed, and other factors to study the risk of energy anomalies in the approach and landing segment of large civilian aircraft [5]. These physics-based or rule-based approaches are designed with the expectation that a normal operation should not violate the standard operation of the system or exceed the boundaries of the nominal dynamics, which relies on a mathematical model of the system [6]. The airspace, on the other hand, is a very complex system with a constantly changing and evolving behavior, which makes it challenging to determine its control equations and limited in practical applications.

Existing tracking and monitoring techniques for aircraft include radar, ADS-B, Aircraft Communications Addressing and Reporting System (ACARS), etc., which makes the sources, quantity and availability of aviation operational data more than ever before [6], especially the mandatory installation of ADS-B. This brings the benefit for data-driven approaches, which can both detect safety risks and improve the efficiency of the overall system. Advances in machine learning techniques have greatly facilitated the use of data-driven approaches to gain insights from operational data, improve aviation system efficiency and enhance safety.

In 2019, Antonio Fernández et al. address the problem of unknown hazard detection during the approach phase by integerating two very powerful machine learning algorithms (HDBSCAN clustering and AutoEncoders) from data science and aviation safety perspective [7]. In 2021, Weizun Zhao et al. proposed an incremental anomaly detection method based on Gaussian mixture model (GMM) to identify common patterns and detect anomalies from flight data [8]. In 2021, Samantha J. Corrado et al. performed a quantitative analysis of spatial and energy anomaly interdependencies of ADS-B trajectory data based on clustering [6]. In 2021, Jiayi Xie et al. conducted a preliminary exploratory spatial data analysis of unstable approach events [9]. Although previous studies have given methods for data analysis and detection of unstable approaches, most of them cut from a single perspective of energy or space, and lacked the combination of example analysis, or excluded the common case of go-around when analyzing abnormal data.

The go-around detection was implemented in previous work [10]. Data-driven and machine-learning techniques were applied to further explore the spatial and energetic connections of flight data, and the outcomes of the clustering were evaluated along with examples of go-around. It is successful in identifying air traffic patterns, establishing energy safety bounds, and quantifying atypical scores based on the total energy. It contributes to a better understanding of the influencing elements and salient features of unstable approaches and go-arounds, and it serves as the foundation for terminal airspace anomaly detection and prediction, which enhances in flight safety.

The remainder of the paper is organized as follows: Sect. 2 introduces the data source and the data pre-processing process. Section 3 identifies air traffic flow from a spatial standpoint. Section 4 analyzes altitude, ground speed, and total energy in order to define energy safety bounds and quantify atypical scores. Finally, a summary and future work outlook are provided.

2   Data Sets and Pre-processing

ADS-B data are an excellent source of position information for studying unstable app-roach and go-around events. The ADS-B data used in this paper comes from Opensky Network [11], a non-profit website that processes and stores ADS-B data from a global network of sensors. Obtain the position report data from September 1, 2020 to August 31, 2021 within a 100*100 square kilometer range centered on ZUCK from the Opensky Network.

The downloaded data contains take-off, landing and transit aircraft, so needs to be pre-processed. First, the position report data is filtered out for barometric altitudes above 10,000ft, as aircraft preparing to land are usually below this altitude. Second, remove invalid transfers with missing significant values and reports with duplicate timestamps. The data is then integrated according to the icao24 address to form flights, applying a multi-level filter to eliminate departures, transits and other flights that are not relevant to this study. Following these operations, the data can be used for go-around detection, while DBSCAN clustering for energy and spatial necessitates resampling of the data to generate a normalized n-dimensional trajectory record. Each trajectory is represented by 50 data points using time-based resampling. Figure 1 depicts the data pre-processing framework.

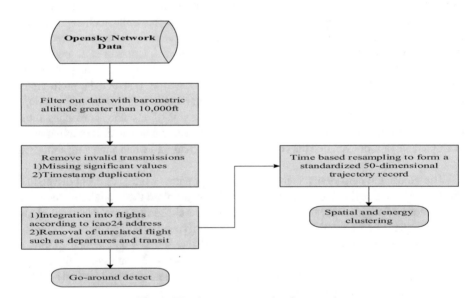

Fig. 1. The data pre-processing framework.

3   Air Traffic Flow and Spatial Anomalies

Spatial clustering is often used to identify air traffic flows and spatial anomaly detection is a byproduct of it. Luis Basora et al. have used the latitude, longitude, and altitude of the entry and exit points of the aircraft through the airspace to represent the entire track, but this method is not ideal in scenarios of high traffic density in the terminal airspace [12]. In this paper, longitude, latitude, and heading are used as clustering features, with heading better representing the inflection point of the trajectory than altitude and highlighting the differences. A random sample of 15 days of data was taken, including approximately 6272 trajectories, each represented by 50 points. Ten data points are sampled at equal intervals from the trajectory data, each point including latitude, longitude and heading information. Due to the large size of the data, using Principal Component Analysis (PCA) to reduce the dimensionality. PCA as a common data analysis method is often used for dimensionality reduction of high-dimensional data, which can extract the main feature components of the data. After dimensionality reduction, get a matrix of 6272 × 5, which is the input of DBSCAN.

DBSCAN relies on two parameters [13], (1) the minimum sample size, which is the minimum number of samples for a cluster, and (2) ε-neighborhood distance threshold, where data points will be in the same cluster if their mutual distance is less than or equal to the specified ε. Euclidean distance is used as the distance function for clustering, which is the most commonly used. Two n-dimensional vector x and y, where $x = (x_1, x_2, \ldots, x_n)$ and $y = (y_1, y_2, \ldots, y_n)$ the Euclidean distance computation proceeds as Eq. (1):

$$D_{ED}(x, y) = \sqrt{\sum_{i=1}^{n} (x_i - y_i)^2} \tag{1}$$

Usually, the minimum sample size is set to 1% of the sample, but the practice has found that few clusters are identified at this point, and the percentage of abnormalities is too high. Eventually, after adjustment, the minimum sample size was set to 0.5% of the sample, at which point the percentage of anomalous trajectories was 8.71%, which is similar to the percentage of anomalies identified in other studies. Eight different clusters were identified, as shown in Fig. 2, with anomalous trajectories in gray.

Each color represents an air traffic flow, representing a different approach path into the terminal airspace. Four go-arounds were detected in the fifteen days, and these trajectories were marked as anomalies in the spatial clustering.

The spatial anomaly detection is realized while identifying air traffic flow, which provides a new perspective for terminal airspace anomaly detection and prediction. At the same time, the trajectory can be re-planned for unstable approaching aircraft based on air traffic flow.

4 Energy Management

4.1 Energy Safety Boundary

Additional metrics must be calculated for energy clustering. The data of the runway center extension is intercepted using air traffic flow and airport runway information, and the corresponding runway entry point is selected based on the aircraft's position and heading. Latitude and longitude are projected onto the UTM coordinate system, and the distance in nautical miles (NM) between each data point and the runway entry point is calculated. Only the final approach segment of the flight is taken into account, ignoring

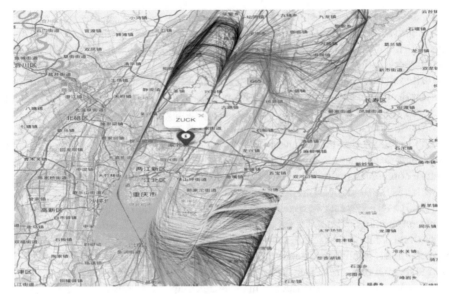

Fig. 2. Spatial clustering results, with different colors representing different air traffic flows.

the aircraft's mass variation, and the energy metrics are expressed in terms of altitude and ground speed [14]. Data points containing distance, altitude, or ground speed are used as inputs to DBSCAN clustering, and the clustering results are more concerned with the shape of the clusters.

Based on the shape of the cluster and the approach procedure of ZUCK, the energy safety boundary is established as Fig. 3. The green in the figure is determined by the shape of the cluster, the orange is warning, the red is critical. The warning and critical values are determined by analyzing the approach procedures and operational flight data. The data in Fig. 3 are from a normal landing aircraft.

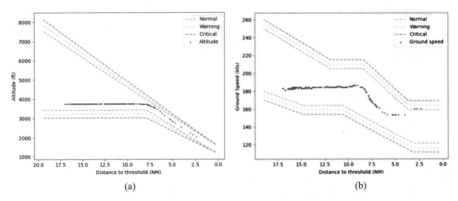

Fig. 3. Flight data of normal landing aircraft with energy safety boundaries. (a) Altitude and (b) Ground speed function of the distance to threshold.

The upper and lower parameters' limits are estimated using historical datasets, which can detect parameter anomalies but are not intuitive enough. To quantify the atypical scores of flight trajectories, the HDBSCAN and GLOSH algorithm are used.

4.2 Atypical Scores

Energy management is the most important challenge for aircraft during the landing phase. As a result, the anomaly of total aircraft energy is used to characterize the atypical approach during the final stages of flight before landing.

The E_T (total aircraft energy) is defined as follows in this study, and since only the landing phase is of concern, the aircraft mass variation is ignored:

$$E_T = E_K + E_P \tag{2}$$

$$E_K = \frac{1}{2}\left(G^2 + V^2\right) \tag{3}$$

$$E_P = g \cdot H \tag{4}$$

where E_K denotes specific kinetic energy, E_P denotes specific potential energy. The ground speed is G, the vertical speed is V, the gravity constant is g, and the altitude is H.

The flight trajectory was divided into multiple trajectory segments using a sliding window, each represented by data points of total energy as a function of distance. The segments were then subjected to PCA (principal component analysis), with the k first principal components retained. Finally, clustering is used to detect and quantify outliers, as well as to calculate outlier scores. The HDBSCAN algorithm is used, which can predict new data points as well as provide outlier scores based on the GLOSH algorithm [15]. The degree of abnormality is quantified on a scale of 0 to 1, with a higher score value indicating a greater distance from normal clustering and a closer proximity to abnormality.

Each movement of the sliding window is assigned a coefficient between 0 and 1. Following testing, the sliding window size is set to 0.3 NM, with an offset factor of 0.05 NM. The average coefficient of all sliding windows containing the point is used to calculate the final atypical score. The algorithm's entire process is depicted in the figure below (Fig. 4).

On October 6, 2020, Hainan Airlines 7409 go-around occurred at ZUCK, and its approach data before go-around was analyzed in relation to the energy safety boundary and atypical scores, as shown in Fig. 5. As shown in the graph, there is a small increase in ground speed around 10 NM, and the atypical scores increases accordingly. At about 7.5 NM, the altitude stopped falling and changed from level to climb, and the atypical scores sharply increased, at which time the aircraft had started to execute the go-around procedure. The red dots in the graph represent atypical scores, and the darker the red, the higher the score.

The aircraft has shown differences from normal landing before go-around, even beyond criticality, which can be useful for the detection and prediction of unstable

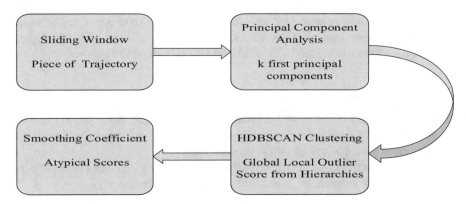

Fig. 4. The process of obtaining atypical scores.

approaches. If the atypical scores consistently exceed the threshold, or if the flight parameters exceed the safety boundary for an extended period of time, the flight is classified as abnormal, and a warning is issued. It is possible to use the energy safety boundary and atypical scores to monitor aircraft approach data to detect anomalies earlier, capture precursors, obtain more time for safety management, and realize the prediction of unstable approach and go-around.

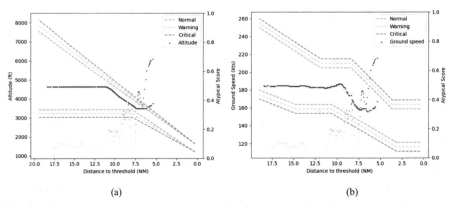

(a)   (b)

Fig. 5. Flight data of the go-around aircraft with energy safety boundaries and atypical scores. (a) Altitude and (b) Ground speed function of the distance to threshold.

5 Conclusions

The clustering algorithm is applied to the terminal airspace in this study, and the application of data-driven methods to aviation data is investigated. Air traffic flow and abnormal trajectories are identified spatially, which aids terminal airspace track planning and management.

For altitude and ground speed, energy safety boundaries were established, and the GLOSH algorithm was used for outlier scoring and validated with go-around data to demonstrate its effectiveness. This method can be used to monitor aircraft approach data, detecting anomalies earlier, catching precursors and gaining more time for safety management.

Future work will build on this to develop a complete tool to monitor the approach and landing phases, capture anomalous precursors, and provide path planning recommendations for unstable approach flights.

Acknowledgments. This work was supported by the Graduate Student Innovation Program-2021YJS019 of the Civil Aviation University of China. Thanks to the OpenSky Network, ADS-B data are freely available for research. The authors would also like to thank the members of the Tianjin Key Laboratory of Advanced Signal Processing for their support of this research.

References

1. Corrado, S., Puranik, T., Pinon, O., Mavris, D.: Trajectory clustering within the terminal airspace utilizing a weighted distance function. In: Proceedings of the 8th OpenSky Symposium pp. 1–10 (2020)
2. Blajev, T., Curtis, W.: Go-around decision-making and execution project. Flight Saf. Found. 3–6 (2017)
3. Haodong, Z.: Aircraft flight anomaly detection algorithm based on flight trajectory. Mod. Comput. (Prof. Ed.) **01**, 27–29 (2018)
4. Proud, S.R.: Go-around detection using crowd-sourced ADS-B position data. Aerospace **7**(2), 16 (2020)
5. Lu, Z., Zhang, S., Dai, R., Huang, M.: Abnormal energy risk criteria of large civil airplanes in approach and landing. Acta Aeronauticaet Astronautica Sinica **42**(6), 624132 (2021)
6. Corrado, S., Puranik, T., et al.: A clustering-based quantitative analysis of the interdependent relationship between spatial and energy anomalies in ADS-B trajectory data. Transp. Res. Part C: Emerg. Technol. **131**, 103331 (2021)
7. Fernández, A., et al.: Flight Data Monitoring (FDM) Unknown Hazards detection during Approach Phase using Clustering Techniques and AutoEncoders (2019)
8. Zhao, W., Li, L., et al.: An incremental clustering method for anomaly detection in flight data. Transp. Res. Part C: Emerg. Technol. **132**, 103406 (2021). https://doi.org/10.1016/j.trc.2021.103406
9. Xie, J., Sun, H., Wang, C., Lu, B.: Analysis of influence factors for unstable approach in fine-grained scale. Geomatics Inf. Sci. Wuhan Univ. **46**(8), 1201–1208 (2021)
10. Xu, Z., Lu, X., Zhang, Z.: Aircraft go-around detection employing open source ADS-B data. In: 2021 IEEE 3rd International Conference on Civil Aviation Safety and Information Technology (ICCASIT), pp. 259–262 (2021). https://doi.org/10.1109/ICCASIT53235.2021.9633714
11. Schafer, M., Strohmeier, M., Lenders, V., et al.: Bringing up OpenSky: a large-scale ADS-B sensor network for research. In: Ipsn-14 International Symposium on Information Processing in Sensor Networks, pp. 83–94. IEEE (2014)
12. Basora, L., et al.: Occupancy Peak Estimation from Sector Geometry and Traffic Flow data (2018)

13. Ester, M., Kriegel, H.-P., Sander, J., Xu, X.: A density-based algorithm for discovering clusters in large spatial databases with noise. In: Second International Conference on Knowledge Discovery and Data Mining, pp. 226–231. AAAI Press (1996)
14. Jarry, G., et al.: Aircraft atypical approach detection using functional principal component analysis (2018)
15. Campello, R.J.G.B., Moulavi, D., Sander, J.: Density-based clustering based on hierarchical density estimates. In: Pei, J., Tseng, V.S., Cao, L., Motoda, H., Xu, G. (eds.) PAKDD 2013. LNCS (LNAI), vol. 7819, pp. 160–172. Springer, Heidelberg (2013). https://doi.org/10.1007/978-3-642-37456-2_14

Research on Pedestrian Detection and Recognition Based on Improved YOLOv6 Algorithm

Zeqiang Sun[1] and Bingcai Chen[1,2(✉)]

[1] College of Computer Science and Technology, Xinjiang Normal University, Urumqi 830054,
China
china@dlut.edu.cn
[2] School of Computer Science and Technology, Dalian University of Technology,
Dalian 116024, Liaoning, China

Abstract. Pedestrian detection is a technology that uses computer vision to determine whether there are pedestrians passing by in the video sequence or pictures, and realizes the positioning of pedestrians. It is an important task in manless driving, automobile intelligence and intelligent monitoring. Aiming at the problems of low efficiency of pedestrian detection and slow running speed on small devices, a YOLOV6-SE (Yolo Look Only Once-SE) model is constructed to detect pedestrian targets。the mAP detected by pedestrians reaches 85.1%, which is 7.3% points higher than that of the original YOLOv6 without adding attention mechanism. The three-layer channel attention module is added to backbone(RepVGG) adopted by YOLOv6, which enables the backbone network to extract more feature information, thus improving the accuracy of the network. After experimental comparison, Good detection performance is achieved.

Keywords: Computer vision · Pedestrian detection · Deep learning · YOLOv6 algorithm · Attention mechanism

1 Introduction

In recent years, with the rapid development of manless driving, intelligent monitoring and intelligent robots, pedestrian detection and tracking technology has also become a research hotspot. As a branch of object detection, pedestrian detection has important application value in security monitoring system, intelligent driving, intelligent transportation and other fields. The pedestrian detection still exist some problems and difficulties, for example, block phenomenon between pedestrians, pedestrian detection, special light conditions and the pedestrian detection for small targets, these are serious influence the pedestrian detection accuracy and real-time performance, therefore, a high precision, fast detection of pedestrian detection algorithm is very necessary.

Traditional pedestrian detection algorithms often have problems such as missed detection and false detection due to the lack of deep network layers. The method based

© The Author(s), under exclusive license to Springer Nature Singapore Pte Ltd. 2023
Q. Liang et al. (Eds.): AIC 2022, LNEE 871, pp. 281–289, 2023.
https://doi.org/10.1007/978-981-99-1256-8_33

on deep learning can improve the accuracy of pedestrian detection and be more adaptive and robust. Usually, the object detection algorithm of deep learning is divided into two stages: object detection algorithm of two stages and object detection algorithm of one stage. The object detection algorithm of two stages first generates the candidate boxes of the object, generates the foreground and background prediction information of the image, and then performs classification and boundary regression on these candidate boxes. Representative algorithms include RCNN [2], Fast RCNN [3], Faster RCNN [4], etc. Because the two-stage target detection algorithm based on candidate box has complex network and slow speed, it is difficult to meet the real-time requirements. Therefore, the one-stage algorithms based on the idea of regression are gradually emerging, representative algorithms include SSD [5], YOLO [6], YOLOv3 [7], etc. Among them, SSD algorithm adopts the fusion of low-level feature information and high-level feature information, and YOLO algorithm adopts the network model based on the idea of regression to divide the image into small squares. The center of the small square is taken as the center point of the candidate box, and then the confidence of each prediction box is obtained. The prediction box with low confidence is filtered out by the threshold, and the final prediction box is obtained by non-maximum suppression.

In order to improve the pedestrian detection effect in the low-light environment, Zhang Mingzhen [8] proposed a dense-YOLO network based underground pedestrian detection model, and through Gamma transform, weighted logarithmic transform and other methods to enhance the light map data, reduce the missed detection rate of underground pedestrian targets. Han Song et al. [9] added a detection layer on the basis of the original YOLOv3 to improve the recognition accuracy of small-sized pedestrian targets. Wang Lihui et al. [10] replaced the YOLOv3 backbone network with GhostNet, retained the multi-scale detection part, and used GIOU to guide the regression task, which greatly improved the detection accuracy and speed.

Aiming at the problems of missing pedestrian target detection caused by pedestrian occlusion and low efficiency of pedestrian target detection, this paper proposes a YOLOv6-SE algorithm. The main work is as follows:

1. The YOLOv6 model is improved, and the RepVGG module, the backbone network of the original YOLOv6 algorithm, is improved into RepVGG-SE [11] network. By adding attention module, the region of interest is located, the attention is focused on the features to be paid attention to, and some unimportant information is suppressed.
2. A more efficient structure (Decoupled Head) was used and designed briefly to decouple the classification and regression branches, so as to further improve the accuracy of the network detection.
3. SIOU loss function is used to not only consider the overlap between the prediction box and the target box, the distance of the center point, aspect ratio and other factors, but also add Angle penalty to effectively reduce the loss of degrees of freedom.

2   YOLO Algorithm and Attention Mechanism

2.1   YOLO

YOLOv1 is the first work of YOLO(You Only Look Once) model, which was proposed by Redmon J Equal in 2016. Since then, the target detection model of YOLO series has gradually become mature. With continuous improvement and optimization, the detection accuracy of YOLO model has been continuously improved, and the detection performance has been getting higher and higher. YOLOv3 algorithm adds FPN(Feature Pyramid Network) on the basis of YOLOv2 algorithm to make the extracted features better fused. YOLOv4 algorithm adopts the backbone Network structure of CSPDarknet53. The Neck part adopts FPN + PAN(Path Aggregation Network), which effectively improves the accuracy of Network prediction. The Head part detects the large, medium and small targets respectively. YOLOv5 algorithm is created and maintained by Ultralytics, which uses many tricks on the basis of YOLOv4 algorithm, including Mosaic data enhancement, Spatial Pyramid Pooling (SPP) structure and so on. YOLOv6 algorithm is a target detection framework developed by Meituan Visual Intelligence Department. Backbone is composed of multiple RepVGGblocks. Backbone not only makes efficient use of hardware, but also has strong representation ability. At the same time, the feature fusion network of RepPAN is designed, the CSPBlock used in YOLOv5 is replaced by RepBlock, and the detection head part adopts a simple and efficient decoupling head structure.

2.2 SE Attention Module

SENet [11]is a new image recognition architecture unveiled by autonomous driving company Momenta in 2017. It models the correlation between feature channels and intensifies importantfeatures to improve accuracy. Its specific operations are as follows:

1. Given an input whose number of feature channels is C, a feature whose number of feature channels is C can be obtained after a series of general transformations such as convolution.
2. 2.Squeeze: Squeeze features along the spatial dimension and turn each two-dimensional feature channel into a real number, which has a global receptive field to some extent and the output dimension matches the input number of feature channels.
3. 3.Excitation: Based on the correlation between the feature channels, each feature channel generates a weight that represents the importance of the feature channel.
4. 4.Reweight: The weight of Excitation output is regarded as the importance of each feature channel. Then, the weight of Excitation output is weighted to the previous feature channel by multiplication, which completes the re-calibration of the original feature in the channel dimension (Fig. 1).

2.3 The Improved Network Structure

YOLOv6 mainly makes many improvements in Backbone, Neck, Head and training strategies. It adopts the hardware-friendly Backbone network design. At the same time, inspired by the design idea of hardware-aware neural network, YOLOv6 designs a more efficient Backbone and Neck structure. Based on RepVGG style, reconfigurable parameterized and more efficient Backbone:EfficientRep and Neck: rep-pan are designed. RepVGG [4] style structure is a kind of multi-branch topology in training. In actual deployment, it can be equivalent to be fused into a single 3x3 convolution, which is a

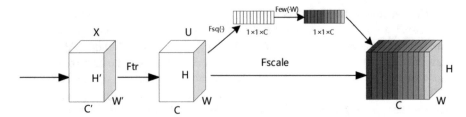

Fig. 1. A Squeeze-and-Excitation Block

reparameterized structure. Through the fused 3x3 convolution structure, the computing power of computationally intensive hardware (such as GPU) can be effectively utilized. Also available is the highly optimized NVIDIA cuDNN and Intel MKL compilation frameworks on Gpus/cpus. Experiments show that, through the above strategy, YOLOv6 reduces the delay on the hardware, significantly improves the accuracy of the algorithm, and makes the detection network faster and stronger.

EfficientRep Backbone: In the aspect of Backbone design, we design an efficient Backbone based on Rep operator. Compared with CSP-backbone adopted by YOLOv5, this Backbone can not only efficiently utilize the computing power of hardware (such as GPU), but also have strong representation ability. The common Conv layer of Backbone with stride = 2 is replaced by the RepConv layer with stride = 2. At the same time, the original CSP-blocks are redesigned as Repblocks, where the first RepConv of the RepBlock will do the channel dimension transformation and alignment. In addition, we optimized the original SPPF for a more efficient SimSPPF.

Rep-pan: In terms of Neck design, in order to make it more efficient in hardware reasoning and achieve a better balance between accuracy and speed, we designed a more effective feature fusion network structure for YOLOv6 based on the hardware sensing neural network design idea. Based on the PAN topology, REP-PAN replaces the CSP-block used in YOLOv5 with RepBlock. At the same time, the operators in the whole Neck are adjusted to achieve efficient reasoning in hardware and maintain good multi-scale feature fusion ability (Fig. 2).

In YOLOv6, we adopt the decoupled detection header structure and simplify its design. The detection head of the original YOLOv5 is realized through the fusion and sharing of classification and regression branches, while the detection head of YOLOX decouples the classification and regression branches and adds two additional 3x3 convolutional layers. Although the detection accuracy is improved, the network delay is increased to a certain extent. Therefore, we simplify the design of decoupling head. At the same time, considering the balance between the representation ability of relevant operators and the computational cost of hardware, we redesign a more efficient decoupling head structure by using Hybrid Channels strategy, which reduces the delay while maintaining the accuracy. Alleviate the extra delay overhead caused by 3x3 convolution in decoupled head. The simplified decoupling head structure is shown in Fig. 3 below.

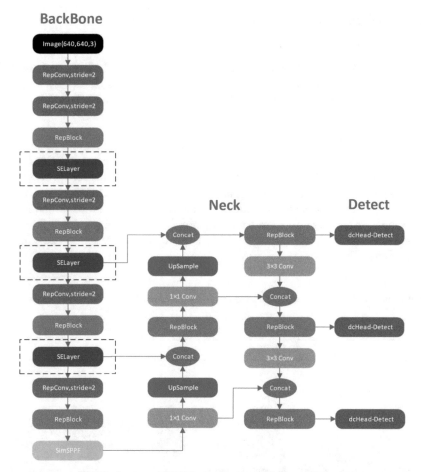

Fig. 2. Improved YOLOv6 network architecture diagram

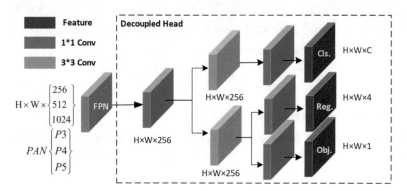

Fig. 3. YOLOv6 simplified decoupled header

2.4   Experimental Results and Analysis

2.4.1   Experimental Platform

The experiment of this paper is completed under Windows platform, and the hardware information configuration is as follows: CPU: Intel(R) Core(TM) I3-10100F CPU@3.60GHz; Graphics processor (GPU):NVIDIA GeForce RTX 2080 TI.

2.4.2   Experimental Data

The data set used in this paper is VOC data set. The original data set is divided into 20 different categories, including 'aeroplane', 'Bicycle', 'bird', etc. In order to better detect pedestrians, the image data containing only pedestrians is extracted from this data set, and a total of 9583 images are screened. The data are divided into 6132 training sets, 1534 validation sets and 1917 test sets. The information of training set, validation set and test set is shown in the following Table 1 the representative images are shown in Fig. 4 below.

Table 1. Number of images in pedestrian data set division, training set, validation set and test set

training sets	validation sets	test sets
6132	1534	1917

Fig. 4. Images of pedestrians in the data set

2.4.3   Evaluation Criteria

MAP value was used as the evaluation index in the experiment.mAP is the mean of the average accuracy of all categories, and its calculation formula is (1).

$$mAP = \frac{\sum_{i=1}^{n} AP_i}{N} \tag{1}$$

where, N is the number of category detection, and AP can comprehensively consider the influence of accuracy and recall rate.PR curve can be obtained by taking accuracy as the vertical axis and recall rate as the horizontal axis. For continuous PR curve, AP in Eq. (2) is

$$AP = \int_0^1 P(R)dR \tag{2}$$

where P is the Precision (3) and R is the Recall (4)

$$Precision = \frac{TP}{TP + FP} \tag{3}$$

$$Recall = \frac{TP}{TP + FN} \tag{4}$$

FPS is used as the standard to measure the speed of pedestrian detection by the model. The larger FPS is, the more images are detected per unit time, and the better real-time performance of the model is achieved. The FPS calculation formula is as shown in formula (5), where NumFigure represents the number of images detected and TotalTime represents the TotalTime spent on detecting these images.

$$FPS = \frac{TotalTime}{NumFigure} \tag{5}$$

2.4.4 Evaluation Setting

As can be seen from the following Table 2, after adding SE attention mechanism to the YOLOv6 backbone network, the mAP detected by pedestrians reaches 85.1%, which is 7.3% points higher than that of the original YOLOv6 without adding attention mechanism, and the improvement effect is quite obvious. Meanwhile, the detection speed FPS value reaches 178. Compared with the original algorithm, the real-time performance is also significantly improved. The use of YOLOv5 algorithm is not good in pedestrian detection. Compared with YOLOv5s version, after increasing the network depth and width, the mAP of YOLOv5l version is increased by 1.8% points, but it brings a serious decline in FPS and is obviously poor in real-time performance.

When trying to replace YOLOv5 backbone network, the original CSPdarknet53 network is replaced by MobileNetv3 network. The mAP decreases by 2.9% points, and the FPS also decreases, so the network performance after replacement is not good. By comparing the improved YOLOv6 algorithm with SSD algorithm and FasterRCNN algorithm, it can be seen that the detection accuracy of SSD algorithm is not as good as that of the proposed algorithm. On the basis of mAP reduction of 1.5% points, FasterRCNN sacrifices a lot of detection speed, and FPS is only 12% of YOLOv6_SE algorithm. Therefore, it does not meet the actual demand in real time. According to the comparison in Table 2, YOLOv6_SE improves the average accuracy mAP and the detection speed is also very fast. This algorithm is suitable for the scene of pedestrian detection.

The detection diagram is shown in Fig. 5 below:

Table 2. Comparison of detection performance of various target detection algorithms

Model	defect recognition accuracy		mAP	GFLOPS	detection speed/fps
	P	R			
YOLOv5s	0.757	0.785	0.796	15.8	93.458
YOLOv5l	0.785	0.773	0.814	107.8	52.356
YOLOv5s_SE	0.776	0.759	0.797	16.5	78.125
YOLOv5_MobileNetv3	0.729	0.763	0.767	38.2	63.291
YOLOv6s	0.780	0.665	0.778	44.19	172.413
SSD	0.531	0.705	0.804	–	112.764
FasterRCNN	0.853	0.756	0.836	–	22.290
YOLOv6s_SE	**0.731**	**0.793**	**0.851**	**44.09**	**178.300**

Fig. 5. Schematic diagram of pedestrian image detection results

References

1. Xiang, S., Wang, L., Jia, C., Jian, Y., Ma, X.: Improve YOLO sheltered pedestrian detection simulation. J. Syst. Simul. 1–15 (2022). https://doi.org/10.16182/j.issn1004731x. Joss. 21–0915
2. Wang, K., Zhou, W.: Pedestrian and cyclist detection based on deep neural network fast R-CNN. Int. J. Adv. Rob. Syst. **16**(2), 1–10 (2019)
3. Girshick, R.: Fast R-CNN. In: Proceedings of the IEEE International Conference on Computer Vision, pp. 1440–1448. IEEE, Los Alamitos (2015)
4. Ren, S., He, K., Girshick, R., et al.: Faster R-CNN: Towards real-time object detection with region proposal networks. IEEE Trans. Pattern Anal. Mach. Intell. **39**(6), 1137–1149 (2017)
5. Liu, W., et al.: SSD: single shot MultiBox detector. In: Leibe, B., Matas, J., Sebe, N., Welling, M. (eds) Computer Vision – ECCV 2016. ECCV 2016. Lecture Notes in Computer Science, vol. 9905, pp. 145-159. Springer, Cham (2016). https://doi.org/10.1007/978-3-319-464 48-0_2

6. Redmon, J, Divvala, S, Girshick, R, et al.: You only look once: unified, real-time object detection. In: Proceedings of the IEEE/CVF Conference on Computer Vision and Pattern Recognition, pp. 779–788. IEEE, Los Alamitos (2016)
7. Redmon, J., Farhadi, A.: YOLOv3: an incremental improvement. In: Proceedings of 2018 IEEE International Conference on Computer Vision and Pattern Recognition, pp. 1–9. IEEE, Washington, DC (2018)
8. Mingzhen, Z.: Underground pedestrian detection model based on dense-yolo network. Ind. Autom. 13(3), 86–90 (2022). https://doi.org/10.13272/j.iSSN.1671-251-x.,17861
9. Song, H., Guojun, M.: Improved multi-scale feature fusion of pedestrian detection algorithm. Light. Control 1–7 (2022). http://kns.cnki.net/kcms/detail/41.1227.TN.20220422.1919.013.html
10. Wang, L., Yang, X., Liu, H., Huang, J.: Pedestrian detection and tracking algorithm based on GhostNet and attention mechanism. Data Acquisit. Process. 37(01), 108–121 (2022)
11. Hu, J., Shen, L., Sun, G.: Squeeze-and-excitation networks. In: 2018 IEEE/CVF Conference on Computer Vision and Pattern Recognition, Salt Lake City, UT, USA, New York, 18–23 June 2018, pp. 7132–7141. IEEE (2018)

Jetson Nano-Based Pedestrian Density Detection in Subway Stations

Cheng Chen and Wei Wang[✉]

Tianjin Key Laboratory of Wireless Mobile Communications and Power Transmission,
Tianjin 300387, China
weiwang@tjnu.edu.cn

Abstract. Computer Vision is one of the popular research in the field of Deep Learning. As a hotspot in the field of computer vision research in recent years, pedestrian detection has a wide application prospect and great economic value in urban transportation, social public safety, national defense security construction, and so on. In this paper, the Yolov4-Tiny algorithm is used as the basic algorithm to implement pedestrian detection in crowded subway stations, and the network structure, detection principle, and training process of the algorithm are studied in detail. The map value achieved by the model after training is 84.42%, and pedestrian images are selected and then tested with good results. In this paper, the trained model is tested on the embedded motherboard Jetson Nano, and the processing speed fps value is about 7, which can meet the requirements of real-time pedestrian detection.

Keywords: Computer vision · Deep learning · Pedestrian detection · Yolov4-Tiny · Jetson Nano

1 Introduction

As one of the main ways for people to travel, the subway has become the primary choice for people with its advantages of speed and convenience. With the rapid increase in the number of people traveling by subway, the crowding of passengers in subway stations has become more and more serious, especially during holidays and the daily morning and evening rush hours. Based on the above problems, the subway station urgently needs a real-time monitoring system for human flow density, and the technical core of this system is the target detection of pedestrians.

With the rapid development of Internet technology, the research and development in the field of computer vision [1] have become more and more mature, and pedestrian target detection [2], as one of the important branches of computer vision, has a wide range of application prospects and economic value. The early traditional target detection algorithms mainly include the frame difference method, optical flow method, background subtraction method [3], etc. A single traditional detection algorithm is not able to achieve the effect we want, and often requires two or more algorithms combined to achieve the demand, which greatly increases the complexity and computational

Q. Liang et al. (Eds.): AIC 2022, LNEE 871, pp. 290–297, 2023.
https://doi.org/10.1007/978-981-99-1256-8_34

power of the algorithm, and also increases the requirements for hardware performance, so the traditional detection algorithm is not suitable for application in practical scenarios. AlexNet network won the 2012 ImageNet Since then, deep learning has reached a rapid development stage, and Convolutional Neural Network [4] (CNN) is also applied to target detection, and target detection based on deep learning is mainly divided into two categories, one is Two-stage [5] target detection algorithm, which mainly generates a series of candidate frames by algorithm first, and then uses The representative Two-stage algorithms include R-CNN, Fast R-CNN, Faster R-CNN, etc. The other category is the One-stage [5] target detection algorithm, which combines the generation of candidate frames and the classification of samples into one step, and its representative algorithms are Yolo series algorithms. Compared with the Two-stage target detection algorithm, the One-stage target detection algorithm has more excellent detection speed and is more suitable for application in practical scenarios.

The rest of the paper is as follows. In Sect. 2, we mainly introduce the basic theoretical knowledge such as convolutional neural networks; In Sect. 3, we introduce the Yolov4-Tiny algorithm and its network structure; In Sect. 4, we migrate the trained pedestrian model to the Jetson Nano motherboard; Sect. 5 is the conclusion.

2 Theoretical Basis

2.1 Convolutional Neural Network

A convolutional neural network is a Multilayer Perceptron (MLP) [6] with two additional parts, the Convolutional Layer and the Pooling Layer, on top of the MLP. A convolutional neural network is generally composed of an input layer, a convolutional layer, a pooling layer, and a fully connected layer. The final result is obtained by extracting the features of the image and dimensionality reduction for images [7] through the computation of convolution and pooling, and by extracting the features of the image through a fully connected layer for classification.

(1) Input layer

When we input an image into the neural network, the image needs to be pre-processed to reduce the impact on training caused by image noise and image overexposure, etc. There are usually three types of processing: homogenization, normalization, and PCA dimensionality reduction [8].

(2) Convolutional layer

The convolution layer is the most important layer of the convolutional neural network, and its main function is to extract the features of the input image. The convolutional calculation mainly involves two parameters: the size of the convolutional kernel and the step size of each move. Therefore, a certain number of pixels are filled in the input to ensure that the size of the output feature map remains unchanged. The output feature size after convolution is calculated as in Eqs. 2, 3, 4.

$$n_{out} \times n_{out} = \left[\frac{N + 2p - n}{s} + 1\right] \times \left[\frac{N + 2p - n}{s} + 1\right] \tag{1}$$

where n_{out} is the output size, N is the input size, p is the number of zero padding, n is the convolution kernel size, and s is the step size of the move.

(3) Pooling layer

Pooling is mainly a process of down-sampling, which can reduce the dimensionality of each feature map and thus retain most of the important information. Common pooling operations include Average Pooling and Maximum Pooling.

(4) Fully-connected layer

The fully connected layer is usually placed at the end of the convolutional neural network and consists of multiple neurons. Its main role is to combine feature information to achieve the classification of image features.

3   Yolov4-Tiny Algorithm-Based Pedestrian Detection Counting

3.1   Introduction of the Yolov4-Tiny Algorithm

Yolov4 algorithm was proposed in 2020, compared with the previous Yolo series algorithm, it has made certain changes from data enhancement, backbone network, activation function, network training, etc., and the detection accuracy and detection speed have been significantly improved, which can be called a quality algorithm with both accuracy and speed. The yolov4-Tiny algorithm is a lightweight version of the Yolov4 algorithm It mainly reduces the number of residual blocks and data stacking in the backbone network structure of the Yolov4 algorithm and keeps the CSP Net [9] and Res Net [9] in the backbone network, so that the lightened model has good detection accuracy, and the network parameters of Yolov4-Tiny are about The network structure of the Yolov4-Tiny algorithm is shown in Fig. 1.

As we can see from the Fig. 1, the Yolov4-Tiny network consists of three parts: the CSPDarknet53-Tiny backbone network, the FPN data enhancement network, and the Yolo Head. The backbone network consists of three DarknetConv2D_Leaky and three Resblok_body stacked, and the Data Enhancement Network has only Convolution and one Up-sampling process, so the network structure of the Yolov4-Tiny algorithm can be very simple, which makes it have a very good detection speed, but there is a loss in the accuracy of small target detection.

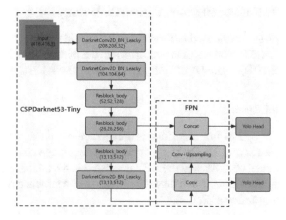

Fig. 1. Yolov4-Tiny algorithm network structure

3.2 Evaluation Criteria

The metrics for evaluating the accuracy of the training model are Average Precision (AP) [10] and Mean Average Precision (mAP) [10], where the formula for calculating mAP is in Eq. 3–1.

$$mAP = \frac{1}{n} \sum_{i=1}^{n} AP_i \qquad (2)$$

where n is the number of target categories to be detected. Since only the category of pedestrians is detected in this paper, the mAP value is the same as the AP value.

The Precision is the proportion of correctly predicted sample frames in the total sample frames, and the Recall is the proportion of correct sample frames in all real frames, which is calculated as follows.

$$\text{Precision} = \frac{TP}{FP + TP} \qquad (3)$$

$$\text{Recall} = \frac{TP}{FN + TP} \qquad (4)$$

TP is the number of models correctly detected and classified to the corresponding species, FP is the number of targets detected wrong by the model, and FN is the number of targets missed by the model.

The Average Precision (AP) is determined by both Precision and Recall, which is calculated as follows.

$$AP = \int_{0}^{1} P(R)dR \qquad (5)$$

where P is the precision value, R is the recall value, and the AP value is equivalent to the integral of Precision to Recall.

3.3 Experimental Environment and Parameter Settings

The deep learning framework used is TensorFlow2.2.0, and the image processing is performed by OpenCV4.2.0. The editor is pycharm, and the python libraries: Numpy1.18.4, Scipy1.4.1, H5py2.10.0, Tqdm4.46.1, Pillow8.2.0, Matplotlib3.2.1.

3.4 Experimental Results

After training, the YOLOv4-tiny algorithm is used to achieve the detection of pedestrian density, and when the Score_Threshold is set to 0.5, the accuracy value can reach 91.23%, the recall rate can reach 65.04%, and the AP value can reach 84.42%. The changes in AP value, accuracy, and recall rate when the model reaches convergence during the training process are shown as follows (Fig. 2).

(a) AP values after trainng   (b) Accuracy values after training   (c) Recall values after training

Fig. 2. Train effect

4   Embedded Jetson Nano-Based Human Flow Density Detection Implementation

4.1 The Choice of Embedded Hardware

Nowadays, there are many kinds of embedded devices on the market, such as Raspberry Pi series, NVIDIA TX2, Xavier, and Jetson Nano, in terms of processing data and computing power, Jetson Nano 4GB is equipped with 128-core NVIDIA Maxwell GPU and Quad-Core The Jetson Nano 4GB is equipped with a 128-core NVIDIA Maxwell GPU and a Quad-Core ARM Cortex-A57MPCore CPU and has good data processing capability. The small size is the most obvious advantage of the Jetson Nano, which is 100 x 80 x 29mm and can be flexibly used in real-world scenarios. In terms of power consumption, the Jetson Nano can be powered by Micro USB with a minimum power of 5W, which is sufficient for simple operations, and up to 10W when powered by the DC port, which allows complex data processing and other operations, and overall the Jetson Nano has low power consumption. It also has good video encoding and decoding capabilities, itself equipped with 16 GB of eMMC5.1 flash memory, and an external hard disk to expand the space while having a variety of interfaces for development, specific parameters and specifications are shown in the table below (Table 1).

Table 1. Jetson Nano hardware parameter specifications

Components	Parameter Specification
CPU	Quad-Core ARM Cortex-A57MPCore
GPU	128-core NVIDIA Maxwell GPU
AI Arithmetic	473 GFLOPS
Video Memory	4GB 64-bit(LPDDR4 25.6GB/s)
Storage	16GB eMMC5.1 Flash Memory
Video Encoder	4K@30I 4 × 1080p@30 I 9 × 720@30 (H.264/H.265)
Video decoder	4K@60I 2 × 4K@30I 8 × 1080p@30 I 18 × 720p@30(H.264/H.265)
Camera interface	2个MIPI CSI-2 DPHY Passage
Connections	Gigabit Ethernet、M.2 Key E
Show	HDMI I DisplayPort
USB	4*USB 3.0
Network	Support USB high-speed network card I Support M.2 dual-band high-speed network card
Power consumption	5-10W

4.2 Environment Configuration

First, connect the Jetson Nano to a Win10 computer, and then burn the boot system in eMMC through an Ubuntu 18.04 virtual machine to prepare for the subsequent boot. The prepared USB stick is then plugged into the computer and the Win32DiskImaager software is used to burn the system on the motherboard. The final access to the power supply and monitor, etc. can be booted and used, at this time the operating system for Linux Ubuntu18.06.

4.3 System Operation Test and Analysis

After configuring the required environment, the trained weight files and so on are migrated to Jetson Nano, and the effect of calculating the number of people is achieved by calculating the number of recognition frames. The scene of the test video is at the escalator in the subway station, and the effect after running the program is as follows (Fig. 3).

The total number of people in the frame is 18, and the number of missed detections is 4. The accuracy of detection is 77.78%, while the speed of detection reaches 7FPS, which can meet the demand for real-time detection.

Fig.3. Detection effect on Jetson Nano

5 Conclusion

This paper mainly realized the detection of passengers in the specified area of subway station by the Yolov4-Tiny algorithm, and migrated the weight file to the embedded board Jetson Nano, after configuring the environment of TensorFlow, OpenCV, python, etc., the accuracy rate is 77.78% in the embedded end of the test, and the recognition speed can reach 7FPS, which can be The actual scene can be detected in real-time by installing a USB or CSI camera on the Jetson Nano motherboard. The algorithm in some specific angle recognition still needs to be improved, the subsequent data set can be reconstructed and increase the number of samples to solve, can also replace the algorithm to Yolov5. In order for the system can be better in the actual scene application, the detection speed of the algorithm still needs to be improved, the subsequent can be considered through TensorRT and other acceleration processes.

Acknowledgement. The work was supported by the Natural Science Foundation of China (61731006, 61971310).

References

1. Weinstein, B.G.: A computer vision for animal ecology. J. Animal Ecol. **87**(3), 533–545 (2018)
2. Zhang, S.D., Shao, Y., Mei, Y., et al.: Using YOLO-based pedestrian detection for monitoring UAV. In: Tenth International Conference on Graphics and Image Processing (ICGIP 2018), pp. 1141–1145. SPIE, 11069 (2019)
3. Al-Heety, A.T.: Moving vehicle detection from video sequences for traffic surveillance system. ITEGAM-JETIA **7**(27), 41–48 (2021)
4. Putra, M.H., Yussof, Z.M., Lim, K.C., et al.: Convolutional neural network for person and car detection using yolo framework. J. Telecommun. Electr. Comput. Eng. (JTEC), **10**(1–7), 67–71 (2018)

5. Lu, X., Li, Q., Li, B., Yan, J.: MimicDet: Bridging the Gap Between One-Stage and Two-Stage Object Detection. In: Vedaldi, A., Bischof, H., Brox, T., Frahm, J.M. (eds) Computer Vision – ECCV 2020. ECCV 2020. Lecture Notes in Computer Science, vol. 12359, pp. 541-557. Springer, Cham (2020). https://doi.org/10.1007/978-3-030-58568-6_32
6. Agrawal, P., Girshick, R., Malik, J.: Analyzing the Performance of Multilayer Neural Networks for Object Recognition. In: Fleet, D., Pajdla, T., Schiele, B., Tuytelaars, T. (eds) Computer Vision – ECCV 2014. ECCV 2014. Lecture Notes in Computer Science, vol. 8695, pp. 329–344. Springer, Cham (2014). https://doi.org/10.1007/978-3-319-10584-0_22
7. Paul, A., Chaki, N.: Dimensionality reduction of hyperspectral images using pooling. Pattern Recogn. Image Anal. 29(1), 72–78 (2019)
8. Wang, D., Gu, J.: VASC: dimension reduction and visualization of single-cell RNA-seq data by deep variational autoencoder. Genom. Proteom. Bioinform. 16(5), 320–331 (2018)
9. Wang, C.Y, Liao, H.Y.M., Wu, Y.H., et al.: CSPNet: a new backbone that can enhance learning capability of CNN. In: Proceedings of the IEEE/CVF Conference on Computer Vision and Pattern Recognition Workshops, pp. 390–391 (2020)
10. Horzyk, A., Ergün, E.: YOLOv3 precision improvement by the weighted centers of confidence selection. In: 2020 International Joint Conference on Neural Networks (IJCNN), pp. 1–8. IEEE (2020)

IVOCT Image Based Coronary Artery Stent Detection Reconstruction

Wei Xia, Tingting Han[(✉)], Jing Gao, and Hongmei Zhong

Tianjin Key Laboratory of Wireless Mobile Communications and Power Transmission, Tianjin Normal University, Tianjin 300387, China
hanting608@163.com

Abstract. Coronary heart disease has become a disease with the highest mortality rate in the world. The main treatment for coronary artery disease is coronary stenting. The detection and reconstruction of stents after coronary stent implantation is currently a difficult problem. In this paper, a reconstruction method for coronary stent detection based on IVOCT images was proposed. This method utilized the continuity of IVOCT image frames and the columnar shadow of the stent to realize two-dimensional and three-dimensional reconstruction of the stent. Median filtering, Otsu method, Canny method, pixel fetching and coordinate conversion were used. It was verified that the reconstruction of stent images could be achieved accurately when the frame spacing of IVOCT images satisfies the sampling theory. The stent recognition reconstruction method could be used in clinical applications to assist physicians in post-stenting status analysis.

Keywords: Stent identification · Stent reconstruction · Boundary segmentation · IVOCT

1 Introduction

Coronary atherosclerotic heart disease resulted from the deposition of lipids in the coronary arteries. Coronary stent implantation had become an important treatment for atherosclerotic heart disease [1]. Coronary arteries were hard to detect due to their stenoses and thin walls. Intravascular Optical Coherence Tomography (IVOCT) systems were widely used in coronary artery detection [2].

Automatic detection and segmentation of stents had been extensively investigated [3]. Nam et al. [4]. Proposed a segmentation method based on fuzzy C-mean (FCM) clustering and wavelet transform to inner lumen contour extraction. Tsantis et al. [5]. Proposed a segmentation technique for automatic lumen extraction and stent structure detection in IVOCT images to quantitatively analyze neoplastic endothelial hyperplasia. Wang et al. [6]. Proposed a greyscale based method that used greyscale and gradient features and directed segmentation of struts using a series of thresholds. However, the brightness and contrast of IVOCT images varied from patient to patient and blood artifacts may exist, making it difficult to apply in practice.

In this paper, a reconstruction method for coronary stent detection based on IVOCT images was proposed. This method utilized the continuity of IVOCT image frames and

Q. Liang et al. (Eds.): AIC 2022, LNEE 871, pp. 298–304, 2023.
https://doi.org/10.1007/978-981-99-1256-8_35

the columnar shadow of the stent to realize the two-dimensional and three-dimensional reconstruction of the stent. Noise points of image were removed by image processing and median filtering. Otsu method and Canny method were used to extract blood vessel boundary. The extraction of the position information of the stent was achieved by summing 100 pixels below the boundary. Coordinate transformation and arrangement of stent position information, reconstruction of 2D and 3D image of stent were based on the continuity of the image frames. The experimental results showed that the method achieved stent detection reconstruction when the IVOCT image acquisition speed satisfied the sampling theory.

2 Method

2.1 Median Filtering

Median filtering was a non-linear method of image noise reduction [7]. Median filtering selected appropriate points to replace the values of contaminated points and protected the sharp edges of the image. Median filtering selected a 3*3 matrix of nine pixel points in the image. We sorted the nine pixels and assigned the centroid of this matrix to the median of these nine pixels. The principle was shown in Fig. 1.

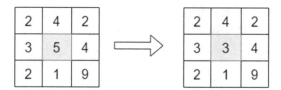

Fig.1. Median filtering principle.

2.2 Otsu Method

The Maximum inter class variance method (Otsu) was an algorithm for determining image segmentation thresholds [8]. The Otsu method automatically selected a global threshold T by counting the histogram characteristics of the entire image. Pixels in the grey level 0-i were foreground pixels, pixels in the grey level i-255 were background pixels and i was the grey level. The global threshold T was the maximum value of the inter-class variance corresponding to i.

The formula for the pixel ratio of foreground to background calculated proportion of foreground and background pixels ω_0 and ω_1. The proportion was computed as:

$$\omega_0 = N_0/(M*N) \tag{1}$$

$$\omega_1 = N_1/(M*N) \tag{2}$$

M and N were the length and width of the image, N_0 was the number of pixels with a grey value greater than the threshold, N_1 was the number of pixels with a grey value less than the threshold. The above parameters satisfy the constant equation:

$$N_0 + N_1 = M * N \tag{3}$$

$$\omega_0 + \omega_1 = 1 \tag{4}$$

The average gray value of the image μ was determined by the foreground pixels and the background pixels. The average gray value was computed as:

$$\mu = \omega_0 * \mu_0 + \omega_1 * \mu_1 \tag{5}$$

μ_0 was the average grey level of foreground pixels, μ_1 was the average grey level of background pixels.

When i was a global threshold, the variance definition equation g achieved a maximum value.

$$g = \omega_0(\mu_0 - \mu)^2 + \omega_1(\mu_1 - \mu)^2 = \omega_0 * \omega_1(\mu_0 - \mu_1)^2 \tag{6}$$

2.3   Canny Edge Detection Algorithm

The Canny edge detection algorithm removed image noise by smoothing the image with a Gaussian filter [9]. The Gaussian equation was computed as:

$$G(x, y) = \frac{1}{2\pi\sigma^2} e^{-\frac{x^2+y^2}{2\sigma^2}} \tag{7}$$

x and y were the coordinates and σ was the standard deviation.

The gradient value of each pixel point in both the positive and negative directions of the gradient $M(x,y)$ were calculated. The gradient equation was computed as:

$$M(x, y) = \sqrt{d_x^2(x, y) + d_y^2(x, y)} \tag{8}$$

$$\theta_M = \arctan(d_y/d_x) \tag{9}$$

θ_M was the direction of the gradient, d_x was the derivative in the horizontal direction and d_y was the derivative in the vertical direction.

The non-maximum suppression technique NMS was used to eliminate edge misdetection. The NMS equation was computed as:

$$M_T(x, y) = \begin{cases} M(x, y), & \text{if } M(x, y) > T \\ 0, & \text{otherwise} \end{cases} \tag{10}$$

A double thresholding algorithm determined the final boundary. The pixels whose gradient values were greater than the high threshold were strong edges, whose gradient values were less than the low threshold were not edges, and whose gradient values were in between were weak edges. The upper and lower threshold bounds were determined automatically by the Canny algorithm.

3 Experimental Results and Discussion

3.1 Experimental Data

The experimental stent images had spacer size of about 500 μm, connector size of about 320 μm and minimum stand size of about 145 μm. The IVOCT image acquisition retraction speeds were 20 mm/s, 10 mm/s and 5 mm/s respectively (corresponding to image frame distances of 200 μm, 100 μm and 50 μm).

The distance between image frames was related to withdrawal speed and rotation speed. The distance of image frames was computed as:

$$L = \frac{S_p}{S_r} \leq \frac{1}{2}R \tag{11}$$

The formula L was the distance between IVOCT image frames, R was the length of stent, S_p was the withdrawal speed. S_r was the rotation speed of the probe which was determined by the scanning frequency of the light source. Therefor, S_p could change the distance between image frames.

3.2 Image Preprocessing

The coronary stent detection image was shown in Fig. 2(a). The imaging catheter, guidewire and protective sheath in the image had an impact on the experimental results when one-dimensional projection was performed for pixel value calculation. The pixel values of imaging catheter, guidewire and protective sheath were assigned to 0. It was necessary to preprocessing the image to avoid errors, as shown in Fig. 2(b).

(a)   (b)

Fig. 2. (a) IVOCT image of a coronary stent. (b) Image with imaging catheter, guidewire and protective sheath removed.

3.3 Stent Identification

The IVOCT system was a frequency domain OCT system based on a swept frequency light source. An optical fiber inside the catheter transmitted and focused near-infrared light onto the vessel wall, and the vessel was imaged by rotating and retracting the fiber. The NIR light could penetrate vessel wall but could not penetrate stent, so there was a columnar shadow area behind the stent.

The image was filtered with median filtering to remove the noise points. The global threshold was obtained by the Otsu method for the noise-reduced image. Canny algorithm was used for the extraction of the vessel boundaries, as shown in Fig. 3. The yellow line showed the vessel boundaries.

Fig. 3. Segmentation of vascular boundaries.

The 100 pixels below the vascular border were summed to clearly distinguish the stent from the tissue region, as shown in Fig. 4. The intensity values of the stent were lower than that of the tissue regions on either side. The areas of low intensity values were arranged in a one-dimensional array to form an unfolding image of the stent.

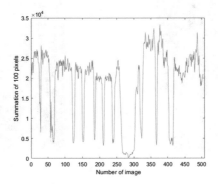

Fig. 4. Summation of 100 pixels below the vascular border.

3.4   Coronary Stent Reconstruction

The coordinate points corresponding to the maximum value of the slope in the stent expansion diagram were extracted. The two-dimensional image of the stent was formed by expanding it in the order of the image frames at one time and the three-dimensional image was generated by converting the stent expansion diagram into a bar chart, as shown in Fig. 5.

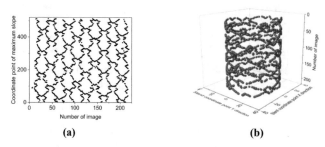

(a)   **(b)**

Fig. 5. (a) 2D reconstruction image. (b) 3D reconstruction image.

3.5 2D Reconstruction of Coronary Stents at Different Distances Between IVOCT Image Frames

IVOCT catheter retraction speeds were 20 mm/s, 10 mm/s and 5 mm/s. The results of the experiment were shown in Fig. 6.

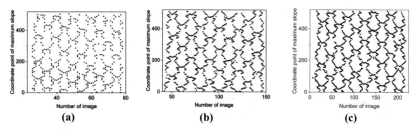

(a)   **(b)**   **(c)**

Fig. 6. (a) 2D image of the stent at 20 mm/s retraction speed. (b) 2D image of the stent at 10 mm/s retraction speed. (c) 2D image of the stent at 5 mm/s retraction speed.

Retraction speeds were 20 mm/s, 10 mm/s and 5 mm/s correspond to image frame distances of 200 μm, 100 μm and 50 μm. Retraction speed of 5 mm/s produced image frame distance less than half of the minimum bracket distance, which satisfied the sampling theory. 50 μm image frame distance allowed for effective bracket 2D image reconstruction. The retraction speeds of 20 mm/s and 10 mm/s didn't satisfy the sampling theory. The support body of the stent and the longer connection body of the stent were reconstructed, but the smaller size connection body couldn't be reconstructed. This experiment proved the stent extraction method proposed in this paper, and also demonstrated the relationship between stent reconstruction and stent retraction speed.

4 Conclusion

In this paper, an algorithm for coronary stent detection and reconstruction based on IVOCT images was proposed. The algorithm used median filtering, the Otsu method, the Canny algorithm and pixel summation to determine stent location information. The algorithm utilized the continuity of IVOCT image frames to achieve 2D image and 3D image reconstruction of stent. Experimental results showed that the algorithm accurately achieved 2D and 3D image reconstruction of the stent when the IVOCT image acquisition satisfied the sampling theory. Subsequent work will focus on the automatic recognition of stent fracture, which will provide effective guidance to doctors for clinical diagnosis stent implantation.

Acknowledgement. This work was supported by the Tianjin Science and Technology Planning Project under Grant No. 20JCYBJC00300 and the National Natural Science Foundation of China under Grant No. 11404240.

References

1. Latus, S., Neidhardt, M., Lutz, M., et al.: Quantitative analysis of 3D artery volume reconstructions using biplane angiography and intravascular OCT imaging. In: Annual International Conference of the IEEE Engineering in Medicine and Biology Society (EMBC), pp. 6004–6007 (2019)
2. Wang, A., Eggermont, J., Reiber, J.H.C., et al.: Fully automated side branch detection in intravascular optical coherence tomography pullback runs. Biomed. Opt. Express **5**(9), 3160–3173 (2014)
3. Ch, A., Gz, B., Yl, C., et al.: Automatic segmentation of bioabsorbable vascular stents in Intravascular optical coherence images using weakly supervised attention network - science direct. Futur. Gener. Comput. Syst. **114**, 427–434 (2021)
4. Nam, H.S., Kim, C.S., Lee, J.J., et al.: Automated detection of vessel lumen and stent struts in intravascular optical coherence tomography to evaluate stent apposition and neointimal coverage. Med. Phys. **43**(4), 1662–1675 (2016)
5. Tsantis, S., Kagadis, G.C., Katsanos, K., et al.: Automatic vessel lumen segmentation and stent strut detection in intravascular optical coherence tomography. Med. Phys. **39**(1), 503–513 (2012)
6. Wang, A., Nakatani, S., Eggermont, J., et al.: Automatic detection of bioresorbable vascular scaffold struts in intravascular optical coherence tomography pullback runs. Biomed. Opt. Express **5**(10), 3589–3602 (2014)
7. Beagum, S.S., Sheeja, S.: A novel restoration technique for the elimination of salt and pepper noise using 8-neighbors based median filter. Adv. Comput. Sci. Technol. **10**(9), 2851–2874 (2017)
8. Essaf, F., Li, Y., Sakho, S., et al.: An improved lung parenchyma segmentation using the maximum inter-class variance method (OTSU). In: International Conference on Computing and Artificial Intelligence, pp. 204–212 (2020)
9. Shah, A.A., Chowdhry, B.S., Memon, T.D., et al.: Real time identification of railway track surface faults using canny edge detector and 2D discrete wavelet transform. Ann. Emerg. Technol. Comput. **4**(2), 53–60 (2020)

Concept Drift Detection and Update Algorithm Based on Online Restricted Boltzmann Machine

Qianwen Zhu, Jinyu Zhou, and Wei Wang[✉]

Tianjin Key Laboratory of Wireless Mobile Communications and Power Transmission, Tianjin 300387, China
weiwang@tjnu.edu.cn

Abstract. In this paper, we address a concept drift detection and update algorithm based on the online restricted Boltzmann machine (O-RBM). We introduce an attention mechanism into RBM, and the parameters of each classifier in the concept drift detection model are updated according to the important information mined by the attention mechanism. The updated model complies with the various data better and judges the types and states of new data online. In the experiments, we compare the performance of the proposed algorithm with the algorithm proposed in our previous work, and the results show that O-RBM plays even better in concept drift detection.

Keywords: Concept Drift · Online Restricted Boltzmann Machine (O-RBM) · Attention Mechanism

1 Introduction

With the rapid development of Internet technology, the data generated by people's production and life explode, which own various forms, large in scale and constantly changing over time, making concept drift occurred more easily. Concept drift is the phenomenon that the information in data stream varies over time or changing environment [1, 2]. In recent years, concept drift detection has become a hot issue.

Liu et al. proposed a drift detection algorithm based on equal intensity K-means space partitioning, which achieves concept drift detection from the perspective of data distribution according to the Pearson's chi-square test data model [3]. Gözüaçık et al. proposed an unsupervised method called D3 which uses a discriminative classifier with a sliding window to detect concept drift by monitoring changes in the feature space [4]. Song et al. considered both temporal and spatial characteristic and proposed a local drift degree (LDD) method that constantly monitors variations in region density [5]. Xu and Wang proposed a combination of dynamic extreme learning machine and threshold detection of drift to classify online data stream. When a drift alert is issued, additional hidden layer nodes are added to the neural network and a new classifier would replace the old one with lower performance once a drift is detected [6]. Although all the above algorithms are good at detecting concept drift, they do not complete uniformity in terms

Q. Liang et al. (Eds.): AIC 2022, LNEE 871, pp. 305–311, 2023.
https://doi.org/10.1007/978-981-99-1256-8_36

of dynamics and efficiency. We have studied concept drift detection algorithms for multi-class classification system based on RBM [7]. In this paper, we introduce an attention mechanism into RBM to update the model parameters online and detect the concept drift.

The rest of the paper is as follows. Section 2 discusses the theoretical knowledge. Section 3 presents the basic framework of the proposed algorithm. Section 4 evaluates the algorithm on different datasets. Section 5 is the conclusion.

2 Basic Theory

2.1 Online Restricted Boltzmann Machine

The multi-class concept drift detection mechanism based on RBM can well realize the detection and judgment of drift types [7]. However, with the passage of time and the continuous inflow of new data, the type of data keeps changed, RBM cannot update the model in time to accommodate the concept change of the data. Therefore, in this paper, we introduce the attention mechanism to obtain the probability distribution via RBM, and then all data blocks are attentional computed according to the data types of the new data blocks. The weights of each data block are renormalized and weighted to get new probability distribution for online detection of concept drift, this process is called online restricted Boltzmann machine (O-RBM). Compared with RBM, O-RBM is able to update model parameters online according to the data characteristics of new data blocks to detect the drift more accurate and timely.

2.2 Attention

The attention mechanism is mostly used in neural network models, learning the weighted information of the input. The core of the attention mechanism is the attention function. There are two kinds of attention functions commonly used: additive attention and dot product attention. The dot product attention can be described as the computational process of mapping a Query and a set of Key-Value pairs to an output, where Query, Key, Value and output are all vectors [8]. The output matrix is computed as

$$Attention(Q, K, V) = Soft \max(\frac{QK^T}{\sqrt{d}}) V \tag{1}$$

where d represents the length of the input data vector.

3 O-RBM-Based Concept Drift Update Algorithm

In the task of data mining, real-time monitoring of data stream is essential. For data stream with concept drift, the classification model should not only ensure the accuracy of the classification, but also match the new concepts. O-RBM updates the model by adding an attention mechanism in RBM. The attention mechanism can select important information from the massive information stream, which is realized by the calculation of weight coefficients. The larger the weight value is, the more attention will be paid to its corresponding data value (*Value*). As new data blocks input, the probability distribution and weight coefficients of each data block can be calculated by RBM. Then the outputs of each classification model in RBM are regarded as the inputs of O-RBM, and the weights of each classifier updated with the attention mechanism to relearn the new data. Thus, the new probability distribution of the input data can be obtained to achieve the online detection and judgement of concept drift. The process of O-RBM-based concept drift detection and update algorithm is shown in Fig. 1. The algorithm owns two parts: data pre-training and concept drift detection. The data is divided into training set and test set in pre-training, and the concept drift detection part consists of three parts: RBM model training, O-RBM model update and concept drift analysis. O-RBM model update mechanism is shown on the right of Fig. 1.

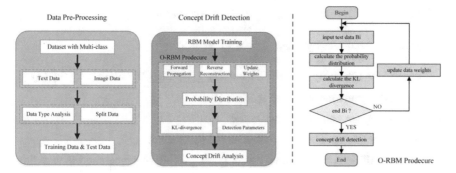

Fig. 1. The procedure of O-RBM-based concept drift update algorithm

The steps of updating the classifier weights based on the attention mechanism are as follows. Firstly, calculate the similarity between *Query* and *Key_i*. In this paper, we use the vector cosine similarity, denoted by Sim_i

$$Sim_i = Similarity(Query, Key\_i) = \frac{QueryKey\_i}{||Query||||Key\_i||} \tag{2}$$

Secondly, Sim_i is numerically transformed to obtain the weighting coefficient of $Value_i$, denoted by a_i a_i

$$a_i = Soft\max(Sim_i) = \frac{e^{Sim_i}}{\sum_{j=1}^{Lx} Sim_j} \tag{3}$$

Finally, the attention value of *Query* is calculated by *Value* and a_i a_i

$$Attention(Query, Source) = \sum_{j=1}^{Lx} a_i Value_i \tag{4}$$

where *Query* corresponds to the weight matrix W_i of the data, *Key_i* calculates the relationship between the current data block B_i and the historical data blocks $B_1, B_2, ..., B_{i-1}$, and *Value*$_i$ is the current data value. The proposed O-RBM assigns different weights to all existing data blocks and the classifier updated with the current data to reduce the effect of historical data on the classifier performance. Algorithm 1 gives the O-RBM-based Concept Drift Update Mechanism.

Algorithm 1: O-RBM-based Concept Drift Update Mechanism

Input: Training data blocks $B_1, B_2, ..., B_{i-1}$ and test multi-class data block $B_i, B_{i+1}, B_{i+2}, ...$

Output: updated parameters of weight

Begin:

1. Train RBM via training data, get W_i and the probability distribution p^i of B_i

2. Calculate the KL-divergence $p_1^i, p_2^i, ..., p_{i-1}^i$ between B_i and $B_1, B_2, ..., B_{i-1}$

3. Obtain Key_i of B_i by $Key\_i = p_{i-1}^i / \sum_{n=1}^{i-1} p_n^i$

4. Calculate the similarity between W_i and Key_i by Eq. (2), obtain a_i by Eq. (3)

5. Update the parameters of RBM by Eq. (4)

6. Repeat 1-5 when next data block comes

7. Complete online update of the RBM model parameters

8. Concept drift detection achieve by comparing KL-divergence and the variation of each data block

End

---

4 Experiments and Analysis

As an extension of our previous work [7], the performance of O-RBM can be estimated by comparing with RBM. In this paper, we investigate the KL-divergence and variance of the probability distributions obtained from the two models to explore whether there is an improvement on O-RBM.

4.1 Dataset

Gaussian Dataset. Create three classes of Gaussian distribution datasets, representing A, B and C respectively. The tested data consists of 10 data blocks with length 50 of each block. There exist mutational drifts in the tested data, where the first five data blocks belong to A, the other five data blocks belong B.

MNIST Dataset [9]. MNIST is a set of image dataset of handwritten digits collated by the National Institute of Standards and Technology (NIST). Taking 250 digits 1 and 2 to form the tested data containing mutation drifts, and digits 1 constitutes the first five data blocks, while digits 2 are the last five data blocks.

Satimage Dataset [10]. The Satimage dataset is a ground-based satellite dataset containing 36 attributes. Among the dataset, six classes A, B, C, D, E and F with 500 samples of are taken for the experiments, the tested data contains asymptotic drifts which generated by another 300 samples of A and 200 samples of F.

4.2 Experimental Results

In this part, we investigated the probability distributions of each class of data blocks obtained from the RBM-based multi-class concept drift detection algorithm [7] and the O-RBM-based concept drift detection and update algorithm on different datasets, respectively. The KL-divergence and variance of the probability distributions of the data blocks are tested and plotted in Figs. 2, 3, and 4.

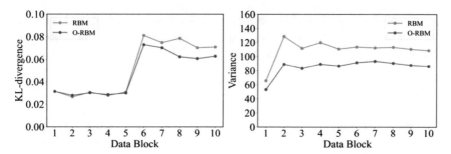

Fig. 2. Comparison of RBM and O-RBM model parameters based on Gaussian dataset

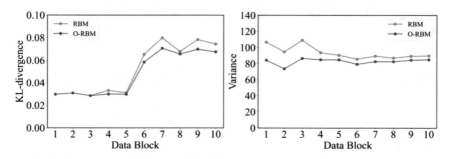

Fig. 3. Comparison of RBM and O-RBM model parameters based on MNIST dataset

From the figures, It can be observed that the KL-divergence of the probability distributions for each data block derived based on RBM are higher than those of O-RBM, which indicates that the variability of the probability distributions derived from RBM is larger. Similarly, the results of the variance obtained by these two methods show that the variance generated by O-RBM are always smaller than those in RBM, which indicates that the fluctuations of the probability distributions in O-RBM are smaller. This is because when there is a concept drift, O-RBM normalizes and updates the weights of all

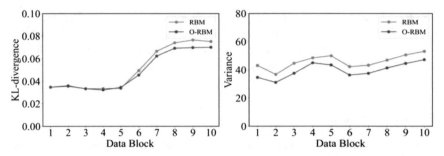

Fig. 4. Comparison of RBM and O-RBM model parameters based on the Satimage dataset

data blocks again based on the data block to reduce the influence of historical data blocks on the model parameters and increase the weights of the new data blocks. In addition, the probability distribution of O-RBM is closer in the data blocks with same concept, so the variance of the probability distribution between the old and new data blocks will be lower than the variance of RBM-based model. The O-RBM owns higher accuracy and more stable classification, which can adapt well to data changes.

5   Conclusion

We studied O-RBM-based concept drift detection and updating algorithm in this paper. An attention mechanism is introduced to achieve concept drift detection and online updating of RBM [7] model. In addition, experiments are conducted on the synthetic dataset and the real datasets, and the KL-divergence and variations of the probability distributions obtained from RBM and O-RBM are compared, respectively. The results show that the O-RBM model proposed in this paper achieves real-time detection of concept drift based on the detection of concept drift, which has higher accuracy and more stable classification results compared to RBM.

Acknowledgement. The work was supported by the Natural Science Foundation of China (61731006, 61971310).

References

1. Wares, S., Isaacs, J., Elyan, E.: Data stream mining: methods and challenges for handling concept drift. SN Appl. Sci. **1**(11), 1–19 (2019). https://doi.org/10.1007/s42452-019-1433-0
2. Iwashita, A.S., Papa, J.P.: An overview on concept drift learning. IEEE Access **7**, 1532–1547 (2018)
3. Liu, A., Lu, J., Zhang, G.: Concept drift detection via equal intensity k-means space partitioning. IEEE Trans. Cybern. **51**(6), 3198–3211 (2021)
4. Gözüaçık, Ö., Büyükçakır, A., Bonab, H., et al.: Unsupervised concept drift detection with a discriminative classifier. In: Proceedings of the 28th ACM International Conference on Information and Knowledge Management, pp. 2365–2368 (2019)

5. Liu, A., Song, Y., Zhang, G., et al.: Regional concept drift detection and density synchronized drift adaptation. In: IJCAI International Joint Conference on Artificial Intelligence (2017)
6. Xu, S., Wang, J.: Dynamic extreme learning machine for data stream classification. Neurocomputing **238**, 433–449 (2017)
7. Zhou, J., Zhu, Q., Shi, R., Wang, W.: Concept drift detection based on restricted Boltzmann machine in multi-class classification system. In: Liang, Q., Wang, W., Mu, J., Liu, X., Na, Z. (eds.) Artificial Intelligence in China. Lecture Notes in Electrical Engineering, vol. 854, pp. 235–241. Springer, Singapore (2022). https://doi.org/10.1007/978-981-16-9423-3_29
8. Vaswani, A., Shazeer, N., Parmar, N., et al.: Attention is all you need. In: Advances in Neural Information Processing Systems (2017)
9. MNIST Dataset. http://yann.lecun.com/exdb/mnist/
10. Satimage dataset. KEEL: a software tool to assess evolutionary algorithms for data mining problems (regression, classi fication, clustering, pattern mining and so on) (ugr.es)

3D Data Augmented Person Re-identification and Edge-based Implementation

Ziyang Bian[1,3], Liang Ma[3], Jianan Li[1(✉)], and Tingfa Xu[1,2(✉)]

[1] Beijing Institute of Technology, Beijing 100081, China
{lijianan,ciom_xtf1}@bit.edu.cn
[2] Beijing Institute of Technology Chongqing Innovation Center,
Chongqing 401120, China
[3] North China Research Institute of Electro-Optics, Beijing 100015, China

Abstract. In recent years, artificial intelligence technology has made breakthrough progress. It is widely used in person re-identification(Re-ID), but its accuracy still needs to be improved. This paper proposes a person re-identification(Re-ID) model based on 3D data augmentation and designs and implements edge computing deployment. The 3D human body model is put into the 3D scene and take automatic virtual photography to obtain the simulation image data. The neural network model is trained through the expanded data set of 3D simulation, and its improvement is significant. Based on this, an onboard scheme of artificial intelligence edge computing based on the Rock chips 3588 was designed to meet the practical needs of person re-identification(Re-ID) applications. This scheme is widely compatible with standard IP cameras based on RTSP/RTMP streams, supports person re-identification of visible and infrared video streams, can control the PTZ through the network interface, and can be updated and iterated remotely through the network.

Keywords: 3D Data Augmentation · Person Re-identification · Edge Computing · Rock chips 3588

1 Introduction

Person re-identification (Re-ID) seeks the same person across multiple non-overlapping surveillance cameras, which has attracted widely study. Person Re-ID has also made significant progress based on the large-scale visible image data set of person Re-ID and convolutional neural networks. Because of its importance in intelligent video surveillance, it has attracted more and more attention in the computer vision field [1–3,8–11]. Recently, the application of the image generation method based on an adversarial method [4,5,12–14] to person Re-ID has attracted much attention. The main challenge of person Re-ID is to deal with the appearance differences caused by lighting, pose changes, and angle changes, as well as the other differences caused by different cameras.

Q. Liang et al. (Eds.): AIC 2022, LNEE 871, pp. 312–319, 2023.
https://doi.org/10.1007/978-981-99-1256-8_37

However, the above tracking methods still have the following problems: (1) The person Re-ID dataset still cannot cover the natural scene, and the accuracy of the training model is low. (2) The current models are mainly run on computers, lacking low-cost edge deployment engineering applications.

2   3D Augmented Person Re-identification

The general data set construction process directly captures all data from the actual physical environment. However, the 3D CG simulation data set construction enhancement technology only collects a small part of data from the real world, including the environment and specific target data. On the one hand, through 3D CG technology and physical simulation algorithms [6], the collected natural environment data is processed to build a 3D simulation scene. On the other hand, the target data of the AI task scanned and collected will be processed to form a three-dimensional digital model consistent with the natural world target. In particular, for AI tasks requiring multidimensional data, multiple image domain data and various sensor data can be further integrated to establish a digital twin model of the target object for use by subsequent system units.

On this basis, computer graphics, physical simulation, and other technologies are comprehensively used to shoot on demand in a three-dimensional virtual environment to generate the simulation image data of the target subject of the expected AI task under various scenes, weather, climate, and lighting conditions at various angles and motion postures. Furthermore, the data can be automatically annotated without manual intervention through the programmed method. The system jointly uses the simulated and natural world data to train the GAN generative countermeasure neural network to improve the generated data's authenticity further. It can obtain the adjustment and conversion of the virtual simulation image data's realistic image style. The final output is a deep neural network artificial intelligence dataset almost identical to real-world image data. Next, we will describe each critical part.

Because the task of person re-identification (Re-ID) focuses on local small-scale human features, only local environment modeling data of specific monitoring areas are needed. Therefore, we selected the local geographic data taken by UAV to conduct 3D modeling and obtain the 3D landform model after data preprocessing. Then, we enrich the geomorphic model and correct the part with poor local texture. We appropriately add and delete the elements such as vegetation and buildings that are not fully three-dimensional and adjust the surface material to meet the requirements of PBR physical natural lighting to obtain the final 3D simulation scene, as shown in Fig. 1.

To collect image data of different pedestrians at different angles, we use the method of photogrammetric modeling to build 3D models of the human body and integrate some existing 3D human models to complete the establishment of the databases, as shown in Fig. 2. We put the 3D human model into the 3D scene and take automatic virtual photography in the digital simulation space to obtain the simulation image data. By adjusting the capture angle, ambient

Fig. 1. Geographical environment 3D model

Fig. 2. 3D human models

light, and other factors of 100 pedestrian subjects' attitude sets, a 6000-piece image data set was formed. Then 10000 pieces of the data set were completed through random noise, hue jitter, rotation, clipping, and other conventional data augmentation methods.

The simulation data set is used to expand the CUHK03 [7] person re-identification (Re-ID) data set. Moreover, the final test person re-identification (Re-ID) data set with the scale of 20000 (the total number of images in the original CUHK03 data set is 13164, and we reserve 3164 images as the test group data, not participating in the neural network model training), as shown in Fig. 3.

Fig. 3. (a) 3D CG simulation dataset construction (b) simulation dataset (c) CUHK03 dataset

3 Edge-based Implementation

As shown in Fig. 4, an AI edge computing on-board solution based on the Rock chips 3588 is designed for low-cost video surveillance. The solution is widely compatible with standard IP cameras based on RTSP/RTMP streams, supports person re-identification of visible and infrared video streams, can control the

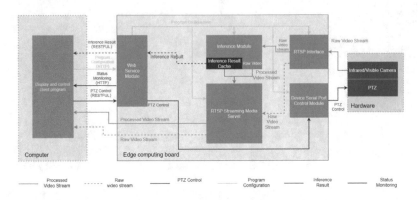

Fig. 4. System Software Dependency Diagram

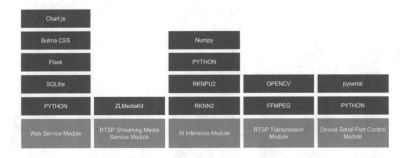

Fig. 5. System logic structure diagram

Fig. 6. The Edge-based Implementation of Person Re-identification

Fig. 7. The Experimental Result of the Edge-based Person Re-identification

PTZ through the network interface, and can be updated and iterated remotely through the network. Function parameters can be configured through the WEB service interface, and the model and function can be updated remotely and iteratively through the network. The system software dependency Diagram is shown in Fig. 5. The design maximizes the system's scalability by abstracting it into

Table 1. Results of the Person Re-identification using 3D Augmented Data

Data set result	3D Augmentation rank-1	3D Augmentation mAP	CUHK03 rank-1	CUHK03 mAP
256 × 128	79.5	74.9	78.1	72.7
224 × 224	80.1	75.3	78.5	74.2
384 × 128	80.8	75.1	79.0	74.1
384 × 192	80.1	74.4	77.8	73.5

independent functional modules, combined through interfaces and configuration information to form specific functions to adapt to changing applications.

4   Result and Analysis

The 3D augmented person Re-identification dataset and CUHK03 dataset are used to train the deep neural network, respectively. Two trained deep neural network models for person Re-identification have been obtained: the 3D augmented model and the CUHK model for short. Finally, the 3D augmented model and the CUHK model was tested and evaluated with the test group data, and the final comparison results were shown in Table 1. The 3D augmented model has significantly surpassed the CUHK model. It is mainly because 3D simulation can supplement the feature dimensions missing in all aspects of the data set. Due to the human cost and time cost limitations, obtaining enough data from the real world to meet the feature dimensions required by the neural network is not easy. The 3D simulation data construction technology can complete the data expansion well and meet the corresponding in-depth neural network training needs.

As shown in Fig. 6 and Fig. 7, we deployed an edge-based Person Re-identification system. The modular design supports iterative upgrades to each functional part, or the introduction of new functional modules, providing the system with maximum backward expansion possibilities. The design provides an intuitive and easy-to-use user interface via the WEB, through which all system functions, parameters, and status can be set uniformly, minimizing the difficulty of using the system.

5   Conclusion

This paper proposes a person re-identification(Re-ID) model based on 3D data augmentation. Based on this, we designed an onboard scheme of artificial intelligence edge computing based on the Rock chips 3588 to meet the practical needs of person re-identification(Re-ID) applications. This paper shows the capabilities of 3D CG simulation data construction technology with the application of artificial intelligence in the CV field. This technology is an interdisciplinary and basic technology that can support various artificial intelligence tasks. With the continuous development of CG technology and artificial intelligence technology, this technology will continue to evolve, constantly improve its data augmentation ability, and expand its application boundaries. In the future, we will continue to enrich and improve the 3D CG simulation data construction technology and further explore its fluid simulation, optical domain conversion, infrared radiation simulation, and other related work. Furthermore, we will strengthen its edge deployment capabilities to use person re-identification technology more widely.

Acknowledgments. This research was funded by the National Key Laboratory Foundation of China grant number TCGZ2020C004 and 202020429036.

References

1. Zheng, L., Yang, Y., Hauptmann, A.G.: Person re-identification: past, present and future, arXiv preprint arXiv:1610.02984 (2016)
2. Karanam, S., Gou, M., Wu, Z., Rates-Borras, A., Camps, O., Radke, R.J.: A systematic evaluation and benchmark for person reidentification: Features, metrics, and datasets. In: IEEE TPAMI (2018)
3. Chen, Y.-C., Zhu, X., Zheng, W.-S., Lai, J.-H.: Person re-identification by camera correlation aware feature augmentation. IEEE TPAMI **40**(2), 392–408 (2018)
4. Ni, H., Song, J., Zhu, X., et al.: Camera-agnostic person re-identification via adversarial disentangling learning (2021)
5. Fu, X., Lai, X.: Unsupervised person re-identification via multi-order cross-view graph adversarial network. IEEE Access **9**, 22264–22273 (2021)
6. Zhou, D.W., Gu, X.X., Wang, Y., et al.: Shader technology based on physical rendering (2022)
7. Wei, L., Rui, Z., Tong, X., et al.: DeepReID: deep filter pairing neural network for person re-identification. In: Computer Vision & Pattern Recognition, IEEE (2014)
8. Zhong, S., Bao, Z., Gong, S., et al.: person reidentification based on pose-invariant feature and B-KNN Reranking. IEEE Trans. Comput. Soc. Syst. **8**(5), 1272–1281 (2021)
9. Ming, Z., Zhu, M., Wang, X., et al.: Deep learning-based person re-identification methods: a survey and outlook of recent works (2021)
10. Zhuang, W., Wen, Y., Zhang, S.: Joint optimization in edge-cloud continuum for federated unsupervised person re-identification (2021)
11. Integrating coarse granularity part-level features with supervised global-level features for person re-identification (2021)
12. Wang, X., Li, S., Liu, M., et al.: Multi-expert adversarial attack detection in person re-identification using context inconsistency. arXiv e-prints (2021)
13. Zhang, C., Wu, L., Wang, Y.: Crossing generative adversarial networks for cross-view person re-identification. Elsevier (2019)
14. Dai, C., Wang, H., Ni, T., et al.: Person re-identification based on deep convolutional generative adversarial network and expanded neighbor reranking. J. Comput. Res. Dev. **2019**(8)
15. Ye, M., Shen, J., Shao, L.: Visible-infrared person re-identification via homogeneous augmented tri-modal learning. IEEE Trans. Inf. Forensics Secur. **16**, 728–739 (2020)

5G Potential Customer Recognition Research Based on Multi-layer Heterogeneous Integration Model

Yuejia Sun[✉], Zhongxian Xu, Ye Guo, Zhihong Zhou, and Lin Lin

China Mobile Research Institute, No.32 Xuanwumen West Street, Xicheng District, Beijing, China
sunyuejia@chinamobile.com

Abstract. With the rapid development of 5G, telecom operators have begun to deploy and promote 5G services. How to accurately recognize 5G potential customers and increase conversion rate of 5G customer has become a very concerned issue for telecom operators. In this paper, a 5G potential customer recognition model based on telecom big data is constructed. Firstly, a influencing factor system is constructed to fully reflect the relevant characteristics of 5G customers. Secondly, feature selection is performed by summing the importance evaluated by multiple tree ensemble models, and it is found that relatively important features include data ARPU, brand type, DOU and its changes, ARPU and its changes, video APP usage and so on. Thirdly, the SMOTE-ENN combined algorithm is applied to improve the performance of the basic model. Then, various machine learning models are used to identify 5G potential customers, and the comparison results show that XGBoost, LightGBM and DNN have better performance. Finally, based on the three best-performing classifiers, a multi-layer heterogeneous integrated model is constructed, which combines the cascade and parallel structures. Experiments show that the MHI model can achieve better performance than the separate classifiers.

Keywords: 5G potential customer · Comprehensive feature selection · SMOTE – ENN · Sample imbalance · XGBoost · LightGBM · DNN · Muti-layer heterogeneous integration

1 Introduction

With the rapid development of 5G technology, 5G has become an important role in driving a new round of digital transformation. Telecom operators have begun to deploy and promote 5G services across the country. Seizing 5G development opportunities has become an important task for telecom operators. Therefore, how to accurately identify 5G customers, recommend 5G related services and improve conversion rate has become issues of great concern to the development of telecom operators.

In recent years, the research on 5G potential customer identification was mainly carried out through the construction of logistic regression, decision tree, tree-based

© The Author(s), under exclusive license to Springer Nature Singapore Pte Ltd. 2023
Q. Liang et al. (Eds.): AIC 2022, LNEE 871, pp. 320–328, 2023.
https://doi.org/10.1007/978-981-99-1256-8_38

ensemble model. In addition, the sample balancing methods have also been focused on for 5G imbalanced data sets. Yiwen Dai [1] used XGBoost algorithm, decision tree and logistic regression algorithm for different levels of 5G package customers, and selected the optimal model for each level of 5G package through recall and precision. Wanying Jin [2] screened out the most important features such as the proportion of 5G users in cities, billing income, and found that the XGBoost, LightGBM and CatBoost worked better. Hong Xu [3] compared three algorithms of sampling and found that the performance of LightGBM with under-sampling was better than other models. Hao Zhang [4] proposed the TABoost algorithm to solve the problem of data imbalance, using the classification hardness distribution to sample the majority class.

2 5G Potential Customer Recognition Model

In this paper, a 5G potential customer identification model is generated based on telecom big data and machine learning models, aiming to identify those customers who are likely to order 5G packages in the future from the existing 4G stock customers, so as to help enterprises quickly migrate customers to 5G. The framework of the model is as follows: firstly, 150 + dimensional features are constructed based on the '1 + 5' influencing factor model, and then processed by comprehensive feature selection and combined sample balance algorithm. Then, 5G customers are identified by logistic regression, decision tree, random forest model, XGBoost model, LightGBM model and deep neural network model, and the optimal, second-best and third-best estimators are selected by comparing the performance results. Finally, a multi-layer heterogeneous integrated model is constructed to improve the performance of the 5G customer comprehensive prediction model.

2.1 Problem Definition and Variable System Construction

5G potential customers are defined as those who are not currently 5G package customers, but may become 5G customers in the next time period. Set T month as the observation period, and T-1 to T-3 months as the observation period. The positive sample set is the group of customers who are 5G customers in month T but are not 5G customers from months T-1 to T-3. The negative sample set is the group of customers who are not 5G customers from month T to month T-4.

5G has the characteristics of high data volume, high speed, and high concurrency. And most 5G packages are integrated products, including data, communications, application products, home services and terminals. Therefore, customers' tendency to migrate to 5G may be related with data usage demand, talk and text demand, app usage demand, home broadband demand, and terminal demand, and may be affected by attributes such as the customer's identity and consumption level. Therefore, a '1 + 5' influencing factor system is constructed based on telecom big data, including one dimension of basic customer attributes, and five dimensions of behaviors, which are data behavior, communication behavior, application behavior, home business behavior, and terminal behavior. Data behavior includes total data volume and network DPI data construction. Communication behavior includes the use of calls, text messages, etc. Application behavior includes the subscription and usage of various application members. Home business behavior includes usage of home broadband, ipTV, etc. Terminal information includes the model, type, and replacement status of the mobile phone. In total, more than 150 dimensions of initial data are extracted, and 4 months of historical data are extracted for each customer (Table 1).

Table 1. 5G potential customer identification variable system

Dimensions	Variables
Attributes	Gender, Age, Region, Occupation, Education Level, Network Age, Brand Type, etc.
Data Behavior	5G packages subscription, DOU, Package data saturation, Package data overage, APP usage data, APP usage times, APP usage minutes(APPs include video, music, reading, live broadcast, etc.)
Communication behavior	ARPU, MOU, Call times, Call minutes, card type, etc.
Home Business behavior	Broadband type, Broadband usage time, Broadband usage data, Broadband TV, Broadband TV boot times, etc.
Application behavior	Application number in package, Application membership, Video members exchange, Music members exchange, etc.
Terminal behavior	Terminal brand, Terminal type, Dual SIM card, 5G mobile phone, Contract phone, etc.

2.2 Feature Extraction and Sample Balancing

Firstly, data preprocessing and feature construction are carried out. Data preprocessing includes repeated data processing, missing data processing, abnormal data processing, and data encoding. Feature construction includes conventional feature extraction and derived feature extraction. Conventional features take the value of T-1 period. Derived features are derived from T-1 to T-3 period values, including summed value, average value, variance, trend value, volatility value, etc. A feature transformation is then performed to normalize the feature values by removing the mean and scaling it to unit variance. The normalization of datasets has a greater impact on recognition models, especially the deep neural network.

Then a comprehensive feature selection method based on tree-based ensemble model is constructed. Firstly, the feature importance is calculated by random forest, XGBoost and LightGBM separately. Then, the importance value of single feature is summed to obtain the comprehensive importance score. Finally, the feature selection results are obtained by matching the comprehensive importance score with the judgment conditions. The feature importance of tree-based model refers to how much each feature contributes to each tree, which is measured by Gini index or out-of-bag (OOB) error rate. In this paper, the importance measurement results are obtained by normalizing the Gini index, as shown in formula 1, where n is the number of decision trees, M_i is the number of nodes of the i_{th} decision tree, C_m is the number of categories of the m_{th} node, C_l and C_r are the number of categories of the new node after branching, p is the proportion of a certain category in the corresponding node.

$$VIM_j^{Gini} = \sum_{i=1}^{N} (\sum_{m \in M_i} (\sum_{c=1}^{C_m} p_{mc}(1-p_{mc}) - \sum_{c=1}^{C_l} p_{lc}(1-p_{lc}) - \sum_{c=1}^{C_r} p_{rc}(1-p_{rc}))) \quad (1)$$

Finally, the combined sample balancing method of SMOTE-ENN is performed, due to the large difference in the number of positive and negative samples. Combined sample balancing is the fusion of over-sampling and under-sampling [5]. For over-sampling, synthetic minority over-sampling technique (SMOTE) is used to synthesize new samples of 5G customers. The synthesis strategy is to randomly select a sample B from its nearest neighbors for each minority sample A, and then randomly select a point on the connection between A and B as a newly synthesized minority class sample. For under-sampling, edited nearest neighbor (ENN) is adopted. For samples of non-5G category, if most of the k-nearest neighbor samples of this sample are different from its own category, then the sample will be deleted, so as to reduce the samples of the non-5G class to achieve balance. In this paper, a SMOTE-ENN combined balancing algorithm is constructed, which first uses SMOTE to expand the samples, and then uses the ENN method to delete the samples in a stalemate state, thereby making the margin between categories larger (Fig. 1).

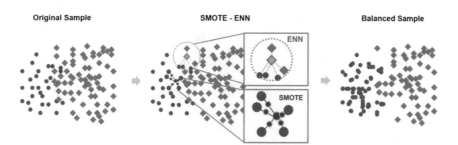

Fig. 1. Principle description of Smote-ENN

2.3 Tree-Based Homogeneous Integrated Model

Random forest is a bagging-based ensemble classifier based on decision tree model, which has the characteristics of random samples and random features. Random sampling is performed in the original sample set to construct N different sample sets and m different features, and then a decision tree model is built for each sub-data set. The results of the decision tree are averaged or voted to obtain the final result. This operation can effectively improve the over-fitting phenomenon of decision tree and make the model have strong generalization ability.

XGBoost is an improved and upgraded algorithm based on Gradient Boosting Decision Tree (GBDT), which is an improved model using the boosting integration strategy.

In each iteration process, a CART is learned to fit the residuals between the last predicted result and the actual result, which eventually generates a strong classifier. The loss function is optimized by changing the error from the first-order Taylor expansion to the second-order Taylor expansion to obtain more accurate results, and the regularization part is added to improve the generalization ability [6].

LightGBM is an efficient implementation of the GBDT algorithm, which has the advantages of higher training efficiency, lower memory usage, higher accuracy, ability to process large-scale data, and support for parallelization. The improvement point is that lightGBM adopts some optimization algorithms, such as leaf-wise growth, histogram algorithm, gradient-based one-side sampling, exclusive feature bunding and so on.

2.4   Deep Neural Network Model

The deep neural network (DNN) model is a multi-layer neural network model based on the expansion of the perceptron, which has multiple hidden layers between the input layer and the output layer. The layers are connected to each other, and the data is continuously transmitted from the previous layer to the next layer [7]. The results of each layer are obtained by linear transformation and nonlinear transformation of the activation function, and finally output by the output layer. In this paper, by parameter tuning, a deep neural network structure is constructed, as shown in Fig. 2, which has 3 hidden layers with relu as the activation function.

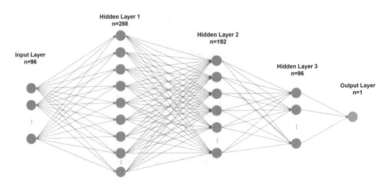

Fig. 2. Deep neural network structure for 5G potential customer recognition

The first stage of the neural network is the forward propagation, which inputs data from the input layer, then performs linear and nonlinear transformations through the hidden layer, and finally reaches the output layer, as shown in formula 2. The second stage of the network is the error back propagation, which aims to continuously optimize the loss function, as shown in formula 3. By finding the negative gradient of the loss function, the weight W and the bias b are updated, and the loss function is continuously reduced, so as to continuously approach the optimal solution.

$$x_j^{(h+1)} = f(\sum_{i=1}^{n} w_{i,j}^{(h)} x_i^{(h)} + b_j^{(h)}) \tag{2}$$

$$L(w, b) = \frac{1}{2} \sum_{j=1}^{n} (\widehat{y_j} - y_j)^2 \tag{3}$$

2.5 Multi-layer Heterogeneous Integrated Model

A heterogeneous integrated model is designed with the cascade structure as the main and the parallel structure as the auxiliary. The integrated strategy is to arrange the optimal estimator as the first layer, the second-optimal estimator as the second layer and the third-optimal estimator as the third layer, followed by a cascade estimator as the last layer. The sample is recognized at each layer by the estimator, and the confidence of its recognition result is judged. If a certain set condition is met, the result is adopted; if not, the sample is input to the next layer for identification. Proceed like this until the sample is fed into the parallel structure of the last layer. The parallel structure integration is to integrate the judgment results of N different types of sub-estimators through voting method (Fig. 3).

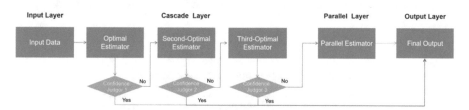

Fig. 3. Multi-layer Heterogeneous Integrated Model

3 Experimental Results and Analysis

3.1 5G Customer Feature Analysis and Sample Balancing Results

Part of the results of the comprehensive feature analysis are shown in Fig. 4. It is found that the features with high importance for 5G potential customer mining include data ARPU, brand type, DOU and its changes, ARPU and its changes, video APP usage, MOU and calling times, network age, contract ordering, broadband type, news APP usage, terminal replacement, etc.

The comparison results of sample balance methods of over-sampling, under-sampling and combined sampling are shown in Fig. 5. It can be seen that the SMOTE-ENN method achieves the best performance.

3.2 Different Classifier Experimental Results

Different classifiers are used to train and test samples, and the comparative results are shown in Table 2. It can be seen that the top three models with the best comprehensive performance are XGBoost, LightGBM and DNN, which will be adopted as the evaluation units of the heterogeneous integration model.

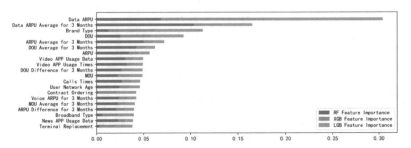

Fig. 4. Experimental results on feature importance of comprehensive feature selection

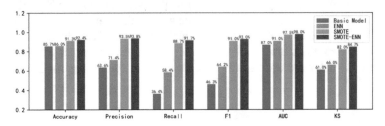

Fig. 5. Comparative results of multiple models

Table 2. Comparative experimental results

Model	Accuracy	Precision	Recall	F1	AUC	KS
Logistical Regression	60.5%	57.0%	87.1%	0.69	0.75	0.41
Decision Tree	85.5%	85.6%	87.2%	0.86	0.85	0.71
Random Forest	89.1%	91.0%	88.1%	0.90	0.96	0.78
XGBoost	92.4%	93.8%	91.7%	0.93	0.98	0.85
LigthGBM	91.4%	93.7%	89.8%	0.92	0.98	0.83
DNN	90.6%	91.4%	90.6%	0.91	0.96	0.81

3.3 Multi-layer Heterogeneous Integrated Model Experimental Results

The structure of multi-layer heterogeneous integration model consists of cascade layer and parallel layer. In the cascade layer, the first estimator is XGBoost, the second estimator is LightGBM, the third estimator is DNN, and the cascade judgment threshold is [0.85, 0.75, 0.9]. The parallel layer is composed of these three layers by voting principle. The final effect evaluation of the MHI model is shown in Fig. 6, from which it can be seen that the MHI model can effectively improve the identification effect of 5G potential customers.

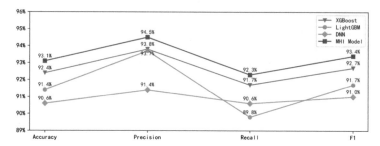

Fig. 6. Multi-layer Heterogeneous Integrated Model

4   Conclusion and Summary

This paper builds a 5G potential customer model based on telecom big data, which improves the performance of the basic classifier by adopting combined sample balancing method and constructing multi-layer heterogeneous integrated model.

- In terms of feature construction, through the 1 + 5 influence factor model and comprehensive feature selection, it is found that the relatively important features of 5G customers include ARPU, data ARPU and dou, brand, Video and news APP usage, MOU and call times, network age, etc.
- In terms of sample balancing, the combined sample balancing algorithm of SMOTE_ENN is performed to improve the over-fitting problem caused by the imbalance of 5G customer data, and achieves a better performance.
- In terms of recognition model, three optimal classifiers are selected as the integrated model units, including XGBoost, LightGBM and DNN. Finally a multi-layer heterogeneous integration model based on cascade and parallel structure is constructed. Experiments show that the MLI model can improve the main effect evaluation value of the recognition model.

References

1. Dai, Y.: Dashuju jishu zai 5G taocan qianke moxing zhong de yanjiu [Application of Big Data Technology in 5G Package Potential Customer Model]. Science & Technology Information, No. 20 (2021)
2. Jin, W.: Jiyu shuju wajue de 5G dianxin yonghu yuce yanjiu [Research on 5G Telecom Customer Prediction based on Data Mining]. Dalian University of Technology, Dalian (2021)
3. Xu, H.: Yunyingshang 5G qianzai kehu shibie yanjiu [Operator 5G Potential Customer Identification Research]. Nanjing Normal University, Nanjing (2021)
4. Zhang, H.: Jiyu jiqi xuexi de 5G qianzai kehu wajue yanjiu [Research on 5G Potential Customer Mining Based on Machine Learning]. Yunnan University of Finance and Economics, Yunnan (2022)
5. Batista, G.E., Prati, R.C., Monard, M.C.: A study of the behavior of several methods for balancing machine learning training data. ACM SIGKDD Explorations Newsl. **6**(1), 20–29 (2004)

6. Chen, T., Guestrin, C: XGBoost: a scalable tree boosting system. In: KDD 2016 Proceedings of the 22nd ACM SIGKDD International Conference on Knowledge Discovery and Data Mining, pp. 785–794 (2016)
7. Tomek, I.: Two modifications of CNN. IEEE Trans. Syst. Man Cybern. **6**, 769–772 (1976)

Dynamic Task Offloading for Air-Terrestrial Integrated Networks: A Learning Approach

Peng Qin[✉], Shuo Wang, HongHao Zhao, Yang Fu, and Miao Wang

State Key Laboratory of Alternate Electrical Power System with Renewable Energy
Sources, School of Electrical and Electronic Engineering,
North China Electric Power University, Beijing 102206, China
125529995@qq.com

Abstract. With the popularization of artificial intelligence, 5G and
other technologies, a number of emerging applications require both
efficient communication and computing service, which poses enormous
challenges to the computing ability and battery capacity of terminal
equipment. Moreover, ground-based 5G system can not provide seamless service especially for hotspot and remote area. To tackle the above
challenges, we minimize the weighting of delay and energy consumption by optimizing the task offloading decision and computing resource
allocation, which, however, is a mixed integer nonlinear programming
(MINLP) issue due to the strong coupling between optimization variables. Therefore, we decompose it into two subproblems and design a
deep reinforcement learning-Based approach to address the first problem with offloading decision-making. For the second computing resource
allocation subproblem,. a Greedy-based solution is proposed. The simulation results indicate that, in comparison to other benchmark approaches,
the proposed method can achieve superior performance.

Keywords: Air-terrestrial integrated networks (ATIN) · Task
offloading · Cost minimization · Machine learning

1 Introduction

WITH the popularization of artificial intelligence, 5G and other technologies,
the explosive growth of data has significantly accelerated the advancement of
mobile network and wireless communication [1]. However, the above-mentioned
wide-range applications all require large amounts of computing resources, which

This work was supported in part by the National Natural Science Foundation
of China (No. 62201212 the Natural Science Foundation of Hebei Province (No.
F2022502017),), the Zhejiang Key Laboratory of Medical Electronics and Digital
Health (No. MEDH202215), and the Fundamental Research Funds for the Central
Universities (No. LAPS21018).

Q. Liang et al. (Eds.): AIC 2022, LNEE 871, pp. 329–339, 2023.
https://doi.org/10.1007/978-981-99-1256-8_39

poses enormous challenges to the battery and computing ability of terminal devices [2]. Cloud computing could offload computing-intensive tasks to cloud servers, which often leads to large transmission delay owing to the considerable distance between equipment and cloud server, making it impossible to satisfy delay-sensitive task like autonomous driving. By shifting computing demand and delay-sensitive tasks to the wireless network edge, Mobile Edge Computing (MEC) improves system quality of service (QoS) [3], which has been extensively researched as a solution to the issue. The power consumption minimization problem of a multi-user MEC system with energy harvesting devices was studied in [4]. It is essential to know how to allocate the MEC server's computing resources effectively because of its limited computing capacity. Moreover, due to the limited terminal battery capacity, energy-saving method needs to be explored to prolong battery life [5]. Last but not least, the ground-based 5G system also can not provide seamless service especially for hotspot and remote area.

The air-terrestrial integrated network (ATIN) is able to address the above-mentioned issues by complementary integration of the air and ground segments [6], which has many advantages, such as seamless coverage, low latency, improved throughput and alleviation of network congestion [7]. Reinforcement learning (RL) was initially created to address the dilemma of making decisions without being aware of global state information (GSI). It can be used to resolve model-free MDP issues and has been extensively studied in [8,9]. In [10], to implement the allocation of computing resources for online learning and boost the overall effectiveness of the system, a dynamic resource allocation scheme based on RL was presented. However, since the state space and action space increases exponentially, it cannot effectively learn high-dimensional information and handle large state space problem, causing the curse of dimensionality. Therefore, Deep RL (DRL) is proposed to take advantage of both the decision-making and learning prediction capabilities offered by RL and deep learning [11].

Inspired by the mentioned technologies and to tackle the above challenges, we propose an ATIN model, which consists of one high altitude platform station (HAPS) and multiple unmanned aerial vehicles (UAVs) equipped with edge servers. Each user has the option of carrying out tasks locally or executing them to the edge. Then, our objective is to optimize the weighting of delay and energy consumption. By giving the delay more consideration, we may pay attention to the delay reduction of low-latency applications. To reduce user equipment's energy consumption, we can pay more attention to the amount of system power consumption. The problem is formulated by jointly considering task offloading decision and computing resource allocation, which, however, is a mixed integer nonlinear programming (MINLP) issue due to the strong coupling between optimization variables. Therefore, we decompose it into two subproblems and design a machine-learning-based approach to solve the first offloading decision making issue. For the second computing resource allocation subproblem, a Greedy-based solution is proposed. We conduct a lot of simulations and compare with some benchmark approaches. The results demonstrate that, our method can produce superior consequences in terms of terminal energy consumption and delay.

2   System Model and Problem Formulation

An ATIN model is proposed as depicted in Fig. 1. Each UAVs and HAPS is equipped with an edge server to provide access and data computing service. Assuming there are 1 HAPS and K UAV, where s_0 and $s_1, \ldots, s_K$ represent HAPS and UAV, respectively. Terminal users $\mathcal{E} = \{e_1, e_2, \ldots, e_I\}$ are randomly distributed on the ground. In this paper, the time slot model is adopted, and there are totally T slots $\mathcal{T} = \{1, 2, \ldots, t, \ldots, T\}$. Considering UAVs' dynamics, users may go beyond the communication range of one server and into another. Considering the terminal device's limited battery capacity and computing capability, tasks that cannot be handled locally will be offloaded to the edge. The ith terminal device's offloading decision can be represented by the vector $\chi = \{x_{i,0}(t), x_{i,1}(t), \ldots, x_{i,K}(t)\}$ Fig. 1.

Fig. 1. System model.

2.1   Communication Model

If a user equipment (UE) e_i chooses edge server s_k for task offloading in slot t , then $x_{i,k}(t) = 1$. For the communication model , we mainly considers large-scale fading [12]. Therefore, the path loss can be determined by

$$L_{i,k,t} = 20\log_{10}\left(\frac{4\pi f_c\sqrt{d_{i,k,t}^2 + r_{i,k,t}^2}}{c}\right) + P_{i,k,t}^{Los}\eta_{i,k,t}^{Los} + \left(1 - P_{i,k,t}^{Los}\right)\eta_{i,k,t}^{NLos} \quad (1)$$

where $r_{i,k,t}$ reflects the distance among the UE and the edge server in horizontal space, $d_{i,k,t}$ is the flying height of the UAV and HAPS, f_c is the carrier frequency,

c is the speed of light. $\eta_{k,n,t}^{Los}$ and $\eta_{k,n,t}^{NLos}$ represent the additional loss generated in addition to the free space path loss of line-of-sight (LoS) links and non-line-of-sight (NLoS) links. The LoS probability is calculated by

$$P_{i,k,t}^{LoS} = \frac{1}{1 + b_1 exp\{-b_2[arctan(\frac{d_{i,k,t}}{r_{i,k,t}}) - b_1]\}} \tag{2}$$

where b_1, b_2 are variables determined by environmental information. Therefore, the UE-edge server link's transmission rate can be calculated by

$$R_{i,k}(t) = B_{i,k}log_2\left(1 + \frac{x_{i,k}(t)P^{TX}10^{-\frac{L_{i,k,t}}{10}}}{\sigma^2}\right) \tag{3}$$

where, $B_{i,k}$ represents the bandwidth, P^{TX} is the UE data transmission power, and σ^2 represents the additive white Gaussian noise power.

2.2   Computing Model

We describe the model for local computing as well as edge computing. The computing task of UE e_i can be represented by a two-tuple vector $X_i = (D_i, F_i)$, where D_i represents the task data size, F_i is the amount of needed CPU cycles. When a task arrives at a particular user e_i in the tth slot, the system should choose whether to compute locally or offload it to the edge.

Local Computing Model. The execution delay and energy consumption for local computing is calculated by

$$T_i^L(t) = \frac{F_i}{f_i^L(t)} \tag{4}$$

$$E_i^L(t) = \mathcal{K}\left(f_i^L(t)\right)^2 F_i \tag{5}$$

where $f_i^L(t)$ represents the allocated local CPU cycles, and $\mathcal{K}$ represents the efficiency coefficient depending on the chip structure [13].

Edge Computing Model. According to the previously described communication model, for e_i offloading tasks to edge server s_k, the corresponding transmission time and computation time can be obtained by formula (6) and (7), respectively

$$T_{i,k}^{E,trans}(t) = \frac{D_i}{R_{i,k}^A(t)} \tag{6}$$

$$T_{i,k}^{E,exe}(t) = \frac{F_i}{f_{i,k}^A(t)} \tag{7}$$

$$T_{i,k}^{E,off}(t) = T_{i,k}^{E,trans}(t) + T_{i,k}^{E,exe}(t) \tag{8}$$

The terminal energy consumption can be calculated by

$$E_i^E(t) = T_{i,k}^{E,trans}(t)P^{TX} \tag{9}$$

The energy consumption of the ith UE for task processing can be expressed by

$$E_i(t) = x_{i,k}(t)E_i^L(t) + \sum_{k=1}^{K} x_{i,k}(t)E_i^E(t) \tag{10}$$

The delay of e_i is

$$T_i(t) = x_{i,k}(t)T_i^L(t) + \sum_{k=1}^{K} x_{i,k}(t)T_{i,k}^{E,off}(t) \tag{11}$$

Note that, after being processed by the edge server, the task data volume is considerably less than it was previously. Therefore, the delay of returning the result from edge server can be ignored.

2.3 Problem Formulation

In this subsection, by taking into account task offloading decision-making and computing resource allocation simultaneously, the problem of minimizing the weighting of delay and energy consumption is expressed as follows

$$\mathbf{P0}: \min_{X,F} \lim_{T \to \infty} \frac{1}{T}\sum_{t=1}^{T}\sum_{i=1}^{I} \varpi_e E_i(t) + \varpi_t T_i(t) \tag{12}$$

$$\text{s.t.} \quad x_{i,k}(t) \in \{0,1\} \quad \forall i \in \mathcal{E}, \forall t \in \mathcal{T} \tag{13a}$$

$$\sum_{k=0}^{K} x_{i,k}(t) = 1 \quad \forall i \in \mathcal{E}, \forall t \in \mathcal{T} \tag{13b}$$

$$\sum_{i=1}^{I} f_{i,k}^A(t) \le f_k^{A,max}(t) \quad \forall i \in \mathcal{E}, \forall t \in \mathcal{T} \tag{13c}$$

$$f_{i,k}^A(t) \ge 0 \quad \forall i \in \mathcal{E}, \forall t \in \mathcal{T} \tag{13d}$$

In (12), ϖ_e and ϖ_t are weights of energy consumption and delay respectively, and $\varpi_e + \varpi_t = 1$. Constraint (13a) and (13b) mean each user can only select at most one edge server for task offloading or process locally in each slot. (13c) and (13d) statement that the MEC server's computing resources are not exceeded by the assigned computing resources.

3   Problem Solution

3.1   DRL-Based Offloading Decision Making

1) **MDP-Based Network Environment.** The task offloading can be modeled as a discrete-time Markov decision process (MDP), expressed as $\{\mathcal{S}, \mathcal{A}, \mathcal{R}\}$, where $\mathcal{S}$ is the state space, $\mathcal{A}$ is the action space, and $\mathcal{R}$ is the value reward function. The sets of state, action, and reward can be described as follows.

State: The system state space for terminal u_k includes its transmission rate and task-related information, expressed as $\mathcal{S}_i(t) = \left\{ R_{i,k}^A(t), D_i(t), F_i \right\}$. Historical state information of $\mathcal{S}_i(1), \ldots, \mathcal{S}_i(t-1)$ and the system's future status is predicted using the existing observations.

Action: In the tth time slot, each terminal either offloads task to edge server or processes locally denoted by variable $x_{i,k}(t)$. Therefore, the action space for terminal u_k can be describe by $\mathcal{A}_i(t) = \{x_{i,0}(t), x_{i,1}(t), \ldots, x_{i,K}(t)\}$.

Reward: In the tth time slot, each agent takes action and gets instant reward. Since minimizing terminal energy usage while maintaining QoS is the optimization goal, the reward function can be designed as

$$R(t) = -\sum_{i=1}^{I} \varpi_e E_i(t) + \varpi_t T_i(t) \tag{14}$$

2) **Problem Solution with Actor-Critic Framework.** Owing to the high-speed mobility of UAVs, the state space grows exponentially, which makes the issue intractable. Therefore, in this subsection, we leverage an actor-critic-based DRL method to make use of both the predicting ability of deep learning, and the decision-making capability of RL. The actor-critic framework contains an actor network and a critic network. The actor network takes appropriate actions occording to the current state and observation, while the value function to evaluate and update the current policy is generated by the critic network.

For our solution, the policy $\pi(a|s, \vartheta) = P(s^t = a|s^t = s, \vartheta^t = \vartheta)$ represented by parameter ϑ, denotes the probability that the system takes an action a in state s under the policy π with ϑ in slot t. We define $J(\vartheta)$ as the value function of state s_0 at the beginning

$$J(\vartheta) \approx V_{\pi_\vartheta}(s_0) = \mathbb{E}_{\pi_\vartheta}\left[\sum_{t=0}^{T} \gamma^t R\left(s^t, a^t\right) \middle| \pi\left(a^t \middle| s^t, \vartheta\right)\right] \tag{15}$$

We could utilize the gradient descent approach to gradually update the policy parameter ϑ of minimizing the value function $J(\vartheta)$, which can be expressed as

$$\vartheta^{t+1} = \vartheta^t - \varphi \nabla J\left(\vartheta^t\right) \tag{16}$$

Since the learning rate is represented as φ, by using the policy gradient approach, we can obtain it. We utilize a new update method instead of the

Algorithm 1. DRL-Based Offloading Algorithm for ATIN

1: **Input:** User and Server information: location, mobility traces, $B_{i,k}, D_i, F_i$
   Output: Task offloading decision: $x_{i,k}(t)$
2: Set the actor network parameter ϑ and the critic network parameter ϖ to their initial values.
3: Randomly initialize actor network $\pi\left(s^t|\vartheta\right)$ and critic network $V\left(s^t, \varpi\right)$
4: **while** episold=1,..., E **do**
5:    Setup simulation environment and generate initial state
6:    **while** time slot $t = 1, \ldots, T$ **do**
7:       Execute action $a_i(t)$ according to the policy $\pi\left(s^t|\vartheta\right)$
8:       Calculate reward $R(t)$ and transfer to next state
9:       Calculate $\mathcal{E} = R(t) + \gamma V\left(s^{t+1}, \varpi\right) - V\left(s^t, \varpi\right)$
10:      Update ϑ according to (17)
11:      Update ϖ according to (18)
12:   **end while**
13: **end while**

discounted return of cost approach to greatly accelerate the agent learning rate. Specifically, we use $V\left(s^t, \varpi\right)$ to estimate the value function of each state space, where ϖ is the update parameter of the critic network.

$$\vartheta^{t+1} = \vartheta^t - \varphi\left(R(t) + \gamma V\left(s^{t+1}, \varpi\right) - V\left(s^t, \varpi\right)\right) \frac{\nabla_\vartheta \pi\left(a^t|s^t, \vartheta\right)}{\pi\left(a^t|s^t, \vartheta\right)} \tag{17}$$

$$\varpi^{t+1} = \varpi^t - \varphi' \nabla_\varpi L(\varpi) \tag{18}$$

where the learning rate for the critic network is represented by φ', and the loss function $L(\varpi)$ is represented as

$$L(\varpi) = \left|R(t) + \gamma V\left(s^{t+1}, \varpi\right) - V\left(s^t, \varpi\right)\right|^2 \tag{19}$$

The proposed reinforcement learning-Based solution is shown by **Algorithm 1**. In lines 1–3, the actor network and critic network are randomly initialized, and the algorithm is configured with E rounds of episodes, each with T time slots. In lines 4–6, the actor network takes action $a_i(t)$ according to the policy $\pi\left(s^t|\vartheta\right)$, and calculates the reward $R(t)$. Then, the system transfers to the next state. We utilize TD-error $\mathcal{E} = R(t) + \gamma V\left(s^{t+1}, \varpi\right) - V\left(s^t, \varpi\right)$ to update the network in line 9, which can greatly speeds up the convergence. Updating critic network and actor network parameters are shown in lines 10–11.

3.2   Greedy-Based Computing Resource Allocation

In this subsection, the CPU cycle frequency allocation for processing task offloaded from e_i in the tth slot is decided by UAV s_k. The fundamental concept of greedy server-side computing resource allocation is to give priority to UE devices . Initialize the computing resources of s_k first, and then the UE device sets N_i with the requirements for offloading computing resources to UAV. Next, determine the total computing resources allotted to e_i. The UE device m_i^* that has the highest goal value will be chosen. Remove e_i^* from the set N_i after that, and then update UAV computing resources. When there are no UEs with unassigned computing resources or no UAVs with computing resources, the iteration of computing resource allocation comes to an end.

4   Performance Analysis and Simulations

Extensive simulations are conducted to verify the method's effectiveness. The simulation platform of this article is Python3.7 and Tensorflow2.0. We consider a 1000m ×1000m area with 50 terminal devices, 3 UAVs and 1 HAPS. Within the region under consideration, the devices are positioned at random. The UAVs circle around a 200-meter-diameter area, and there is roughly a 120° angle between two UAV. Each UAV has a 100-meter flying height and a 300-meter communication coverage range, respectively. The maximum available computing resources $f_0^{A,max}(t)$ and $f_k^{A,max}(t)$ are uniformly distributed within [164, 236], and [30.5, 42.6] GHz, respectively. In Table I, the specific simulation settings are presented. The neural network chooses ReLU as the activation function, each layer has 64 neurons, the learning rate δ is 0.0001, and the discount coefficient γ is 0.9. We compare with three benchmark methods:

(a) UCB Algorithm: UCB based offload strategy [15] is adopted for computing offloading, and the same method as this paper is adopted for computing resource allocation.
(b) Random Method: The user equipment randomly offloads tasks to UAVs or HAPS, and the edge server randomly allocates computing resources.

Table 1. Simulation parameters

Parameter	Value	Parameter	Value
T	100	τ	$0.1\,\mathrm{s}$
I	50	K	3
P^{TX}	23 dbm	f_c	0.1 GHZ
$d_{i,0,t}, d_{i,k,t}$	$150\,\mathrm{m}, 100\,\mathrm{m}$	σ^2	-104 dBm
$\eta_{i,j}^{LoS}, \eta_{i,j}^{NLoS}$	0.1,21	b_1, b_2	4.88,0.43
$B_{i,k}$	1 Mhz	$f_i^L(t)$	$5 \times 10^8 cycles/bit$
$\mathcal{K}$	$1 \times 10^{-27} J/Hz^3/s$	D_i	$5 \times 10^5 bits$

The convergence performance for learning-based task offloading is displayed in Fig. 2. According to the definition of the reward function, the larger the reward value is, the lower UE system costs. As can be seen, the proposed approach

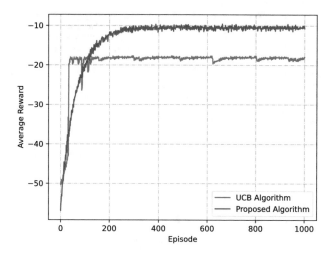

Fig. 2. Convergence performance.

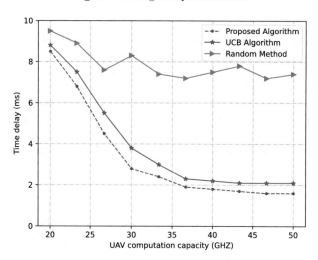

Fig. 3. Impact of the computing capacity on time delay.

converges after approximately 300 episodes. The adopted actor-critic solution, which employs the TD error-based network updating method and the capacity of the deep neural network to offer approximation to complex functions, is what causes the better performance. This demonstrates the potency of our proposed DRL-based offloading decision-making method.

In Fig. 3, we contrast the UE delay versus different UAV edge server side's maximal computing capacity. As can be observed, our propose method offloads more tasks to the edge side as the maximum processing capability of UAVs

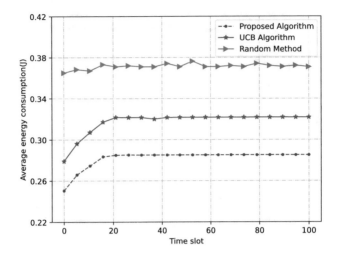

Fig. 4. Average energy consumption.

increases, which can lower the time delay for UEs. The results also show that our proposed approach outperforms other algorithms.

The average energy consumption of UEs over time is shown in Fig. 4. As can be observed, compared to other methods, our approach consistently performs well. This is because learning-based method jointly optimizes offloading decison and resource allocation for UEs, thus it can significantly reduce the average energy consumption. Meanwhile due to that the random algorithm cannot learn the optimal offloading decision, the average energy consumption is much higher.

5   Conclusion

In this paper, we propose an ATIN model consisting of multiple UAVs and one HAPS equipped with edge servers. By jointly optimizing the decision to offload terminal tasks and the computing resource allocation. Due to that the formulated issue is an MINLP problem, it is divided into two subproblems. DRL and Greedy-based solutions are designed to solve the first offloading decision making issue and the second computing resource allocation subproblem, respectively. Various simulations confirm the effectiveness of our method.

References

1. Liu, Y., Peng, M., Shou, G., Chen, Y., Chen, S.: Toward edge intelligence: multi-access edge computing for 5g and internet of things. IEEE Internet Things J. **7**(8), 6722–6747 (2020)
2. Qin, P., Fu, Y., Feng, X., Zhao, X., Wang, S., Zhou, Z.: Energy-efficient resource allocation for parked-cars-based cellular-V2V heterogeneous networks. IEEE Internet of Things J. **9**(4), 3046–3061 (2022)

3. Qin, P., Fu, Y., Tang, G., Zhao, X., Geng, S.: Learning-based energy efficient task offloading for vehicular collaborative edge computing. IEEE Trans. Veh. Technol. (2022). https://doi.org/10.1109/TVT.2022.3171344
4. Zhang, G., Chen, Y., Shen, Z., Wang, L.: Distributed energy management for multiuser mobile-edge computing systems with energy harvesting devices and QoS constraints. IEEE Internet Things J. **6**(3), 4035–4048 (2019)
5. Wu, Y., et al.: Secrecy-driven resource management for vehicular computation offloading networks. IEEE Netw. **32**(3), 84–91 (2018)
6. Qin, P., Zhu, Y., Zhao, X., Feng, X., Liu, J., Zhou, Z.: Joint 3D-location planning and resource allocation for XAPS-enabled C-NOMA in 6G heterogeneous Internet of things. IEEE Trans. Veh. Technol. **70**(10), 10594–10609 (2021)
7. Qin, P., Fu, Y., Zhao, X., Wu, K., Liu, J., Wang, M.: Optimal task offloading and resource allocation for C-NOMA heterogeneous air-ground integrated power internet of things networks. IEEE Trans. Wireless Commun. (2022). https://doi.org/10.1109/TWC.2022.3175472
8. Silver, D., et al.: Mastering the game of go with deep neural networks and tree search. Nature **529**, 484–489 (2016)
9. Cheng, N., et al.: Space/aerial-assisted computing offloading for IoT applications: a learning-based approach. IEEE J. Sel. Areas Commun. **37**(5), 1117–1129 (2019)
10. Liu, S., Liu, H., Zheng, K.: A reinforcement learning-based resource allocation scheme for cloud robotics. IEEE Access **6**, 17215–17222 (2018). Sun, Y., Zhou, S., Xu, J.: EMM: energy-aware mobility management for mobile edge computing in ultra dense networks. IEEE J. Sel. Areas Commun. **35**(11), 2637–2646 (2017)
11. Zhu, S., Gui, L., Cheng, N., Zhang, Q., Sun, F., Lang, X.: UAV-enabled computation migration for complex missions: a reinforcement learning approach. IET Commun. **14**(15), 2472–2480 (2020)
12. Qin, P., Wang, M., Zhao, X., Geng, S.: Content service oriented resource allocation for space-air-ground integrated 6G networks: a three-sided cyclic matching approach. IEEE Internet Things J. (2022). https://doi.org/10.1109/JIOT.2022.3203793
13. Shang, B., Liu, L.: Mobile-edge computing in the sky: energy optimization for air-ground integrated networks. IEEE Internet Things J. **7**(8), 7443–7456 (2020)
14. Liu, J., Zhao, X., Qin, P., Geng, S., Meng, S.: Joint dynamic task offloading and resource scheduling for WPT enabled Space-Air-Ground Power Internet of Things. IEEE Trans. Netw. Sci. Eng. **9**(2), 660–677 (2022)
15. Sun, Y., Zhou, S., Xu, J.: EMM: Energy-aware mobility management for mobile edge computing in ultra dense networks. IEEE J. Sel. Areas Commun. **35**(11), 2637–2646 (2017)

Distributed Dynamic Spectrum Access for D2D Communications Underlying Cellular Networks Using Deep Reinforcement Learning

Zhifeng Jiang[1], Liang Han[1,2(✉)], and Xiaocheng Wang[1,2]

[1] College of Electronic and Communication Engineering, Tianjin Normal University, Tianjin 300387, China
hanliang@tjnu.edu.cn
[2] Tianjin Key Laboratory of Wireless Mobile Communications and Power Transmission, Tianjin Normal University, Tianjin 300387, China

Abstract. In this paper, we investigate a deep Q-network (DQN)-based method for applying a dynamic spectrum access model to device-to-device (D2D) communications underlying cellular networks. Dynamic spectrum access (DSA) devices have two critical concerns, namely avoiding interference to primary users (PUs) and interference coordination with other secondary users (SUs). We consider that the issues faced by DSA users are also applicable to the D2D communication underlying cellular network. Therefore, we propose a distributed dynamic spectrum access scheme based on deep reinforcement learning (DRL). It enables each D2D user to learn a reliable spectrum access policy through imperfect spectrum sensing without knowledge of system prior information, avoiding collisions with cellular users and other D2D users and maximizing system throughput. Finally, the simulation results demonstrate the effectiveness of our proposed dynamic spectrum access scheme.

Keywords: Device-to-device (D2D) communication · Distributed dynamic spectrum access (DSA) · Deep reinforcement learning (DRL)

1 Introduction

Radio spectrum resources are an essential resource. According to a white paper published by Cisco on global mobile data traffic forecasts for 2017–2022, global mobile data traffic will grow sevenfold between 2017 and 2022 [1]. However, related studies have revealed a phenomenon that many spectrum resources are not used effectively [2,3]. D2D communication technology is considered a feasible solution to the problem of poor spectrum resources, with the advantages of improved spectrum efficiency and reduced communication delays [4,5]. Furthermore, considering the limitations of traditional static spectrum allocation policies, dynamic spectrum access techniques have also been proposed to improve spectrum efficiency [6]. In D2D communication technology, cellular users are

© The Author(s), under exclusive license to Springer Nature Singapore Pte Ltd. 2023
Q. Liang et al. (Eds.): AIC 2022, LNEE 871, pp. 340–348, 2023.
https://doi.org/10.1007/978-981-99-1256-8_40

subject to severe interference when D2D users share the same spectrum as cellular users and strong interference between D2D users can also seriously affect the quality of communication [4,7,8]. Similarly, dynamic spectrum access also faces two fundamental problems, namely interference coordination between DSA users and interference suppression for primary users [9].

Previous research has proposed a number of schemes for spectrum allocation between D2D users and cellular users [10,11]. These studies investigated the reuse of cellular user resources by D2D communication users in a non-orthogonal spectrum allocation. In [10], the authors proposed a distributed Q-learning-based spectrum allocation scheme to maximize D2D user system throughput and maintain the QoS requirements for cellular users. In [11], the authors proposed a distributed DRL-based spectrum allocation scheme to address the issue of interference and resource allocation between D2D and cellular users, with the aim of maximizing system throughput. However, little existing research has considered the use of distributed spectrum access schemes to avoid conflicts between D2D communication users and cellular users as well as other D2D users.

In this paper, we consider an uplink scenario of a D2D underlying cellular communication network, and to address the collision problem between DUEs and CUEs, we propose a DRL-based distributed dynamic spectrum access scheme. In particular, we introduce the concept of a "reusable area" [12], where D2D users can choose the number of reusable CUEs based on the range of "reusable areas". According to the DRL theory, we enable each agent to learn the optimal access policy only through imperfect spectrum sensing without knowing the system a priori information, increasing the system throughput while avoiding collisions with other DUEs and CUEs.

2 System Model

We consider a dynamic spectrum access scenario in the uplink of a D2D underlying cellular network. As shown in Fig. 1, the system model includes N cellular users (CUEs) denoted by $\mathbb{N} = \{1, 2, \ldots, N\}$ and K pairs of D2D users denoted by $\mathbb{K} = \{1, 2, \ldots, K\}$, each D2D pair consists of a set of transmitters (DTx) and receivers (DRx). d_{ii} denotes the distance between DTx and DRx, d_{jk} denotes the distance between the CUEs and the BS, and d_{ik} denotes the distance between DTx and the BS. We assume that the system has N channels and that each CUE transmits on a unique channel, thus avoiding interference between CUEs.

2.1 Channel Model

We adopt the WINNER II channel model to calculate the path loss generated by the signal propagation in space [15], which is described as a distance-dependent function

$$PL(d, f_c) = \overline{PL} + B \log_{10}(d[m]) + C \log_{10}\left(\frac{f_c(GH_z)}{5}\right), \tag{1}$$

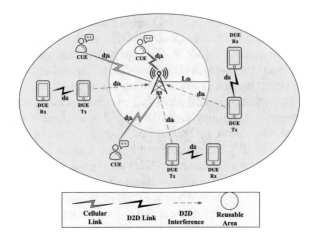

Fig. 1. Uplink scenario for D2D communications underlying cellular networks.

where f_c denotes the carrier frequency, $\overline{PL}$, B and C denote the unit distance loss reference, the path loss exponent and the path loss frequency dependence, respectively. For simplicity, we assume the existence of a strong line of sight (LOS) path between the signal transmission links. Therefore, our model can use the Rician channel model for channel modeling [13], which can be expressed as

$$h = \sqrt{\frac{\kappa}{\kappa+1}}\sigma e^{j\theta} + \sqrt{\frac{1}{\kappa+1}}CN\left(0, \sigma^2\right) \tag{2}$$

where $\sigma^2 = 10^{-\frac{\overline{PL}+B\cdot\log_{10}(d[m])+C\cdot\log_{10}\left(\frac{f_c[GHz]}{5}\right)}{10}}$ is determined by path loss, κ means the $\kappa - factor$, defined as the power ratio of the LOS component to the scattering component, θ denotes the phase and takes the value of a uniform distribution between 0 and 1, $CN\left(\cdot\right)$ denotes a circularly symmetric complex Gaussian random variable.

2.2   Uplink Signal Model

For the uplink scenario, when DUEs and CUEs transmit in the same time slot, DUEs can cause harmful interference to CUEs. Hence, the instantaneous signal to interference plus noise ratio (SINR) received by the BS from the CUEs can be expressed as

$$SINR_j = \frac{P_c \cdot |h_{jk}|^2}{P_d \cdot |h_{ik}|^2 + B \cdot N_0} \tag{3}$$

where P_c and P_d represent the transmit power of the CUE and the DTx, respectively. $|h_{jk}|^2$ denotes the channel gain of the cellular link, and $|h_{ik}|^2$ denotes the channel gain from the ith D2D transmitter to the BS, , which can be derived according to (2). B and N_0 represent the channel bandwidth and noise spectral density, respectively. Furthermore, we assume in this model that each channel can be used by at most one D2D pair.

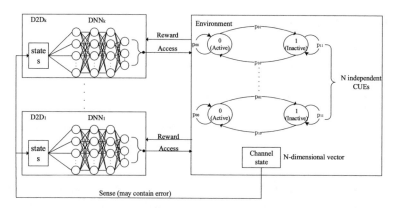

Fig. 2. Dynamic spectrum access framework based on D2D underlying cellular networks.

3 DRL-based Dynamic Spectrum Access Scheme

3.1 Deep Reinforcement Learning

A reinforcement learning model contains three components: possible states in the environment, possible actions that the agent may take based on a policy π, and a feedback reward function that the agent receives after making an action. These three components are defined as s_t, a_t, and r_{t+1}. The goal of the agent is to learn an optimal policy π^* to maximize the cumulative discount reward $R_t = \sum_{i=0}^{\infty} \gamma^i r_{t+1+i}$, where $\gamma \in [0,1]$ represents the discount factor. Q-values are updated with the following rules:

$$Q\left(s_t, a_t\right) = Q\left(s_t, a_t\right) + \\ \alpha\left(r_{t+1} + \gamma \max_a Q\left(s_{t+1}, a\right) - Q\left(s_t, a_t\right)\right), \tag{4}$$

where $\alpha \in (0,1]$ is the learning rate. Furthermore, the policy function π is updated by means of the ϵ-greedy algorithm.

DRL uses deep neural network (DNN) to approximate Q values (DQN), i.e. $Q(s_t, a_t; \boldsymbol{\theta}) \approx Q(s_t, a_t)$, where $\boldsymbol{\theta}$ is the network weights. In DQN, the TD algorithm is mostly used to calculate the loss function,

$$Loss(\theta) = E[(y_t - Q(s_t, a_t; \boldsymbol{\theta}))^2] \tag{5}$$

where y_t represents the target Q-value and is defined as

$$y_t = r_{t+1} + \gamma \max_a Q\left(s_{t+1}, a; \boldsymbol{\theta}\right) \tag{6}$$

After that, the agent can minimize the loss function by the gradient descent algorithm as follows:

$$\boldsymbol{\theta_{t+1}} = \boldsymbol{\theta_t} + \alpha E\left[(y_t - Q\left(s_t, a_t; \boldsymbol{\theta}\right)) \nabla Q\left(s_t, a_t; \boldsymbol{\theta}\right)\right] \tag{7}$$

where α is the learning rate.

3.2   Uplink Dynamic Spectrum Access Framework

In our proposed scheme, as shown in Fig. 2, we have designed the distributed dynamic spectrum access as a deep reinforcement learning model. In Sect. 2, we show that the system model consists of N channels and K D2D users, where each channel is occupied by a CUE. Therefore, we describe the channel states as *Active* and *Inactive*, denoted by 0 and 1 respectively, where *Active* indicates that the channel is occupied by a CUE and *Inactive* indicates that the channel is not occupied by a CUE. In addition, the detailed definitions of "state", "action" and "reward" are as follows.

State: At the beginning of each time slot, each D2D pair will sense the channel state in the environment, which may contain errors. Subsequently, the agent uses the sensed results as input data for the neural network to be trained. Therefore, the state space $\mathbf{S}^k(t)$ of each D2D pair is defined as $\mathbf{S}^k(t) = [s_1^k(t), \ldots, s_n^k(t)]$, where $\mathbf{S}^k(t)$ denotes an N-dimensional vector, k denotes the kth D2D pair, n denotes the number of channels and $s_n^k(t)$ denotes the state of the channel (*Active* or *Inactive*).

Action: The agent decides whether to access and which wireless channel to access based on the spectrum access policy. Hence, the action space A can be defined as $A \in \{0, 1, \ldots, N\}$, where $a_t = 0$ means that the agent does not access the channel and $a_t = N$ means that the agent accesses the nth channel.

Rewards: According to the situations that the agent may face after making an action choice, the following reward function setting scheme is developed.

1. The D2D pair collides with the CUE. This indicates that the D2D pair accesses the channel where the CUE is located when the cellular link resources cannot be reused. In Sect. 2, we mention the concept of warning signals. Therefore, we give a penalty value of -4 as the result of a warning signal being received by the agent. For convenience, we define this case as $\mathbb{C}$.
2. A collision between D2D users. This case indicates that different D2D users are accessing the same channel. We set the reward value for this case to 0 and define this case as $\mathbb{D}$.
3. The D2D pairs do not access any channel. We set the reward value for this case to 1. Similarly, we define this case as $\mathbb{I}$.
4. The D2D pair successfully accesses the channel. The reward value for this case should be set to the maximum. We considered the normalized $S\hat{I}NR_j$ and applied it to our reward function setting, described as $1 + \log_2(1 + S\hat{I}NR_j)$. We define this case as $\mathbb{S}$.

In summary, the reward function for the kth D2D pair on the nth channel can be described as

$$r_{t+1}^k = \begin{cases} -4, & \text{Case } \mathbb{C} \\ 0, & \text{Case } \mathbb{D} \\ 1, & \text{Case } \mathbb{I} \\ 1 + \log_2(1 + S\hat{I}NR_j), & \text{Case } \mathbb{S} \end{cases} \qquad (8)$$

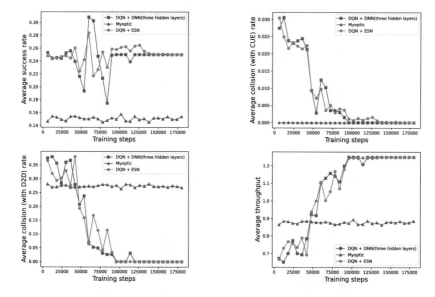

Fig. 3. Performance evaluation in non-orthogonal scenarios ($P = \frac{1}{2}R$).

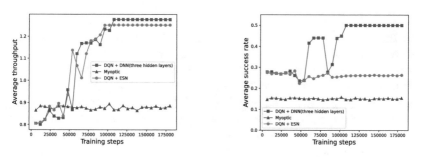

Fig. 4. Performance evaluation in non-orthogonal scenarios ($P = R$).

4 Performance Evaluation

In this section, we evaluate the performance of the algorithmic framework proposed in the scheme. Specifically, we compare the algorithm used in the scheme with the Myopic algorithm [14] based on a priori information about the system and the DQN+ESN [13] to verify the performance.

In our scheme, we consider a cellular cell scenario with a radius of 100m. The locations of the D2D pair and the CUE in this cell are randomly generated and we specify that the communication distance between the transmitter and the receiver of the D2D pair is randomly generated in the range of 20 m and 40 m. Therefore, the location of the CUE may fall within the "reusable area" that we have defined. We express the percentage of the reusable area in the cell by $P = \frac{\pi \cdot L_{th}^2}{\pi \cdot R^2}$, where the choice of the threshold distance L_{th} is determined by the value of P. The specific simulation parameters are shown in Table 1.

Table 1. Simulation parameters

Parameter	Value
Cell radius	100 m
Carrier frequency	2 GHz
Bandwidth	2 MHz
D2D transmit power P_d	23 dBm
CUE transmit power P_C	23 dBm
Noise power density N_0	−174 dBm/Hz
Path loss reference $\overline{PL}$	41
Path loss exponent B	22.7
Path loss frequency dependencer C	20
κ factor	8
Learning rate α	0.01
Discount factor γ	0.5
Exploration ϵ	$0.7 \longrightarrow 0$
Spectrum sensing error probability	$(0, 0.2)$

In this scenario, we set the number of D2D users to 4 and the number of CUEs to 2. The positions of the CUEs are randomly generated, and by calculating the distances from the CUEs to the BS, we can obtain the distances to be 95.75 m and 30.36 m. Therefore, we first consider the case of $P = \frac{1}{2}R$, after which we can calculate the corresponding threshold distance L_{th} of $50\sqrt{2}$ m. According to the range of reusable zones corresponding to the threshold distance, the D2D pair can reuse one CUE resource. The simulation results are shown in Fig. 3, where the performance obtained using the DQN-based method is significantly better than using the Myopic algorithm. In particular, the Myopic algorithm selects the action with the greatest immediate reward and therefore it performs well in avoiding collisions with the CUE. However, simulation results show that using the DQN-based method after training also maximizes throughput while achieving collision avoidance. In this scenario, our scheme performs approximately the same as DQN+ESN. Additionally, we consider the non-orthogonal access scenario when $P = R$. In this scenario, the reusable area covers the entire cellular cell. Therefore, all CUEs in the cell can be reused by the D2D pair. The simulation result is shown in Fig. 4, which shows that our scheme achieves the theoretical maximum success rate and has better throughput.

5   Conclusion

In this paper, we investigated the case of dynamic spectrum access in the uplink of a D2D underlying cellular network. Specifically, we proposed a distributed dynamic spectrum access scheme under imperfect spectrum sensing conditions, which aims to allow D2D users to learn an optimal spectrum access policy to maximize throughput without knowing a priori information. Besides, we introduced the concept of a reusable area, where the D2D user can choose the number of reusable CUEs based on the coverage of that area. Simulation results show that our scheme can avoid collisions while maximizing the system throughput.

Acknowledgment. This work was supported by the National Natural Science Foundation of China (62001327, 61701345), Natural Science Foundation of Tianjin (18JCZDJC31900).

References

1. Forecast, G., et al.: Cisco visual networking index: global mobile data traffic forecast update, 2017–2022. Update **2017**, 2022 (2019)
2. Yin, S., Chen, D., Zhang, Q., Liu, M., Li, S.: Mining spectrum usage data: a large-scale spectrum measurement study. IEEE Trans. Mob. Comput. **11**(6), 1033–1046 (2012)
3. McHenry, M.A., Tenhula, P.A., McCloskey, D., Roberson, D.A., Hood, C.S.: Chicago spectrum occupancy measurements & analysis and a long-term studies proposal. In: Proceedings of the first International Workshop on Technology and Policy for Accessing Spectrum, August 2006
4. Asadi, A., Wang, Q., Mancuso, V.: A survey on device-to-device communication in cellular networks. IEEE Commun. Surv. Tutorials **16**(4), 1801–1819 (2014)
5. Ansari, R.I., Chrysostomou, C., Hassan, S.A., Guizani, M., Mumtaz, S., Rodriguez, J., Rodrigues, J.J.: 5g d2d networks: techniques, challenges, and future prospects. IEEE Syst. J. **12**(4), 3970–3984 (2018)
6. Kolodzy, P., Avoidance, I.: Spectrum policy task force. Federal Commun. Comm. Washington, DC, Rep. ET Docket **40**(4), 147–158 (2002)
7. Shah, S.W.H., Mian, A.N., Crowcroft, J.: Statistical qos guarantees for licensed-unlicensed spectrum interoperable d2d communication. IEEE Access **8**, 27 277–27 290 (2020)
8. Kai, Y., Wang, J., Zhu, H., Wang, J.: Resource allocation and performance analysis of cellular-assisted OFDMA device-to-device communications. IEEE Trans. Wirel. Commun. **18**(1), 416–431 (2019)
9. Song, H., Liu, L., Ashdown, J., Yi, Y.: A deep reinforcement learning framework for spectrum management in dynamic spectrum access. IEEE Internet Things J. **8**(14), 11 208–11 218 (2021)
10. Zia, K., Javed, N., Sial, M.N., Ahmed, S., Pirzada, A.A., Pervez, F.: A distributed multi-agent RL-based autonomous spectrum allocation scheme in d2d enabled multi-tier hetnets. IEEE Access **7**, 6733–6745 (2019)
11. Gong, P.-Y., Wang, C.-H, Sheu, J.-P., Yang, D.-N.: Distributed drl-based resource allocation for multicast d2d communications. In: 2021 IEEE Global Communications Conference (GLOBECOM), pp. 01–06. December 2021
12. Huang, J., Yang, Y., He, G., Xiao, Y., Liu, J.: Deep reinforcement learning-based dynamic spectrum access for d2d communication underlay cellular networks. IEEE Commun. Lett. **25**(8), 2614–2618 (2021)
13. Chang, H.-H., Song, H., Yi, Y., Zhang, J., He, H., Liu, L.: Distributive dynamic spectrum access through deep reinforcement learning: a reservoir computing-based approach. IEEE Internet Things J. **6**(2), 1938–1948 (2019)

14. Zhao, Q., Krishnamachari, B., Liu, K.: On myopic sensing for multi-channel opportunistic access: structure, optimality, and performance. IEEE Trans. Wirel. Commun. **7**(12), 5431–5440 (2008)
15. Meinilä, J., Kyösti, P., Jämsä, T., Hentilä, L.: Winner ii channel models. In: Radio Technologies and Concepts for IMT-Advanced, February 2008

Maximizing Energy-Efficiency in Wireless Communication Systems Based on Deep Learning

Kaiyang Dong[1], Liang Han[1,2](✉), and Yupeng Li[1,2]

[1] College of Electronic and Communication Engineering, Tianjin Normal University, Tianjin 300387, China
hanliang@tjnu.edu.cn

[2] Tianjin Key Laboratory of Wireless Mobile Communications and Power Transmission, Tianjin Normal University, Tianjin 300387, China

Abstract. In recent years, many power allocation algorithms to maximize energy efficiency (EE) have emerged in wireless communication systems (WCS), but these traditional power allocation algorithms have high computational complexity. The advanced deep learning technique proposed in this paper is shown to solve the transmission power control problem in wireless networks to optimize EE. From a machine learning perspective, the conventional power allocation algorithms can be viewed as a nonlinear mapping between channel gains among users and the optimal power allocation scheme, and deep neural network (DNN) can be trained to learn this nonlinear mapping. Based on this, a DNN-based power allocation method is proposed, and the specific structure of the DNN and the system model of the DNN method are introduced to maximize the EE among users in WCS. The results show that the performance of the proposed method using DNN is essentially the same as that achieved by the conventional algorithm, but the computational time is greatly reduced.

Keywords: Machine learning · deep learning · DNN · power allocation algorithm · energy efficiency

1 Introduction

With the emergence of new wireless technologies and the explosive growth of mobile devices, mobile users have an increasing demand for high data rates [1, 2].

Currently, most of the algorithms on resource allocation in wireless communication systems (WCS) are concerned with the spectral efficiency (SE) problem [3, 4]. With the rapid growth of high rate, the energy consumption is also growing at an alarming rate. Thus, energy efficiency (EE) is becoming more important in wireless communications. Usually, when the optimal SE is achieved, the optimal EE is not necessarily achieved. Therefore, it is equally important to study EE-SE balancing in WCS, and many power allocation algorithms to optimize EE have been proposed in the last few years. In [5], the joint power control and spectrum resource allocation problems in SWIPT-based energy

© The Author(s), under exclusive license to Springer Nature Singapore Pte Ltd. 2023
Q. Liang et al. (Eds.): AIC 2022, LNEE 871, pp. 349–357, 2023.
https://doi.org/10.1007/978-981-99-1256-8_41

harvesting D2D underlying networks are addressed. The joint optimization problem is formulated as a two-dimensional matching of D2D pairs with cellular user equipment (CUE), and a preference building algorithm based on Dinkelbach algorithms and Lagrange pairwise decomposition is proposed. In [6], the global optimal solution with exponential complexity of the number of network links is obtained by combining the fractional programming theory.

Researchers have also recently started to study the application of deep learning (DL) in WCS and have achieved good results, most notably in terms of reduced computational complexity compared to traditional iterative algorithms. For example, an encoder for optical wireless communications was implemented using DNN in [7]. Compared to the traditional resource allocation using some hypothetical data to analyze the system model to derive the optimal policy, the DL approach can also use the hypothetical data to obtain the optimal policy. In [8], the authors used the channel implementation as the input, and the power obtained after the WMMSE algorithm as the label of the neural network, and trained a deep neural network. The deep learning algorithm achieved essentially the same results as the WMMSE algorithm, but with reduced online computation time. In [9], channel estimation and signal detection in orthogonal frequency division multiplexing systems were implemented with deep learning algorithms. In [10], in order to maximize the EE, an optimized transmit power scheme can be obtained by DL without deriving many complex and tedious mathematical formulas, reducing the time complexity.

2   System Model and Iterative Algorithm for EE Maximization

We consider a wireless network modeled by a Gaussian interference channel with M communication links, each with a receiver and a transmitter. The interference from other link is treated as noise, which is a common type of ad hoc networks. The interference channel model for 4 links is illustrated in Fig. 1.

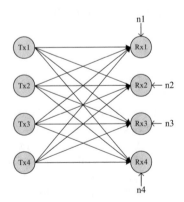

Fig. 1. Interference channel model for 4 communication links

In this section we introduce the power allocation problem for EE maximization in WCS. The EE maximization problem can be expressed as:

$$\max \eta_{EE} = \frac{\sum\limits_{i=1}^{M} \log_2\left(1 + \frac{P_i g_{ii}}{\sum\limits_{j=1,j\neq i}^{M} P_j g_{ij} + n_i}\right)}{\sum\limits_{i=1}^{M}(\xi_i P_i + P_i^c)} \tag{1}$$

$$s.t. \ C1: R_i \geq R^{\min}, \forall i$$
$$C2: 0 \leq P_i \leq P_i^{\max}, \forall i$$

where P_i denotes the transmitter power of link i, g_{ij} denotes the channel gain between the transmitter of link j and the receiver of link i, n_i is the noise power of the receiver of link i, ξ_i denotes the inefficiency factor of the constant power amplifier of the transmitter of link i, P_i^c denotes the circuit power consumption, $R^{\min}$ denotes the minimum rate requirement of the link, and C2 ensures that the transmitter power is non-negative and cannot exceed the maximum value.

For the objective function in (1), it is a non-convex optimization and NP-hard problem. According to [11], the above problem is equivalent to the following problem:

$$\max_{P\in\Omega}\left\{\sum_{i=1}^{M}\log_2\left(1 + \frac{P_i g_{ii}}{\sum\limits_{j=1,j\neq i}^{M} P_j g_{ij} + n_i}\right) - \eta_{EE}^{opt}\left[\sum_{i=1}^{M}(\xi_i P_i + P_i^c)\right]\right\} = 0 \tag{2}$$

where Ω is the feasible domain for P, and $\eta_{EE}^{opt} = \max\limits_{P\in\Omega} \eta_{EE}$. We can use Dinkelbach algorithm to solve this problem. Denote the EE after k-th iteration as η_{EE}^k, the objective function can be rewritten as

$$\sum_{i=1}^{M}\log_2\left(1 + \frac{P_i g_{ii}}{\sum\limits_{j=1,j\neq i}^{M} P_j g_{ij} + n_i}\right) - \eta_{EE}^k\left[\sum_{i=1}^{M}(\xi_i P_i + P_i^c)\right] = f(P) - h(P) \tag{3}$$

where

$$f(P) = \sum_{i=1}^{M}\log_2\left(\sum_{j=1}^{M} P_j g_{ij} + n_i\right) - \eta_{EE}^k\left[\sum_{i=1}^{M}(\xi_i P_i + P_i^c)\right] \tag{4}$$

and

$$h(P) = \sum_{i=1}^{M}\log_2\left(\sum_{j=1,j\neq i}^{M} P_j g_{ij} + n_i\right) \tag{5}$$

where $P = [P_1, P_2, ..., P_M]^T$ and superscript $(\cdot)^T$ denotes the transpose operator. The $f(P) - h(P)$ in Eq. (2) is the difference of two concave functions, which is a standard

DC function. Again, since $h(\boldsymbol{P})$ is differentiable, the CCCP algorithm is used to handle it. Using the first-order Taylor expansion at $\boldsymbol{P}^{(k)}$, $h(\boldsymbol{P})$ can be expanded as $h(\boldsymbol{P}^{(k)}) + \nabla h^T(\boldsymbol{P}^{(k)})(\boldsymbol{P} - \boldsymbol{P}^{(k)})$, where $\nabla h(\boldsymbol{P}^{(k)})$ is the gradient of $h(\boldsymbol{P})$ at the point $\boldsymbol{P}^{(k)}$. Therefore, the optimal problem can be seen as the convex optimization problem which can be solved efficiently by the interior-point method. The whole procedure above can be expressed as Algorithm 1.

Algorithm 1 Optimization Algorithm for EE

1: Set $k = 0$, choose a feasible $\boldsymbol{P}^{(0)} \in \Omega_1$, $EE^{(k)} = \eta_{EE}^{(k)}\big|_{\boldsymbol{P}=\boldsymbol{P}^{(k)}}$, $\delta > 0$.

2: **repeat**

3: $\quad \boldsymbol{P}^{(k+1)} = \arg\max_{p \in \Omega1}\{f(\boldsymbol{P}) - [h(\boldsymbol{P}^{(k)}) + \nabla h^T(\boldsymbol{P}^{(k)})(\boldsymbol{P} - \boldsymbol{P}^{(k)})]\}$;

4: $\quad$ Set $EE^{(k+1)} = \eta_{EE}^{(k+1)}\big|_{\boldsymbol{P}=\boldsymbol{P}^{(k+1)}}$;

5: $\quad$ Set $k = k + 1$;

6: **until** $\left|\max_{p \in \Omega1}\{f(\boldsymbol{P}) - [h(\boldsymbol{P}^{(k)}) + \nabla h^T(\boldsymbol{P}^{(k)})(\boldsymbol{P} - \boldsymbol{P}^{(k)})]\}\right| < \delta$.

7: **return** $\boldsymbol{P}$.

---

3 Deep Neural Network Based Method

In this section, we will design a DNN to achieve power allocation in the WCS network to maximize the energy efficiency of the system.

3.1 DNN System Architecture

We need to verify that DNN can perform real-time power allocation to users in WCS, and we will compare the performance at different distances between transmitters and receivers. Our approach is to fix the distance between each communication link, denoted as d, and the same distance between each communication link, for different groups of d_i distances is corresponding to different optimal power allocations. For different groups, we compare the power allocation corresponding to different distance groups after changing the distance between the transmitters and receivers of the links, but need to ensure that the communication links of the same groups are the same. In this way, it is possible to compare the power allocation schemes of conventional algorithms and DNN-based methods corresponding to each set of distances. Thus, the system architecture of the DNN method for transmitter and receiver power allocation in WCS is shown in Fig. 2 below.

3.2 Data Generation and Pre-processing

For each d_i distance group we generate 10000 training data. There are 5 d_i distance groups, so a total of 50000 training set data need to be generated, and each d_i distance

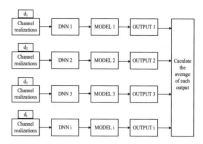

Fig. 2. The structure of the deep neural network

also needs to generate 1000 validation sets separately. We use the cross-validation method and randomly partition the entire training set, with 90% of the data used for training and 10% for validation. Eventually, before we input the generated data to the DNN, we need to normalize the data so that all the data is compressed to between 0 and 1, and also the output of the DNN is between 0 and 1.

3.3 Network Architecture

Our proposed DNN scheme, which consists of one input layer, three hidden layers, and one output layer, has the specific DNN structure shown in Fig. 3. The number of nodes in the input layer is 16, each hidden layer consists of a linear fully connected layer and a BN layer, the number of nodes in the three hidden layers are 1024, 1024 and 512, and the number of nodes in the output layer is 4. We use ReLU as the activation function of the hidden layer to improve the nonlinear fitting ability of the DNN. The ReLU function can be written as

$$relu(x) = \max(0, x) \tag{6}$$

Since the output of the neural network is between 0 and 1, sigmoid is chosen as the activation function of the output layer. Sigmoid function can be expressed as

$$sigmoid(x) = \frac{1}{1 + e^{-x}} \tag{7}$$

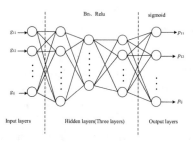

Fig. 3. The structure of the deep neural network

3.4   Training Stage

For the loss function, we use the mean square error between the power assigned by the Dinkelbach algorithm and the power assigned by the DNN prediction, which can be expressed as

$$loss = \frac{\sum |P_{dinkelbach} - P_{dnn}|^2}{n} \tag{8}$$

where $P_{dinkelbach}$ denotes the optimal power allocation of the Dinkelbach algorithm, P_{dnn} denotes the power predicted to be allocated by the DNN method, and n denotes the size of the data set. We use Adam optimizer as an optimization algorithm to make the training of the neural network faster. In addition, since we use the ReLU activation function in the hidden layer, the parameters are initialized using Kaiming initialization, which prevents the activation output from exploding or disappearing and improves the performance of the DNN. We use the loss value of the validation set corresponding to the distance $d_i = 40$ between transmitter and receiver as the basis for selection, and the loss of the validation set during training is shown in Fig. 4a and Fig. 4b.

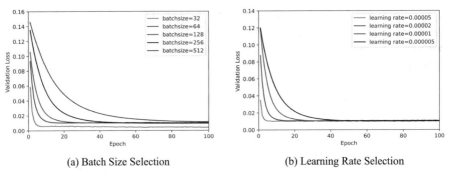

(a) Batch Size Selection   (b) Learning Rate Selection

Fig. 4.   Validation Loss

4   Results and Analysis

4.1   Simulation Environment

To implement our proposed DNN approach, we use python 3.6.1 and pytorch 1.10.2 to build the proposed DNN network structure. The server running the model uses an Intel(R) Xeon(R) W-2223 CPU with 32GB of memory size. The main purpose of this subsection is to experimentally study the power allocation for maximizing EE for a WCS with four pairs of transmitters and receivers, and to compare the performance of the DNN method with that of the conventional Dinkelbach algorithm.

4.2 Results

We set the learning rate to 0.00001 and the batch size to 128. The settings of the simulation parameters are shown in Table 1 below. The EE performance of the Dinkelbach algorithm and the EE performance of the DNN method in the communication scenario of this paper are shown in Fig. 5 below.

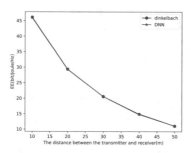

Fig. 5. Comparison of DNN method and Dinkelbach algorithm in EE performance

To prove that the performance of DNN method and Dinkelbach is very similar in this communication scenario, the accuracy table of DNN method is listed, in Table 2. Table 3 show the computational time of the two methods, which can be convenient to compare the computational complexity of the two methods.

Table 1. Simulation parameters.

Parameters	Value
The distance between the transmitter and receiver of the different links	200 m
The noise power	-66 dBm
The constant circuit power consumption (p_i^c)	23 dBm
The maximum transmit power of transmitter	20 dBm
Constant power amplifier inefficiency factor of the transmitter of link i (ξ_i)	4
Path loss exponent	4

Table 2. Performance comparison for the two methods.

Accuracy (DNN/Dinkelbach)			
d_i (m)	DNN	Dinkelbach	Accuracy
10	45.9650	46.1041	99.70%
20	29.3182	29.3713	99.48%
30	20.5676	20.5714	99.98%
40	14.7497	14.8732	99.17%
50	11.0231	11.0404	99.84%

Table 3. Computational time for the two methods.

Computation Time (s)		
d_i (m)	DNN	Dinkelbach
10	0.000028	10.9323
20	0.000028	10.5032
30	0.000025	9.6310
40	0.000027	8.3441
50	0.000033	7.8890

4.3 Results Analyze

From the comparison of the simulation results in Fig. 5 we can see that each point basically overlaps, which means that the performance of the DNN method is basically the same as that of the traditional algorithm Dinkelbach. From Table 2 we can see that basically the accuracy of each point can reach more than 99% of the traditional Dinkelbach algorithm. We can see from Table 3 that the time consumed by the DNN method is very short for different distances, while the Dinkelbach algorithm takes a very long time.

5 Conclusion

This study aimed to design a DNN-based method to solve the power allocation problem in WCS networks where the optimization objective function was EE. The conventional algorithm Dinkelbach used to solve the EE maximization was introduced. Then we presented our DNN approach, introduced the structure of the DNN, and the whole system model. Then, we started to prepare a large amount of data for training the DNN model by channel modeling to obtain a large amount of channel gain information, which was then fed into the traditional Dinkelbach algorithm to maximize EE and saved the optimal power allocation scheme, thus generating the dataset we need. Then, we input the dataset into the DNN network structure for training and use MSE as the loss function.

At the end of the training phase, we compared the performance of the DNN method and the Dinkelbach algorithm.

In the future, we can use unsupervised learning to allow deep learning-based methods to outperform even traditional algorithms. It is truly possible to achieve power allocation in real wireless communication systems with excellent performance and low computational time.

Acknowledgment. This work was supported by the National Natural Science Foundation of China (61901301, 61701345), and the Natural Science Foundation of Tianjin (18JCZDJC31900).

References

1. Wu, Q., Li, G.Y., Chen, W., Ng, D.W.K., Schober, R.: An overview of sustainable green 5G networks. IEEE Wireless Commun. **24**(4), 72–80 (2017)
2. Pan, C., Elkashlan, M., Wang, J., Yuan, J., Hanzo, L.: User-centric C-RAN architecture for ultra-dense 5G networks: challenges and methodologies. IEEE Commun. Mag. **56**(6), 14–20 (2018)
3. Sun, Y., Peng, M., Poor, H.V.: A distributed approach to improving spectral efficiency in uplink device-to-device-enabled cloud radio access networks. IEEE Trans. Commun. **66**(12), 6511–6526 (2018)
4. Han, L., Zhang, Y., Li, Y., Zhang, X.: Spectrum-efficient transmission mode selection for full-duplex-enabled two-way D2D communications. IEEE Access **8**, 115982–115991 (2020)
5. Zhou, Z., Gao, C., Xu, C., Chen, T., Zhang, D., Mumtaz, S.: Energy-efficient stable matching for resource allocation in energy harvesting-based device-to-device communications. IEEE Access **5**, 15184–15196 (2017)
6. Zappone, A., Björnson, E., Sanguinetti, L., Jorswieck, E.: Globally optimal energy-efficient power control and receiver design in wireless networks. IEEE Trans. Sig. Process. **65**(11), 2844–2859 (2017)
7. Lee, H., et al.: Deep learning framework for wireless systems: applications to optical wireless communications. IEEE Commun. Mag. **57**(3), 35–41 (2019)
8. Sun, H., Chen, X., Shi, Q., Hong, M., Xiao, F., Sidiropoulos, N.D.: Learning to optimize: Training deep neural networks for wireless resource management. In: Proceedings of the IEEE 18th International Workshop on Signal Processing Advances in Wireless Communications (SPAWC), Sapporo, Japan, pp. 1–6 (2017)
9. Ye, H., Li, G.Y., Juang, B.H.: Power of deep learning for channel estimation and signal detection in OFDM systems. IEEE Wireless Commun. Lett. **7**(1), 114–117 (2018)
10. Lee, W., Kim, M., Cho, D.H.: Deep power control: transmit power control scheme based on convolutional neural network. IEEE Commun. Lett. **22**(6), 1276–1279 (2018)
11. Dinkelbach, W.: On nonlinear fractional programming. Manage. Sci. **13**(7), 492–498 (1967)

Dynamics Analysis and Optimisation of Television Jumpers's Safety Clamp

Renjun Wang, Zhongrong Gou$^{(\boxtimes)}$, Rutao Wang, and Yu Tian

Zhonghuan Information College, Tianjin University of Technology, Tianjin, China
gouzhongrong2008@126.com

Abstract. A television jumper is a large entertainment project in which visitors ride a jumper to watch mages that correspond in real time, and is a high speed, heavy duty lift. This paper uses virtual prototype technology to analyse and optimise the safety clamp of a television jumper. A 3D solid model of the safety caliper is carried out using the software Pro/E. Then, appropriate constraints and loads are added to create the virtual prototype of the safety clamp. The virtual prototype was simulated to obtain the braking performance curve of the safety clamp, and the braking performance of the safety caliper was analysed. This paper gives the core algorithm GISTIFF. The parametric optimisation method provided by ADAMS is used to optimise the preload and contact force material of the safety clamp, providing a reference method for the commissioning and installation of the safety clamp as well as improving the design safety.

Keywords: Safety Clamp · Kinetic Analysis · Virtual Prototype · GISTIFF rigid integral algorithm · Parameter Optimisation

1 Introduction

In recent years, the amusement industry has entered a new phase of rapid development with the gradual emergence of the effects of the government's policy to stimulate domestic demand. However, we must also be aware that major safety incidents do occur in our amusement facilities. This paper gives the core algorithm GISTIFF and investigates the dynamics of Television Jumpers's Safety Clamp and thus verifies the safety of the project [8, 9]. It is used to provide guidance for installation and commissioning by means of realistic optimisation of reasonable parameters to ensure the safety of the project.

2 The Establishment of the Three-Dimensional Model of the Safety Clamp

Using Pro/E software, a 3D model of a safety clamp can be easily created [7]. A three-dimensional solid model of the safety clamp is shown in Fig. 1.

Q. Liang et al. (Eds.): AIC 2022, LNEE 871, pp. 358–364, 2023.
https://doi.org/10.1007/978-981-99-1256-8_42

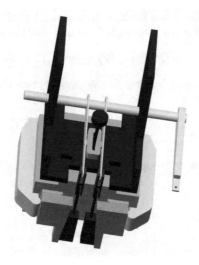

Fig. 1. Three-dimensional solid model of the safety clamp

3 ADAMS-Based Dynamics Simulation of the Safety Clamp

In the analysis of the braking performance of the safety clamp, it is necessary to assume that the braking force provided by the safety clamp is constant throughout the braking process, i.e. the braking acceleration of the safety clamp is assumed to be constant during the braking process. For progressive calipers the braking acceleration needs to be controlled between 0.2 g and 1.0 g [11–13].

3.1 Calculation of the Braking Acceleration

For the lift there are several common failure states, corresponding to various states of braking acceleration analysis is discussed as follows.

The traction machine stops rotating. When a passenger platform descends at excessive speed, the traction machine is braked and operation stops [3–5]. The mechanical equation is shown in Eq. (1).

$$4F = \alpha * G/2 * g - (P + Q + W) * g = (\alpha * G/2 + P + Q + W) * a \qquad (1)$$

The acceleration of the passenger-carrying platform can be derived as in Eq. (2) as.

$$a = \frac{8F + \alpha * G * g - 2(P + Q + W) * g}{\alpha * G + 2P + 2Q + 2W} \qquad (2)$$

where W is the weight of the hoisting wire rope; G is the counterweight weight. F is the braking force per safety clamp; a is the traction factor, i.e. the ratio of wire rope tension on both sides of the traction sheave; P is the weight of the passenger carrying platform; Q is the weight of the load.

The traction machine is still in drive operation. The traction effect on the traction sheave is the opposite of the above, with the mechanical equation shown in Eq. (3)

$$4\alpha * F + G/2 * g - \alpha * (P + Q + W) * g = (G/2 + \alpha * (P + Q + W)) * a \quad (3)$$

The acceleration during braking of the passenger-carrying platform is therefore derived as shown in Eq. (4).

$$a = \frac{8\alpha F + [G - 2\alpha(P + Q + W)] * g}{G + 2\alpha(P + Q + W)} \quad (4)$$

The acceleration of the passenger carrying platform when the safety clamp is actuated is shown in Eq. (5).

$$a = \frac{4F - (P + Q) * g}{P + Q} \quad (5)$$

4 Dynamics Analysis and Optimisation of Television Jumpers's Safety Clamp

4.1 Calculation of Braking Distance

As uniform acceleration is assumed during braking, the formula for calcu-lating the braking distance is shown in Eq. (6) (7) [10].

$$S_{min} = \frac{v_p^2}{2a_{max}} \quad (6)$$

$$S_{max} = \frac{v_p^2}{2a_{min}} + A \quad (7)$$

where v_p is the safety clamp speed; a_{max} is the maximum acceleration; a_{min} is the minimum acceleration; and A is the distance the passenger-carrying platform.

4.2 Analysis of Simulation Results

Using the software ADAMS to simulate the virtual prototype, the distance simulation curve, velocity simulation curve, acceleration simulation curve and the impact force simulation curve of the safety clamp wedge are ob-tained [6]. The following conclusions can be drawn from the analysis of the four curves:

It can be concluded that the overall force on the caliper is smooth. Overall, the current caliper parameters can basically meet the needs of safe work.

5 Safety Clamp Parameterised Design and Optimisation

In this paper, the preload force, the stiffness of the cylindrical coil spring and the friction coefficient between the guide and the wedge are used as variables to be optimised in order to find the best solution within a predetermined range.

5.1 GISTIFF Rigid Integration Algorithm

The GSTIFF integrator (INTEGRATOR) is the default integrator for ADAMS. Its main solution steps are: prediction, iterative correction, analysis of the integration error, and optimisation of the integration step and the order of the integration polynomial.

Prediction: The GSTIFF integrator predicts the state vectors y_{n+1} and y_{n+1} at t_{n+1} using the Taylor series and the so-called implicit backward differential Gear integral polynomial at tn, respectively. Equation (10) is called the implicit backward differential Gear integral polynomial, which is derived from a Newtonian post-interpolation polynomial of highest order 6.

$$y_{n+1} = y_n + \frac{\partial y_n}{\partial t} h + \frac{1}{2!} \frac{\partial y_n}{\partial t^2} h^2 + \cdots \tag{8}$$

$$y_{n+1} = \frac{1}{h\beta_0} \left[y_{n+1} - \sum_{i=1}^{k} \alpha_i y_{n-i+1} \right] \tag{9}$$

where h time step $h = t_{n+1} - t_n$. y_{n+1}, y_{n+1} are the approximations of $y(t), y(t)$ at $t = t_{n+1}$; β_0, α_i are the values of the coefficients of the Gear integration procedure.

Iterative correction: The state vector y_{n+1}, y_{n+1} values are substituted into the system equation. The system of first order differential equations is transformed into a system of nonlinear algebraic equations. If $g(y_{n+1}, y_{n+1}, t_{n+1}) = 0$ then the value of the state vector y_{n+1} is the state at t_{n+1}, if $g(y_{n+1}, y_{n+1}, t_{n+1}) \neq 0$, then the next step is to use a modified Newton-Raphson iterative correction to find the state that satisfies the system equation.

Expanding Eq. (8) at $t = t_{n+1}$, we get:

$$\begin{cases} F(q_{n+1}, u_{n+1}, \dot{u}_{n+1}, \lambda_{n+1}, t_{n+1}) = 0 \\ G(u_{n+1}, \dot{q}_{n+1}) = u_{n+1} - \frac{-1}{h\beta_0} \left[q_{n+1} - \sum_{i=1}^{k} \alpha q_{n-i+1} \right] = 0 \\ \Phi(q_{n+1}, t_{n+1}) = 0 \end{cases} \tag{10}$$

The iterative correction formula is Eq. (11):

$$\begin{cases} F_k + \frac{\partial F}{\partial q} \Delta q_k + \frac{\partial F}{\partial u} \Delta u_k + \frac{\partial F}{\partial \dot{u}} \Delta \dot{u}_k + \frac{\partial F}{\partial \lambda} \Delta \lambda_k = 0 \\ G_k + \frac{\partial G}{\partial q} \Delta q_k + \frac{\partial G}{\partial u} \Delta u_k = 0 \\ \Phi_k + \frac{\partial \Phi}{\partial q} \Delta q_k = 0 \end{cases} \tag{11}$$

where k denotes the kth iteration:

$$\Delta q_k = q_{k+1} - q_k; \ \Delta u_k = u_{k+1} - u_k; \ \Delta \lambda_k = \lambda_{k+1} - \lambda_k \tag{12}$$

From Eq. (9):

$$\Delta \dot{u}_k = -(1/h\beta_0) \Delta u_k \tag{13}$$

From Eq. (10):

$$\partial G/\partial q = (1/h\beta_0) I \quad \partial G/\partial u = I \tag{14}$$

Substitute Eq. (14) (15) into (12):

$$
\left[
\begin{array}{ccc}
\frac{\partial F}{\partial q} & \left[\frac{\partial F}{\partial q} - \frac{1}{h\beta_0}\frac{\partial F}{\partial \dot{u}}\right] & \left[\frac{\partial \Phi}{\partial q}\right]^T \\
\left[\frac{1}{h\beta_0}\right]I & I & 0 \\
\frac{\partial \Phi}{\partial q} & 0 & 0
\end{array}
\right]_k
$$

$$
\left\{
\begin{array}{c}
\Delta q \\
\Delta u \\
\Delta \lambda
\end{array}
\right\} =
\left\{
\begin{array}{c}
-F \\
-G \\
-\Phi
\end{array}
\right\}_k
$$

(15)

where the coefficients on the left-hand side of the equation are the Jacobi matrix of the system, $\partial F/\partial q$ system stiffness matrix, $\partial F/\partial u$ system damping matrix, $\partial F/\partial u$ system mass matrix mass matrix.

Equation (15) shows that the problem of solving a non-linear algebraic system of equations has been transformed into a problem of solving a linear system of equations. By decomposing the system Jacobi matrix, q_{k+1}, u_{k+1}, λ_{k+1}, q_{k+1}, u_{k+1}, λ_{k+1} can be calculated, The correction step is repeated until the state vector satisfies the convergence condition, such that The system equation $g(y_{n+1}, y_{n+1}, t_{n+1}) = 0$ holds approximately.

Checking the integration error and optimising the integration step and polynomial order: The principle of the GISTIFF integrator optimisation is to keep the order as constant as possible. The GISTIFF integrator is optimised by keeping the order as constant as possible and giving priority to order reduction if the order is to be changed, thus avoiding frequent. The GISTIFF integrator is optimised by keeping the order as constant as possible and giving priority to order reduction if necessary.

The previous estimation, correction, checking of integration errors, optimisation of integration steps and polynomial steps are repeated. The previous process of estimating, correcting, checking for integration errors, optimising integration steps and polynomial orders is repeated until the solving time reaches the specified time.

5.2 Optimisation Settings

Firstly, three optimisation objectives need to be defined [1, 2]: the shortest possible braking distance; the greatest possible braking force; and the smallest possible impact force. Secondly, the definition of the constraint function is completed. Finally the selection of the optimisation variables is completed.

5.3 Optimisation of Safety Clamp Braking Performance

In Fig. 2, the change curve before the caliper optimisation is shown as the solid red line and the change curve after the caliper optimisation is shown as the dashed blue line. By comparing the curves of braking distance, braking speed and impact force before and after optimisation, it can be concluded that the braking time is reduced by approximately 0.08 s and the braking distance by 79 mm. The rate of change of the braking speed is significantly larger after optimisation, i.e. the braking force is significantly increased after optimisation. The impact of the optimised wedge on the guideway is significantly reduced. In summary, the optimised safety caliper performance has been significantly improved.

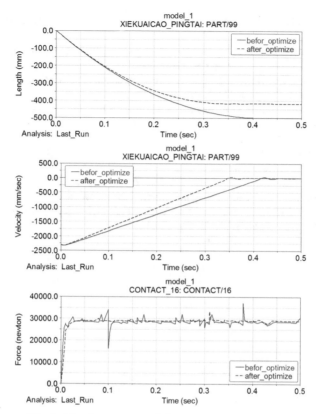

Fig. 2. The change curve of distance-speed-impact force

6   Conclusion

In this paper, a three-dimensional model of the safety caliper is constructed, and the dynamics of the safety caliper is analysed using conventional calculations and the software ADAMAS. The spring preload, spring stiffness and wedge friction coefficient are optimised by means of post-over-parametric design to improve the braking performance of the caliper.

References

1. Arnold, H.: Safety-type clamp for electric plug and socket (1956)
2. China, P.: Modeling for dynamic analysis of elevator mechanical system. Machine Design and Research (1999)
3. Erich, A.: Axial thrust safety clamp - has u=shaped locking piece with arms, and outward bent pieces (1990)
4. Mizuno, S., Fujita, Y., Togashi, N.: Elevator system (2004)
5. Guo, W.: The working principle and fault analysis of lift safety clamp. Mod. Manufact. Technol. Equipment **11**, 2 (2016)

6. Group, F.T.R.: New safety clamp feature for wedge wire filter range. Food Trade Review (3), 78 (2008)
7. Hans-Hermann, S.: Emergency operating device for device driven by permanent magnet synchronous motor e.g. passenger elevator (1999)
8. Jin, X., Li, C., Li, Z., Wei, S.: Optimal design of key components of safety clamps for traction lifts. Mechanical Design and Manufacture (2020)
9. Lin, Z.: Analysis of linkage device of elevator speed limiter safety clamp. Modern Manufacturing Technology and Equipment (2017)
10. Merrick Iii, W.S.: Safety clamp for walking beam compressor (2010)
11. Ren, Y., Zhang, L., Li, Q., Zhang, C., Li, X.: A clamping mechanism for a progressive safety clamp. CN212581269U (2021)
12. Wang Qian, W., Xiao, L.S., Zhao, L., Xia, P.: Simulation of dynamics and dynamic characteristics of a new safety clamp pulling mechanism. Mech. Des. **39**(4), 9 (2022)
13. Xu, H., Liu, S., Wang, Y., Li, X., Chen, A.: A lift car high-speed heavy-duty safety clamp and assembly method. CN110857201A (2020)

Financial Time Series Forecast of Temporal Convolutional Network Based on Feature Extraction by Variational Mode Decomposition

Mengting Zhao[✉]

Yunnan University, Kunming 650500, China
zhaomengting@mail.ynu.edu.cn

Abstract. The analysis and forecasting of non-stationary financial data with artificial intelligence techniques are useful for comprehending the future economic climate and have become a hot research topic. In this study, we propose a novel approach combining variational mode decomposition (VMD) and temporal convolutional network (TCN) for single-dimensional financial time series forecasting. First, VMD is used to reduce the influence of random noise on the prediction model. After that, we utilize TCN to learn the temporal dependence of subsequences. Additionally, we also compare the effectiveness of VMD and EMD for feature extraction of financial time series. The impact of the modal quantity K's value is also evaluated when VMD is used to analyze financial data. We validate the proposed method in a broad experimental analysis over 3 publicly available datasets. Experimental results show that our method achieves significant improvements in the prediction accuracy of financial data.

Keywords: variational mode decomposition · financial time series · temporal convolutional network · feature extraction

1 Introduction

Financial data is an inseparable part of everyday life. The analysis and prediction of financial time series have always been a hot research topic. However, the nonlinear and non-stationary properties of financial data series have made forecasting more complex and challenging. Therefore, several different methods have been proposed for a long time to forecast financial time series.

Methods for financial time forecasting consist primarily of traditional statistical techniques and machine learning techniques [1]. Traditional machine learning methods, such as support vector machines and deep learning methods [2]. Deep learning models have achieved superior performance than traditional machine learning [3]. For the application of neural networks in the forecasting of financial time series, they can be divided into three categories [1]. The first category is also acknowledged as the technical analysis method, the method takes the destination financial time series itself as input directly. The second group, the fundamental

© The Author(s), under exclusive license to Springer Nature Singapore Pte Ltd. 2023
Q. Liang et al. (Eds.): AIC 2022, LNEE 871, pp. 365–374, 2023.
https://doi.org/10.1007/978-981-99-1256-8_43

analysis model, utilizes additional external market information data as inputs in addition to those used by the first group. This kind of method can incorporate learning from external information. The third approach is investigated as the signal decomposition algorithm model.

Although RNN-like networks are the most prevalent solution for time series problems today, S. Bai et al. [4] argued that training RNNs over long time horizons is notoriously hard. Some convolutional networks can achieve better performance than RNNs. In the meantime, it avoids the typical drawbacks of recursive models, such as the gradient explosion issue and the lack of memory retention [4]. The architecture proposed by S. Bai et al. is called Temporal Convolutional Network (TCN) [4]. Later, TCNs were applied in diverse fields of chemical and electric power [5]. Moreover, TCNs have a longer memory than RNNs. This characteristic makes TCNs more appropriate for financial data [6].

In time series forecasting, the adaption of appropriate decomposition techniques can effectively improve the efficiency [7]. The literature [8] enhanced the forecasting results extremely considerably after introducing empirical mode decomposition (EMD) into the hybrid model of ARIMA and artificial neural network model. However, the previous VMD decomposition class algorithms generally decompose the subseries for modeling prediction and then add up the predicted values of each subseries as the final result [9]. This method requires the estimation of the number of decomposed modes prior to producing a more effective decomposition [9]. In actuality, however, there is no standardized method for determining the number of VMD modes [1]. Moreover, the excessive number of subsequences in the traditional decomposition class algorithm induces a reduction in computational efficiency, because each sequence needs to be modeled independently [10].

In this paper, a new VMD-TCN method is proposed for predicting financial time series. The proposed method utilizes both the original data and the subseries generated by VMD as the input of TCN to forecast future data. In our method, the decomposed subsequences are utilized as feature inputs instead of modeling the subsequences separately from the previous decomposition class prediction methods. Additionally, by employing the original data and the subsequences together as the input of the network, the analysis of the decomposed residuals in the conventional method can be omitted.

2   Methodology

Before understanding the model of this paper, the first two subsections first introduce the variational mode decomposition and the time-series convolutional network, on which the model of this paper is constructed. This is followed by the model of this paper.

2.1   Variational Mode Decomposition

Variational Mode Decomposition (VMD) [11] is a new method for time-frequency analysis. It can decompose the signal into multiple single-component amplitude-modulated frequency signals at one time. The endpoint effect and spurious component problems encountered in the iterative process are avoided. This method can effectively handle nonlinear and non-stationary signals. Compared with empirical mode decomposition (EMD) and other methods, this method has better noise immunity and can effectively avoid the problems of modal aliasing.

For the input signal, the VMD algorithm thus generates the constrained variational problem as

$$\min_{\{u_k\}\{w_k\}} \left\{ \sum_k ||\partial_t \left[\delta(t) + \frac{j}{\pi t} \right] * u_k(t) e^{-jw_k t}||_2^2 \right\}, \tag{1}$$

$$s.t. \sum_k u_k = f \tag{2}$$

where f denotes the original input signal, $u_K = u_1, u_2, \ldots U_K$ and $w_K = w_1, w_2, \ldots, w_k$ are the modes and their corresponding center frequencies, " $*$" represented the convolution operator. ∂ denotes the derivative of time concerning the function. $\delta(t)$ denotes the Dirac distribution.

The above literature models the summation reconstruction of subsequences separately to obtain the prediction results, failing to take into account the residuals between the original sequence and the sum of subsequences in practice [9]. The results of such summation reconstruction do not make effective use of the residuals.

$$FT(t) = \sum_{k=1}^{K} IMF_k(t) + R(t) \tag{3}$$

2.2 Temporal Convolutional Network

Convolutional Neural Network (CNN) is a widely used neural network that can extract local data features by convolutional computation [12]. Temporal convolutional network (TCN) [5] adds the following special structures to CNN to make the model more suitable for sequence learning and avoid the gradient problem.

1) Causal convolution: the direction of information transfer between layers is restricted to a single direction, which is consistent with the requirement that only historical information can be obtained for temporal tasks.

2) Inflated convolution: Inflated convolution uses interval sampling, which allows the network to obtain a larger field of perception with fewer layers. For the filter f: the input sequence $X = (x_1, x_2, \ldots, x_t)$ dilated convolution calculation at s is shown in Eq. 4, where $s - d \cdot i$ represents the direction of the past.

$$F(s) = (X_d^* f)(s) = \sum_{(i=0)}^{(k-1)} f(i) \cdot X_{(s-d \cdot i)} \tag{4}$$

3) Residual connectivity: The output of the residual connection is expressed as a linear summation of the nonlinear functions of the input and the input,

which enables the transfer of information across layers. The rectified linear unit (ReLU) is used as the activation function, which is defined as

$$f(x) = max(0, x) \tag{5}$$

2.3 Proposed Method

The overall framework of the model in this paper is shown in Fig. 1 below. After the necessary data preprocessing, the financial time series is decomposed into K subseries with different frequency domain characteristics using variational mode decomposition. After that, the original sequence and the subseries are simultaneously used as the input of the temporal convolutional network. The features are extracted across time steps using TCN while preserving the long memory property, and the final prediction results are obtained. The Variational Mode Decomposition is more robust in the analysis of noisy signals compared to other techniques such as the commonly used empirical modal decomposition. In this paper, a temporal convolutional network is used for feature extraction and machine learning prediction, which can operate more efficiently using convolution and can learn and converge quickly when the sample size is large.

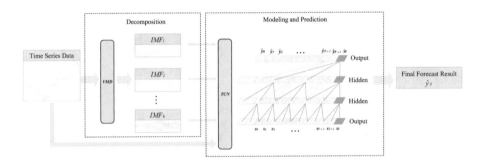

Fig. 1. Overall graphical abstract of the proposed technique, VMD-TCN

In the variational mode decomposition, the main parameters include the number of modes K and the penalty parameter α. A large value of K leads to over-decomposition and makes it difficult to extract effective information from the subsequence. Conversely, too small a K will under-decompose. It cannot effectively distinguish the molecular sequence from the original sequence, and it is difficult to utilize the features of the subsequence. The penalty parameter α taken too large will cause the loss of band information, and vice versa, will information redundancy. In this model, for the two most important parameters K and α of VMD, the number of modes K needs to ensure that the neighboring center frequencies are not too close to each other. The penalty parameter α is set to the default value of 2000 based on experiments.

After decomposing the original data into subseries, TCN is used to predict the financial time series. As shown in the Fig. 2 below Unlike the common decomposition class algorithms that sum the subseries separately as the final prediction result, this model uses both the original data and the decomposed subseries as the input to the neural network to make better use of the original data. In other words, the prediction is based on feature extraction.

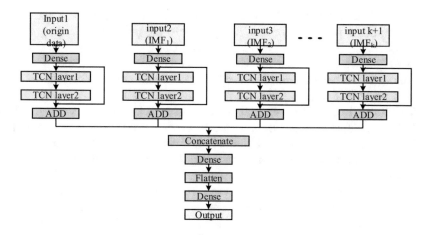

Fig. 2. TCN with IMF input in this paper

The learning rate of TCN is 0.001, the number of training rounds is 100, and the Dropout is 0.1. The convolutional kernel size is 24, and 128 filters are used in the convolutional layer. To improve the convergence efficiency of the network, the training samples are normalized by the standard deviation. Also in this paper, TCN uses an amsgrad optimizer instead of the Adam method of automatic parameter tuning. This method increases the impact of rare but information-rich features. It has a smaller learning rate than Adams and rmsprop, while the training is smoother.

3   Experiment

In this paper, three empirical datasets are used to validate the model validity of this paper. The first dataset is Google stock data, obtained from stooq. The second dataset and the third dataset are obtained from macrotrends, both from publicly available exchange rates. The datasets are divided into training and test sets according to 8:2. That is, the first 80% of the data is used for training and the second 20% for testing.

3.1   Evaluation Metrics and Benchmark Experiments

Root mean squared error (RMSE), (Mean Absolute Error, MAE), (Mean squared error, MSE), and (R-Square, R2) were used to evaluate the experimental results.

In this paper, multiple models are used for comparison to confirm the effectiveness of the proposed models. Multiple input single output TCN (MISO-TCN) is to use other relevant data as the input of TCN at the same time to predict the relevant results.

3.2   Google Stock Experiments

This dataset contains Google stock data for the period 2018.3.1–2021.9.1. It includes opening price, high price, low price, closing price, and trading volume. In the following, the closing price is predicted using the method of this paper.

Table 1 shown below compares the evaluation indexes of this paper model and other models. To better show the performance of this model, different values of K are taken as 3, 5, and 8 respectively in the experiment.

From Table 1, it can be seen that the best effect is the comparison test MISO-TCN. The model stands out because of the other inputs (open price, high market) However, in practice, the problem of predicting financial data often does not provide other data with a strong correlation. Therefore, it is limited in practical applications.

Without considering MISO-TCN, the model proposed in this paper performs the best among several models for a single-dimensional financial series. When $K = 5$, this model is the best. Compared with VMD-TCN ($K = 5$) and TCN, the four indicators of this model are significantly reduced. The RMSE, MAPE, and MAE are reduced to about 60% of the original one, and the MSE is reduced to 35%. The R-squared significantly improved from 0.80 to 0.92. The next most effective was VMD-TCN ($K = 8$). After that, the EMD-TCN model effect is similar to the VMD-TCN ($K = 3$) effect.

Table 1. Experimental results of Google closing

Model	RMSE	MAPE	MAE	MSE	R2
VMD-TCN ($K = 3$)	0.117188	0.434427	0.215841	0.06064	0.869139
VMD-TCN ($K = 5$)	0.180828	0.078919	0.151606	0.032699	0.941598
VMD-TCN ($K = 8$)	0.217865	0.102660	0.187141	0.047465	0.897571
TCN	0.302048	0.131297	0.253222	0.091233	0.80312
EMD-TCN	0.252677	0.115354	0.220394	0.063845	0.862221
MISO-TCN	0.076957	0.303679	0.055100	0.005922	0.993254
LSTM	0.383153	0.291253	0.146806	0.146806	0.827541

The results of this model are excellent for different values of K. Figure 3 below shows the visualization of the prediction effect of google closing price using this model. The middle subplot shows that the red dashed line fits best with the real value of the blue solid line for K = 5. This is followed by K = 8, K = 3.

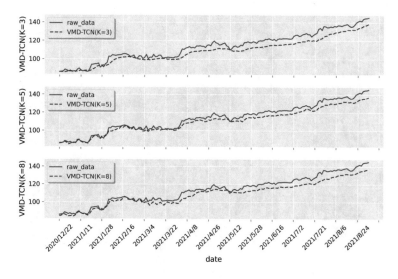

Fig. 3. Comparison of Google stock closing price forecast results

3.3   GBP/CNY Exchange Rate Experiments

The second dataset contains the exchange rates of six currencies against the RMB. The period is from 2016/1/4 to 2018/12/31. There are 794 data items. This subsection is about the experiment of the pound exchange rate. The exchange rate of the pound to yuan has a weak correlation with the exchange rate of the US dollar and the exchange rate of the Japanese yen.

The results of the evaluation index comparison between this paper model and other models are shown in Table 2 below. It can be seen that this model has an outstanding effect. The top three experimental effects are VMD-TCN (K = 5), VMD-TCN (K = 8), and EMD-TCN. The feature extraction financial time series prediction incorporating the decomposition algorithm has good results. In particular, the RMSE, MAPE, MAE, and MSE are significantly decreased with VMD-TCN (K = 5), and the R-squared is improved from 0.85 to 0.94.

Table 2. Experimental results of pound exchange

Model	RMSE	MAPE	MAE	MSE	R2
VMD-TCN(K = 3)	0.130995	1.434427	0.100589	0.01716	0.903419
VMD-TCN(K = 5)	0.101864	0.997054	0.075735	0.010376	0.941598
VMD-TCN(K = 8)	0.111037	0.908559	0.085388	0.012329	0.930607
TCN	0.163247	1.222499	0.127152	0.02665	0.850006
EMD-TCN	0.112974	0.893247	0.089297	0.012763	0.928164
MISO-TCN	0.169357	1.402675	0.129961	0.028682	0.838569
LSTM	0.161134	1.365867	0.110384	0.022841	0.832269

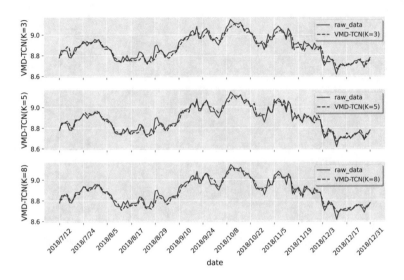

Fig. 4. Comparison of GBP forecast results

When K takes different values, the effect of this model is visualized in Fig. 4. In particular, the middle subplot (VMD-TCN K = 5) shows that the red predicted values and the blue true values fit perfectly within a certain range. The third subplot shows good predictions for $K = 8$ as well. At fluctuations, there is a slight shift.

3.4   USD/CNY Exchange Rate Experiment

In this subsection, the dollar exchange rate from the previous Subsect. 3.3 exchange rate dataset is chosen as the prediction series. Correlation analysis shows that the US dollar to yuan exchange rate shows a weak correlation with the Australian and Japanese yen exchange rates.

As Table 3 shows the results of the USD exchange rate forecasting. It can be seen that the model in this paper works best. The top three experimental effects are VMD-TCN (K = 5), VMD-TCN (K = 3), and VMD-TCN (K = 8), respectively. In particular, VMD-TCN (K = 5) is optimal for various evaluation indexes. The difference is smaller for K taking the value of 3 or 8. Meanwhile, the improvement effect is not obvious when compared with ordinary TCN. The effect of EMD-TCN is not ideal. Other weak correlation sequences are introduced in the MISO-TCN experiment, and yet the experimental results are not as effective as the model in this paper.

As shown in Fig. 5 is a visualization of the comparative effect of this paper's model in the dollar exchange rate experiment. It can be seen that the overall trend is more consistent between the predicted values in the red dashed line and the true values in blue. The intermediate subplot (VMD-TCN $K = 5$) is a better fit in terms of detail. There is also a better fit in the case of fluctuations. The

Table 3. Experimental results of dollar exchange rate

Model	RMSE	MAPE	MAE	MSE	R2
VMD-TCN(K = 3)	0.109082	0.109082	0.070959	0.009252	0.908585
VMD-TCN(K = 5)	0.082254	0.094792	0.068933	0.006766	0.933152
VMD-TCN(K = 8)	0.096888	0.119198	0.078898	0.009386	0.907265
TCN	0.097239	0.098653	0.068375	0.009456	0.906576
EMD-TCN	0.112268	0.190075	0.095988	0.012604	0.875468
MISO-TCN	0.113591	0.142477	0.081513	0.012903	0.872515
LSTM	0.111976	0.221924	0.083021	0.020399	0.869049

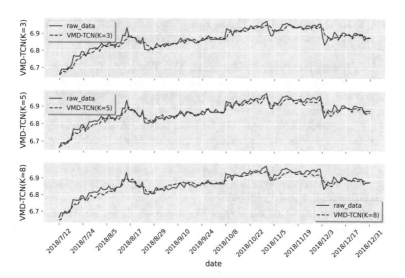

Fig. 5. Comparison of USD exchange rate forecast results

other two subplots show a small bias in the predictions when $K = 8K = 3$ in small fluctuations.

4 Summary

In this paper, we present a new model combining VMD and TCN for forecasting financial time series. Different from previous decomposition-based forecasting methods, in our method, both the subsequences and the original data are used as input features of our model. The proposed model does not require analysis of residuals and could improve efficiency by predicting subsequences jointly. Experimental results demonstrate the effectiveness of the proposed representation on financial time series prediction outperforming the other method on three benchmark datasets.

References

1. Huang, Y., Gao, Y., Gan, Y., Ye, M.: A new financial data forecasting model using genetic algorithm and long short-term memory network. Neurocomputing **425**, 207–218 (2021)
2. Carmona, R.: The influence of economic research on financial mathematics: evidence from the last 25 years. Financ. Stochast. **26**(1), 85–101 (2022)
3. Sezer, O.B., Gudelek, M.U., Ozbayoglu, A.M.: Financial time series forecasting with deep learning: a systematic literature review: 2005–2019. Appl. Soft Comput. **90**, 106181 (2020)
4. Bai, S., Kolter, J.Z., Koltun, V.: An empirical evaluation of generic convolutional and recurrent networks for sequence modeling. arXiv preprint arXiv:1803.01271 (2018)
5. Samal, K.K.R., Panda, A.K., Babu, K.S., Das, S.K.: Multi-output TCN autoencoder for long-term pollution forecasting for multiple sites. Urban Clim. **39**, 100943 (2021)
6. Guo, Yu., Zhang, S., Yang, J., Guanghui, Yu., Wang, Y.: Dual memory scale network for multi-step time series forecasting in thermal environment of aquaculture facility: a case study of recirculating aquaculture water temperature. Expert Syst. Appl. **208**, 118218 (2022)
7. Antwi, E., Gyamfi, E.N., Kyei, K., Gill, R., Adam, A.M.: Determinants of commodity futures prices: decomposition approach. Math. Probl. Eng. **2021**, 1–24 (2021)
8. Büyükşahin, Ü.Ça., Ertekin, Ş.: Improving forecasting accuracy of time series data using a new Arima-Ann hybrid method and empirical mode decomposition. Neurocomputing **361**, 151–163 (2019)
9. Yang, Y., Wang, Z., Gao, Y., Jinran, W., Zhao, S., Ding, Z.: An effective dimensionality reduction approach for short-term load forecasting. Electric Power Syst. Res. **210**, 108150 (2022)
10. Zhou, F., Huang, Z., Zhang, C.: Carbon price forecasting based on CEEMDAN and LSTM. Appl. Energy **311**, 118601 (2022)
11. Dragomiretskiy, K., Zosso, D.: Variational mode decomposition. IEEE Trans. Sig. Process. **62**(3), 531–544 (2013)
12. Alzubaidi, L., et al.: Review of deep learning: concepts, CNN architectures, challenges, applications, future directions. J. Big Data **8**(1), 1–74 (2021). https://doi.org/10.1186/s40537-021-00444-8

User Selection and Resource Allocation for Satellite-Based Multi-task Federated Learning System

Mingyu Zhang[1], Xiaoyong Wu[2], Zhilong Zhang[1(✉)], Danpu Liu[1], and Fangfang Yin[3]

[1] Beijing Laboratory of Advanced Information Network, Beijing Key Laboratory of Network System Architecture and Convergence, Beijing University of Posts and Telecommunications, Beijing 100876, China
zhilong.zhang@outlook.com
[2] Shanghai Radio Equipment Research Institute, Shanghai 201109, China
[3] State Key Laboratory of Media Convergence and Communication, School of Information and Communication, Communication University of China, Beijing 100024, China

Abstract. In view of the development of low-orbit satellite constellation and the increasing demand for computing ability and privacy protection of space-based applications, a satellite-based multi-task federated learning system is designed. Specifically, with consideration of both the fluctuations of satellite-to-ground communication links and local training impacts on global models, a user selection strategy is established. On this basis, the functional relationship between task convergence speed and wireless communication factors is analyzed, and the KM optimal matching algorithm is adopted to dynamically allocate communication resources. Simulation results show that our proposed scheme outperforms baseline methods in both total weighted error rate and accuracy, and achieves the fastest convergence rate at the same time.

Keywords: Satellite computation · Wireless communication · Multi-task Federated learning

1 Introduction

Terrestrial mobile communication networks are difficult or unable to provide ubiquitous services to areas such as desert, deep sea and outer space [1]. Satellite communication network is a feasible supplement to terrestrial networks to achieve global coverage and seamless connection to support various computing tasks such as artificial intelligence (AI). Considering the distributed massive data processing in remote areas, the satellite communication network combined with the federated learning framework can be used to achieve globally distributed training. On the one hand, as a general distributed learning framework, federated learning can achieve the final convergence state iteratively through local training and global aggregation [2]. It can avoids the risk of privacy disclosure by uploading local training parameters in each iteration. On the other hand, satellites can provide global seamless services as federated learning aggregation global nodes, especially in some terrestrial networks, such as remote neighborhoods and ships. It can perfectly fill the service gaps.

Based on the above considerations, satellite-based federated learning system can effectively solve the problem of incomplete coverage of ground communication and data privacy. From this perspective, the works in [3–6] take into account the resource allocation, and capacity management in satellite communication, on this basis, the authors in [7] use the deep learning model on the satellite to preprocess the original observation data and only transmit the relevant information to the ground. For the combination of wireless communication factors and federated learning, the authors in [9] propose an iterative algorithm to reduce energy consumption. However, most of the existing studies do not fully consider user selection strategy under the satellite-based federated learning system. In addition, there is a multi-task environment in reality, that is, each user has multiple personalized services to be processed, few studies have considered this situation.

In this paper, by leveraging the advantage of global seamless coverage of satellite, we propose a joint client selection and resource allocation strategy for multi-task federated learning system. In each iteration, the satellite is selected independently as the global aggregation node for global training. The iterations must be repeated until convergence.

2   System Model

2.1   Task Model

The system contains a set $\mathcal{M} = \{1, 2, ..., M\}$ of remote clients, each of which has computing ability and stores the data sets of tasks. Considering the factors of data privacy and the limits of communications, the original data does not share between clients. There exists a set $\mathcal{S} = \{1, 2, ..., S\}$ of low-orbit satellites which can provide service of global model aggregation.

The set $\mathcal{J} = \{1, 2, ..., J\}$ is the category of tasks to be trained, and each client at least has one learning task. The structure of federated learning is shown in Fig. 1. The specific learning process is as follows: in each iteration, 1) the clients update the parameters according to the local model; 2) the most suitable satellite is selected as the global aggregation node to complete the global aggregation; 3) select the appropriate clients to upload their local gradients to the selected satellite for aggregation; 4) the satellite updates the global model; 5) the global model is sent to the clients. Iterate this process until the algorithm converges. In the client selection stage of our proposed system, specifically, first to filter the optimal clients by optimization problem by considering the communication factors, and then to filter the tasks through the influence of the local training results on global convergence.

2.2   Communication Model

In each iteration of federated learning, if the client m transmits the local training result of task j to the selected satellite s through satellite-to-ground channel, then according to

Shannon's formula, the transmission rate can be approximately expressed as:

$$c_{smj}^U\left(r_{smj}, P_{smj}\right) = \sum_{n=1}^N r_{smj,n} B_n^U \log_2\left(1 + \frac{P_{smj} h_{smn}^U}{I_n^U + B_n^U N_0}\right), \tag{1}$$

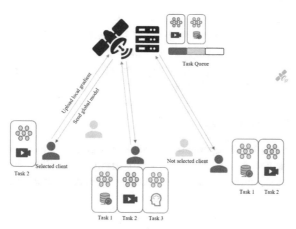

Fig. 1. Structure of satellite-based FL system

where $r_{smj} = \left[r_{smj,1}, \ldots, r_{smj,N}\right]$ is subcarrier allocation matrix, $r_{smj,n}$ represents the n-th subcarrier is allocated to the client m to transmit the task j to satellite s, with a value of 0 or 1. For a subcarrier, it only can be assigned to one user for one task, and it need to be met $\sum_{s=1}^S \sum_{m=1}^M \sum_{j=1}^J r_{smj} = N$. P_{smj} is the transmission power of client m transmit task j to satellite s, h_{smn}^U is the channel gain of uplink, I_n^U is the interference of n-th subcarrier, B_n^U is the uplink transmission bandwidth of n-th subcarrier, and N_0 is the noise power spectral density.

For global model distribution, the downlink transmission rate is:

$$c_{sm}^D = B^D \log_2\left(1 + \frac{P_s h_{sm}^D}{I^D + B^D N_0}\right), \tag{2}$$

where B^D represents the downlink transmission bandwidth, P_s is the transmission power of satellite s, and I^D is the interference of downlink transmission. h_{sm}^D is the average channel gain of downlink.

The delays of uplink and downlink both contain transmission delay and propagation delay, which are expressed as follows, respectively:

$$l_{smj}^U\left(r_{smj}, P_{smj}\right) = \frac{Z(w_{mj})}{c_{smj}^U(r_{smj}, P_{smj})} + t_{sm}, \tag{3}$$

$$l_{smj}^D = \frac{Z(g_j)}{c_{sm}^D} + t_{sm}, \forall j \in \mathcal{J}, \tag{4}$$

where w_{mj} is the local training model parameters of client m with task j, and $Z(w_{mj})$ is the gradient result size of client m with task j. g_j is the global model of task j, and $Z(g_j)$ is the size of global gradient result. t_{sm} is propagation delay of satellite s to client m.

The energy consumption of client m with task j is:

$$e_{smj}(\mathbf{r}_{smj}, P_{smj}) = \varsigma \omega_{mj} \vartheta^2 Z(w_{mj}) + P_{smj} l_{smj}^U(\mathbf{r}_{smj}, P_{smj}), \tag{5}$$

where ς is energy consumption coefficient of each client, ϑ is the CPU frequency calculated by the client, ω_{mj} is the number of CPU cycles per bit required by the client to perform the task j, $\varsigma \omega_{mj} \vartheta^2 Z(w_{mj})$ represents the energy consumption of client m to train the task j, and $P_{smj} l_{smj}^U(\mathbf{r}_{smj}, P_{smj})$ represents the energy consumption caused by the transmitting the local gradient result.

2.3   Federated Learning Model

For the selection of global aggregation satellite node, the satellite with the shortest average distance is selected as the global aggregation node. When the satellite begins global training, only the tasks that participate in federated learning are parametrically averaged, that is:

$$g_j(\mathbf{a}, \mathbf{P}, \mathbf{R}) = \frac{\sum_{m=1}^M K_{mj} a_{mj} w_{mj}}{\sum_{m=1}^M K_{mj} a_{mj}}, \tag{6}$$

where $\mathbf{a} = \begin{bmatrix} a_{00} & \cdots & a_{0J} \\ \vdots & \ddots & \vdots \\ a_{M0} & \cdots & a_{MJ} \end{bmatrix}$ indicates the task is selected or not, with a value of 0 or 1. $\mathbf{P}$ is transmission power vector with P_{smj} to indicate the transmission power of client m transmit task j to satellite s, and $\mathbf{R}$ is subcarrier vector with r_{smj} to indicate the subcarrier allocation.

For the selection of ground clients, first of all, because the instability of communication transmission is taken into account in the proposed model, we set γ_T and γ_E for communication limits, and through optimization problem solving to roughly filter clients. The further filtering is based on local gradient results.

2.4   Optimization Problem

Affected by the combination of communication instability and imbalance of data sample, this model finally needs to minimize the weighted sum of loss functions for all tasks, and the optimization problem can be established as follows:

$$\min_{\mathbf{a}, \mathbf{P}, \mathbf{R}} \sum_{j=1}^J \sum_{m=1}^M \rho_{mj} f_{mj}(g_j(\mathbf{a}, \mathbf{P}, \mathbf{R})) \tag{7}$$

$$\text{s.t.} \sum_{n=1}^N r_{mj,n} = a_{mj}, \forall m \in \mathcal{M}, \forall j \in \mathcal{J} \tag{8}$$

$$l_{smj}^U(\mathbf{r}_{smj}, P_{smj}) + l_{smj}^D \leq \gamma_T, \forall m \in \mathcal{M}, \forall j \in \mathcal{J}, \forall s \in \mathcal{S} \tag{9}$$

$$\sum_{j=1}^{J} e_{smj}\left(r_{smj}, P_{smj}\right) \leq \gamma_E, \forall m \in \mathcal{M}, \forall j \in \mathcal{J}, \forall s \in \mathcal{S} \tag{10}$$

$$a_{mj} \in \{0, 1\}, \forall m \in \mathcal{M}, \forall j \in \mathcal{J}, n = 1, \ldots, R \tag{11}$$

$$r_{mj,n} \in \{0, 1\}, \forall m \in \mathcal{M}, \forall j \in \mathcal{J}, n = 1, \ldots, R \tag{12}$$

$$0 \leq \sum_{j=1}^{J} P_{smj} \leq P_{max}, \forall m \in \mathcal{M}, \forall j \in \mathcal{J}, \forall s \in \mathcal{S} \tag{13}$$

where f_{mj} is the loss function of task j in client m, and different tasks correspond to different loss functions, such as taking variance in prediction tasks. ρ_{mj} is the weight of task j of client m, g_j is the global model of task j, γ_T and γ_E indicate the time requirements and energy requirements, respectively.

3   Optimization of Task Selection and Resource Allocation

In order to solve the above problems, the subcarrier allocation need to be optimized to enable the current satellite to aggregate the optimal sample data in each iteration.

First of all, we analyze the linear correlation between weights and influencing factors. To determine the weight ρ_{mj} by the impact factors of packet loss rate, task category weight, task data size, delay, energy consumption and so on.

Then, we analyze the expected convergence rate of federated learning system, the upper bound of convergence can be given [10], then we can obtain the relationship between convergence rate and wireless communication factors, and the optimization problem can be expressed as

$$\min_{R} \sum_{m=1}^{M} \sum_{j=1}^{J} \rho_{mj} \left(1 - \sum_{n=1}^{N} r_{smj,n}\right), \tag{14}$$

Finally, KM optimal matching algorithm is adopted to solve the linear function problem with nonlinear constraints to complete the subcarrier allocation.

After the clients filtering, considering that the imbalance of data distribution will affect the global convergence rate, we change the selection probability of tasks for further filtering:

$$\Delta p_{mj}^{t} = p_{mj}^{t-1} \cdot \min \left[\left(\frac{Q}{\max\limits_{m \in \mathcal{M}, j \in \mathcal{J}} Q - \min\limits_{m \in \mathcal{M}, j \in \mathcal{J}} Q} + \beta \frac{K_{mj}}{\sum_{m \in \mathcal{M}, j \in \mathcal{J}} K_{mj}} \right)^{\partial}, 1 \right], \tag{15}$$

where Δp_{mj}^{t} is the selection probability increment, p_{mj}^{t-1} is the selection probability of the last iteration, $Q = \langle \nabla F_{mj}(w^t), \nabla F(w^t) \rangle$ represents the effect of local training on global convergence. $\nabla F_{mj}(w^t)$ is the gradient result of client m of task j, k_{mj} is the size of data of task j on client m, and β, ∂ are influence factors.

4 Simulation Results

Simulation parameters are shown in Table 1. We consider a typical Iridium system consisting of LEO satellites with an orbital altitude of 600 km to simulate the LEO constellation. The small-scale fading over Ka-band is set as Rician fading with the K parameter is 7, and the satellite network adopts the free-space path loss model. We use three basic tasks: classification, regression and prediction to represent actual tasks, such as remote sensing image classification tasks, agricultural income prediction in remote areas, environmental information fitting and so on. We define three situations for comparison: 1) considering the influence of wireless communication factors without considering the effect of local training gradient; 2) considering the effect of local training gradient result on global convergence without considering the effect of communication; 3) neither is considered. The simulation parameters are set as follows:

Table 1. Simulation parameters

Parameters	Value
The categories of tasks (J)	3
Energy requirement (γ_E)	0.005 J
The size of data of task j on client $m(K_{mj})$	1000–8000
The downlink transmission bandwidth (B^D)	50 MHz
The uplink transmission bandwidth (B^U)	30 MHz
The CPU frequency (ϑ)	10^9
The energy consumption coefficient (ς)	10^{-27}
Delay requirement (γ_T)	500 ms
The number of clients (M)	6–8
The propagation delay between satellite s and client $m(t_{sm})$	[0–5] * 1e−3 ms
The number of satellites (S)	4
The transmission power of satellite $s(P_s)$	43 dBm

From Fig. 2, since the definition of loss function is different for different tasks, all error rates are normalized as vertical axes here. We can see that the algorithm proposed in this paper can achieve the minimum error rate and is more stable after convergence. When taking communication factors and local training result into consideration, each iteration will select appropriate delay-constrained tasks to improve the training efficiency, while the negative effects of individual extreme error data can be filtered to speed up the convergence.

Figure 3, (a), (b) and (c) are classification, regression and prediction tasks respectively. It can be seen that the algorithm proposed in this paper achieves the minimum error rate in the three tasks, and the convergence is the most stable.

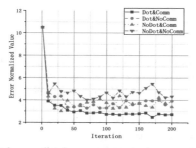

Fig. 2. Convergence of total normalized error parameter by comparing different affecting task selection factors

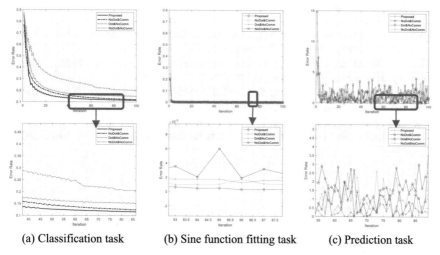

(a) Classification task   (b) Sine function fitting task   (c) Prediction task

Fig. 3. Convergence of all kinds of tasks by comparing different affecting task selection factors

5   Conclusion

In this paper, we combine the federated learning system with the satellite network, and establish the user selection strategy, on this basis, we formulate the resource allocation methods. In this model, user tasks selection is considered from two aspects: wireless communication factors and local task data training results. With the goal of minimizing the total weighted sum of all tasks, the KM optimal matching algorithm is used to realize the optimal allocation of wireless communication resources. The experimental results show that the model proposed in this paper achieves the best accuracy.

Acknowledgement. This work was supported by the Shanghai Science and Technology Commission Research Project under Grant 20511106700, the Beijing Natural Science Foundation under Grant No. L202003, and the Open Research Project of the State Key Laboratory of Media Convergence and Communication, Communication University of China (SKLMCC2021KF009).

References

1. Internet of Things — Number of Connected Devices Worldwide 2015–2025. https://www.sta tista.com/statistics/471264/iot-number-of-connecteddevices-worldwide/
2. Chen, M., et al.: Distributed learning in wireless networks: recent progress and future challenges. IEEE J. Sel. Areas Commun. **39**(12), 3579–3605 (2021)
3. Deng, B., Jiang, C., Yao, H., Guo, S., Zhao, S.: The next generation heterogeneous satellite communication networks: integration of resource management and deep reinforcement learning. IEEE Wirel. Commun., 105–111 (2020)
4. Jiang, C., Zhu, X.: Reinforcement learning based capacity management in multi-layer satellite networks. IEEE Trans. Wirel. Commun., 4685–4699 (2020)
5. Zhu, X., Jiang, C., Kuang, L., Zhao, Z., Guo, S.: Two-layer game based resource allocation in cloud based integrated terrestrial satellite networks. IEEE Trans. Cognit. Commun. Netw., 509–522 (2020)
6. Deng, B., Jiang, C., Guo, S.: Energy minimization of resource allocation in cloud-based satellite communication networks. IEEE Commun. Lett., 2353–2356 (2019)
7. Giuffrida, G., et al.: Cloud scout: a deep neural network for on-board cloud detection on hyperspectral images. Remote Sens., 2205 (2020)
8. Yang, Z., Chen, M., Saad, W., Hong, C.S., Shikh-Bahaei, M.: Energy efficient federated learning over wireless communication networks. IEEE Trans. Wirel. Commun. **20**(3), 1935–1949 (2021)
9. Chen, M., Yang, Z., Saad, W., Yin, C., et al.: A joint learning and communications framework for federated learning over wireless networks. IEEE Trans. Wirel. Commun., 269–283 (2021)

Industrial Time-Series Signal Anomaly Detection Based on G-LSTM-AE Model

Mengru Hu[✉] and Pengcheng Xia

School of Electronic and Optical Engineering, Nanjing University of Science and Technology, Nanjing, China
mengru.hu@njust.edu.cn

Abstract. In practical industrial scenarios, efficient anomaly detection is important for the development of industrial system safety and maintenance. In this paper, an industrial time-series signal anomaly detection algorithm based on Gaussian confidence interval and long short-term memory auto-encoder (G-LSTM-AE) is proposed to improve the accuracy and slow detection efficiency of traditional industrial time-series signal anomaly detection. The proposed algorithm is consisted of four steps: time-series data preprocessing, G-LSTM-AE model constructing, Gaussian confidence interval solving, and time-series signal real-time detection. To verify the effectiveness of our proposed algorithm, we conduct the experiments on automatic guided vehicle (AGV) dataset and Skoltech Anomaly Benchmark (SKAB) dataset, respectively. Compared with the existing unsupervised time-series signal anomaly detection algorithm, the accuracy and recall rate of the proposed G-LSTM-AE based algorithm achieves 95% (3–5% improvement). Meanwhile, the Gaussian confidence interval anomaly analysis method improves the reliability and practicality of anomaly detection in industrial scenarios.

Keywords: Industrial time-series signals · Anomaly detection · Gaussian confidence intervals · LSTM-AE · Unsupervised learning

1 Introduction

With the Internet of Things (IoT) technology development, the industrial manufacturing systems safety issues become an important research topic [1]. Considering that most of the industrial data are time-series signal, the study of anomaly detection based on time-series signal has gradually become a hot issue [2,3]. A variety of methods for anomaly detection based on time-series data have been proposed. For example, [4,5] proposed a time series anomaly detection algorithm based on GANs model, which achieved some effect but increased the complexity of the optimization model. [6] proposed a support vector machines (SVM) for anomaly detection and [7] proposed the isolation forest (iForest) based time series anomaly algorithm. Although these supervised methods have made some progress on time series anomaly detection, a large amount of labels are required, which is not feasible for industrial scenarios.

© The Author(s), under exclusive license to Springer Nature Singapore Pte Ltd. 2023
Q. Liang et al. (Eds.): AIC 2022, LNEE 871, pp. 383–391, 2023.
https://doi.org/10.1007/978-981-99-1256-8_45

On the other hand, the Auto-encoder (AE) and its variants are widely used unsupervised methods [8] to learn data features using reconstruction principles. [9] proposed an AE algorithm for anomaly detection on turbo data, and [10] used Variational Auto- Encoder (VAE) algorithm to reconstruct the handwritten dataset (MNIST). Although these AE-based algorithms don't need any labels, they have shown poor performance on industrial time-series signal anomaly detection. Recently, [11–13] stated the LSTM models have significant advantages for learning long-term information of sequences.

In this paper, we combine the LSTM network which shows the advantage of learning sequence features and AE network which is sensitive to nonlinear data, and proposed an industrial time-series signal anomaly detection algorithm based on Gaussian confidence interval (GCI) and LSTM-AE model. Our contributions are as follows:

1) We proposed a G-LSTM-AE based framework to detect anomalies in industrial periodic time-series signals.
2) A GCI-based threshold method is proposed to detect anomalies. Compared with the traditional threshold settings, GCI can reduce time of adjusting model and increase confidence of threshold.
3) The experimental results on an industrial real-time measurement dataset of AGV and a public dataset SKAB show that the accuracy and recall of the proposed algorithm can reach more than 95%.

2   Anomaly Detection Based on G-LSTM-AE Model

2.1   LSTM-AE Network

In this paper, we propose a LSTM-AE network to learn periodic time series features. The LSTM-AE structure is depicted in Fig. 1, where the encoder and decoder of the AE network are replaced with LSTM cells. X is the input of the network and X^R is the reconstructed output.

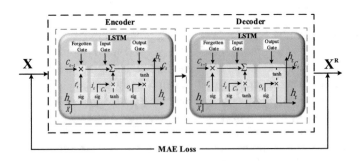

Fig. 1. LSTM-AE Structure

2.2 Gaussian Confidence Interval

Gaussian distribution, also known as normal distribution, is essentially a probability model. If a random variable X obeys a normal distribution with expectation μ and standard deviation σ, can be written as $X \sim \mathcal{N}\left(\mu, \sigma^2\right)$. If $\mu = 0$ and $\sigma = 1$, we denote $F(x)$ as $\phi(x)$, given by

$$\phi(x) = \frac{1}{\sqrt{2\pi}} \int_{-\infty}^{x} e^{-\frac{t^2}{2}} dt. \tag{1}$$

Hence, the probability that a sample X locates between x_1 and x_2 is given by

$$P(x_1 < X < x_2) = \phi\left(\frac{x_2 - \mu}{\sigma}\right) - \phi\left(\frac{x_1 - \mu}{\sigma}\right). \tag{2}$$

The confidence interval indicates the extent which the true value of a sample falls around the predicted value. Combined with the standard deviation of the Gaussian distribution, we apply the result of dichotomous classification to obtain the confidence interval estimating the result of classification prediction, denoted as GCI. According to (1), we have $\phi(2) = 0.9772$, and $P\{|X - \mu| < 2\sigma\} = 2\phi(2) - 1 = 0.9544$.

In this paper, the GCI is set to approximately 95% of the samples centered on μ in the Gaussian distribution. The formula is as follows

$$GCI|_{95\%} = (\mu - 2\sigma, \mu + 2\sigma). \tag{3}$$

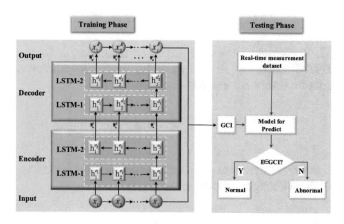

Fig. 2. Anomaly Detection Algorithm Framework on G-LSTM-AE Model

2.3 Detection Framework Based G-LSTM-AE

Based on the LSTM-AE model and GCI-based thresholding method, we proposes an anomaly detection algorithm based on G-LSTM-AE. Figure 2 presents the algorithm framework. The algorithm consists of training phase and testing phase. In the training phase, training LSTM-AE model to get GCI. In the testing phase, the real-time measurement data are input to calculate the reconstruction error. If the reconstruction error is within GCI, the test data are regarded as normal or abnormal.

2.4 Industrial Time-Series Signal Anomaly Detection on G-LSTM-AE Model

Data Pre-processing. The measured real-time-series data of an enterprise are collected as $X = \{X_1, X_2, X_3, \ldots, X_N\}$, where $X_i = \{X_{i_1}, X_{i_2}, X_{i_3}, \ldots, X_{i_L}\}$, $X_i \in X$. Before the construction of our proposed model, the dataset is divided into training dataset, validation dataset and test dataset, respectively. Then, convert them to tensor form, denoted as $\widehat{X}, \check{X}, \ddot{X}$.

Train Model. Train the LSTM-AE model with input X, and output X^R. To evaluate the performance of the model, we chose the average absolute error as loss function, given by

$$L\_1 = \frac{1}{L} \sum_{j=1}^{L} |X_{i,j} - X_{i,j}^R|, \tag{4}$$

where L_1 is the loss function of X_i, $X_{i,j} \in X_i$.

Obtain GCI. To calculate the GCI, input all trian data $\widehat{X}$ into the predicted network of LSTM-AE model so as to acquire the corresponding outputs $\widehat{X}^R$. The reconstruction errors are computed by

$$E = \frac{1}{L} \sum_{j=1}^{L} |\widehat{X}_{i,j} - \widehat{X}_{i,j}^R|. \tag{5}$$

Modeling all the reconstruction errors as Gaussian distribution to obtain 95% GCI of the time-series signal according to (3).

Anomaly Detecting. The data collected in real time is fed into the trained model to obtain the reconstruction error. If it is within the GCI, it is a normal data point, otherwise it is an abnormal data point. The GCI abnormality discrimination is shown in Fig. 3.

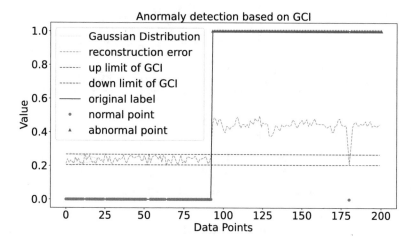

Fig. 3. Identification Abnormalities Process Based on GCI

3   Experiments and Dissuasions

3.1   Experimental Datasets

The anomaly detection model involved in this paper is based on the measured dataset of an industrial AGV and the public dataset SKAB for algorithm performance validation.

The voltage periodic time series signal of AGV is selected for the experiment and characteristics are shown in Fig. 4. There are three types of time-series voltage signals in the public dataset SKAB: the voltage data of closing the device outlet, which is denoted as V1, the voltage data of closing the device inlet, denoted V2, and the voltage data of the device disturbance, denoted as V3. The number and abnormal rates of datasets are shown in Table 1.

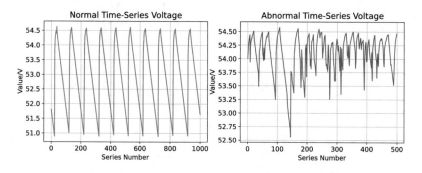

Fig. 4. Voltage Characteristics

Table 1. Statistics of the datasets

	AGV	*SKAB-V1*	*SKAB-V2*	*SKAB-V3*
Number sequences	166000	18160	4312	14929
Abnormal rate	25.7%	53.2%	54.3%	35.1%

3.2   Experimental Settings

In our experiments, the proposed G-LSTM-AE based detection is compared with traditional unsupervised learning anomaly detection algorithms such as AE and VAE. The specific parameters of the algorithm are shown in Table 2.

Table 2. Parameters table

Models	*Parameters*	*Values*
G-LSTM-AE	Layer numbers	2
	Nerve nodes numbers	2–256
	Learning rate	8e–3
	Epoch numbers	100
	Loss function	L_1
	tOptimization function	Adam
	Confidence level	95%
	Window length	L
AE	Input layer of nerve nodes numbers	6
	Output layer of nerve nodes numbers	6
	Learning rate	1e–3
	Epoch numbers	100
	Loss function	MAE
	Optimization function	Adam
VAE	Input layer of nerve nodes numbers	6
	Hidden layer of nerve nodes numbers	8
	Output layer of nerve nodes numbers	12
	Learning rate	1e–4
	Epoch numbers	100
	Loss function	MAE
	Optimization function	Adam

The accuracy rate, recall rate, false alarm rate, time-consuming ratio T_{Rate} of tuning out the best threshold and threshold applicability G_{Rate} are used as evaluation metrics in this paper. The calculation formula is as follows:

$$Accuracy = \frac{TP+TN}{TP+FN+FP+TN}, Recall = \frac{TP}{TP+FN},$$

$$FA = \frac{FP}{FP+TN} + \frac{FN}{TP+FN}, \qquad (6)$$

$$T\_Rate = \frac{T}{5}, G\_Rate = \frac{L_G}{L_E}, \qquad (7)$$

where TP is the number of true positive detections, FN is the number of false negative detections, FP is the number of false positive detections, and TN is the number of true negative detections. From (3), 5 is used as the baseline in terms of the adjustment interval of 1%. T is the total times which tune out the optimal threshold. L_E is the length of reconstruction error and L_G is the length of GCI. It indicates that the larger accuracy, recall rate and the threshold applicability, the smaller false alarm rate and the time-consuming ratio in a experimental.

3.3   Experimental Results

Based on the AGV, their reconstruction errors are shown in Fig. 5. The mean value of MAE distribution for normal data is 0.24, while the mean value of MAE distribution for abnormal data is 0.43. Meanwhile, there is almost no crossover between larger value interval of MAE for normal data and smaller value interval of MAE for abnormal data. In other words, the appropriate threshold interval can accurately distinguish normal from abnormal data. However, the MAE of AE or VAE intervals are strongly overlap. Obviously, the G-LSTM-AE model proposed in this paper can achieve excellent abnormal time-series signal detection.

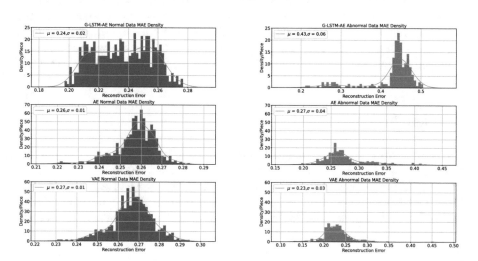

Fig. 5. Reconstruction Error Distribution

Table 3. Results of G-LSTM-AE, AE, VAE anomaly detection models

Different Models	*Accuracy*	*Recall*	*FA*
G-LSTM-AE model on AGV dataset	96.0%	95.7%	5.3%
G-LSTM-AE model on SKAB-V1 dataset	95.6%	95.2%	5.0%
G-LSTM-AE model on SKAB-V2 dataset	98.3%	96.5%	3.2%
G-LSTM-AE model on SKAB-V3 dataset	95.0%	94.3%	5.8%
AE Model on AGV dataset	56.5%	45.3%	12.5%
AE Model on SKAB-V dataset	45.6%	67.2%	22.5%
VAE Model on AGV dataset	68.4%	93.2%	100%
VAE Model on SKAB-V dataset	55.8%	83.7%	89.4%

Table 4. Anomaly detection results for different setting threshold methods

Different Models	*Accuracy*	*Recall*	*FA*	*G_Rate*	*G_Rate*
G-LSTM-AE model on AGV dataset	96.0%	95.7%	5.3%	60%	95%
LSTM-AE model on AGV dataset	97.2%	96.0%	3.2%	>100%	5%

Table 3 shows results of three algorithms, which of AE and VAE anomaly detection based on SKAB-V1, SKAB-V2, and SKAB-V3 are calculated by the average. The G-LSTM-AE anomaly detection algorithm shows that the accuracy and recall rates can reach more than 95%, while the false alarm case is negligible. On the other hand, the AE anomaly detection algorithm has low accuracy and the VAE anomaly detection algorithm can only determine the anomalous data under misjudgment, which have no practical significance. Table 4 compares the two methods of setting thresholds. One thing is that their accuracy and recall rate are as high as 95%. Another thing is that the GCI has advantages on short time-consuming and wide applicability, which mildly fluctuates in metrics but can be neglected.

4 Conclusion

This paper has proposed an algorithm for detecting time-series signal anomalies based on G-LSTM-AE network. Experiments show that this method is effective on detecting industrial time-series abnormalities with an accuracy of more than 95%. The effectiveness of the proposed method is verified by comparing it with state-of-the-art unsupervised learning anomaly detection methods. In the future, we can consider more data features to further improve the performance of the proposed G-LSTM-AE detection algorithm.

References

1. Wong, T., Luo, Z.: Recurrent auto-encoder model for large-scale industrial sensor signal analysis. In: Pimenidis, E., Jayne, C. (eds.) EANN 2018. CCIS, vol. 893, pp. 203–216. Springer, Cham (2018). https://doi.org/10.1007/978-3-319-98204-5_17

2. Xu, L.D., He, W., Li, S.: Internet of things in industries: a survey. IEEE Trans. Ind. Inf. **10**(4), 2233–2243 (2014)

3. Marjani, M., Nasarnddin, F., Gani, A., et al.: Big IoT data analytics: architecture, opportunities, and open research challenges. IEEE Access **5**(99), 5247–5261 (2017)

4. Li, D., et al.: Anomaly detection with generative adversarial networks for multivariate time series. arXiv preprint arXiv:1809.04758 (2018)

5. Li, D., et al.: MAD-GAN: multivariate anomaly detection for time series data with generative adversarial networks. arXiv preprint arXiv:1901.04997 (2019)

6. Ma, J., Perkins, S.: Time-series novelty detection using one-class support vector machines. In: Proceedings of IJCNN, vol. 3, pp. 1741–1745. IEEE (2003)

7. Liu, T.F., Ting, M.K., Zhou, Z.-H.: Isolation-based anomaly detection. ACM Trans. Knowl. Discovery Data **6**(1), 1–39 (2012)

8. Huang, C., et al.: Self-supervision-augmented deep autoencoder for unsupervised visual anomaly detection. IEEE Trans. Cybern. **52**(12), 13834–13847 (2021). https://doi.org/10.1109/TCYB.2021.3127716

9. Renström, N., Bangalore, P., Highcock, E.: System-wide anomaly detection in wind turbines using deep autoencoders. Renewable Energy **157**, 647–659 (2020)

10. KIngma, D.P.: Auto-encoding variational bayes. arXiv: 1312.6114 (2013)

11. Hochreiter, S., Schmidhuber, J.: Long short-term memory. Neural Comput. **9**(8), 1735–1780 (1997)

12. Sutskever, I., Vinyals, O., Le, Q.V.: Sequence to sequence learning with neural networks. In: Advances in Neural Information Processing Systems, pp. 3104–3112 (2014)

13. Miwa, M., Bansal, M.: End-to-end relation extraction using LSTMs on sequences and tree structures. In: Proceedings of the 54th Annual Meeting of the Association for Computational Linguistics (Volume 1: Long Papers), pp. 1105–1116 (2016)

14. Rumelhart, D.E., Hinton, G.E., Williams, R.J.: Learning representations by back-propagating errors. Nature **323**(6088), 533–536 (1986)

Construction and Analysis of Scientific Research Knowledge Graph in the Field of Hydrogen Energy Technology

Min Zhang[1,2], Rui Yang[1,2(✉)], and Jingling Xu[1,2]

[1] Wuhan Library of Chinese Academy of Sciences, Wuhan 430071, China
yangr@mail.whlib.ac.cn
[2] Hubei Key Laboratory of Big Data in Science and Technology, Wuhan Library of Chinese Academy of Sciences, Wuhan 430071, China

Abstract. In order to effectively reveal the various scientific research entities and semantic relationships among them in the field of hydrogen energy technology, the top-down construction method was used to explore the construction process of the scientific research knowledge graph in the field of hydrogen energy technology. Ontology was used to construct the schema layer of the knowledge graph, after knowledge extraction and knowledge fusion in the data layer, the knowledge was stored in the Neo4j graph database. The knowledge graph constructed included 345300 entities of 8 types and 2167484 entity relationships of 12 types. Through the construction of the knowledge graph, the complex knowledge system visual analysis of various scientific research entities and their relationships can be effectively realized, it can provide support for researchers to grasp the whole research situation in the field of hydrogen energy technology.

Keywords: hydrogen energy technology field · knowledge graph · knowledge extraction · knowledge fusion · Neo4j

1 Introduction

As a clean, efficient, safe and sustainable energy, hydrogen energy is an important alternative to fossil energy to achieve carbon neutrality [1]. With the increase of global climate pressure and the acceleration of energy transition, the development of hydrogen energy has been attached great importance, and more and more countries and regions have promoted the development of hydrogen energy to the strategic level of energy. In July 2020, the European Commission released the EU Hydrogen Strategy [2]; In December 2020, the U.S. Department of Energy released the Department of Energy Hydrogen Program Plan [3]; In March 2022, the National Development and Reform Commission and the National Energy Administration jointly released the Medium-and Long-term Plan for the Development of Hydrogen Energy Industry (2021–2035) [4].

In order to meet the great demand of high quality development of hydrogen energy, this paper taked the relevant research papers in the field of hydrogen energy technology

Q. Liang et al. (Eds.): AIC 2022, LNEE 871, pp. 392–402, 2023.
https://doi.org/10.1007/978-981-99-1256-8_46

in Web of Science database as the data source, and adopted the top-down method to construct the scientific research knowledge graph in the field of hydrogen energy technology. The knowledge graph included 8 types of scientific research entities and 12 types of entity relationships. Based on the constructed knowledge graph, various analysis were carried out.

2 Research Design

2.1 Knowledge Graph Framework Design

There are two ways to construct knowledge graph: top-down and bottom-up. In this paper, the top-down approach was adopted to construct the scientific research knowledge graph in the field of hydrogen energy technology [5]. The research framework, as shown in Fig. 1.

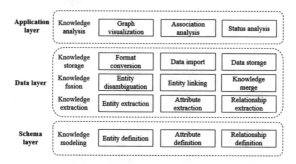

Fig. 1. Research framework of the knowledge graph

Firstly, the schema layer of knowledge graph was constructed, and the entities, entity attributes and entity relationships were defined. Then, in the data layer, according to the data model defined in the schema layer, the entities, entity attributes and entity relationships were extracted, and the knowledge fusion was carried out by the technologies of entity disambiguation, entity linking and knowledge merge, the entities and entity relationships data after knowledge fusion were imported into graph database in a certain format. Finally, various analysis were carried out on the basis of the constructed knowledge graph data in the application layer.

2.2 Data Source

By means of database resource analysis, literature investigation and expert consultation, the paper retrievals in the field of hydrogen energy technology were constructed, and the retrieval time range was from 2012 to 2021. After eliminating the repetitive and inconsistent papers, 46562 valid papers in the field of hydrogen energy technology were obtained as the main data research objects.

3   Knowledge Graph Construction

3.1   Knowledge Graph Schema Layer Design

Schema layer is the conceptual model and logical foundation of knowledge gragh. Ontology is often used to construct the schema layer of knowledge graph to constrain the data [6]. In this paper, the ontology modeling tool Protege was used to construct the ontology model of the scientific research knowledge graph in the field of hydrogen energy technology, as shown in Fig. 2.

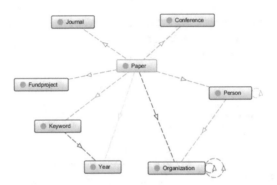

Fig. 2. Ontology model of the knowledge graph

8 types of scientific research entities were defined, including "Paper", "Person", "Organization", "Fundproject", "Journal", "Conference", "Subject" and "Year". 12 entity relationships of the scientific research knowledge graph were defined, as shown in Table 1.

3.2   Knowledge Graph Data Layer Construction

3.2.1   Knowledge Extraction

Knowledge extraction needs to complete the extraction of entities, attributes of entities and relationships among entities according to the data model defined in the schema layer of knowledge graph, and the obtained raw data is parsed into triples containing entities and their relationships, form a preliminary knowledge representation. By combing and analyzing the information of each field of the paper data, the entity extraction models were constructed by using the methods of text preprocessing, string processing, named entity recognition, etc., the models automatically extracted different types of entities and their attributes according to different extraction rules [7–9]. The paper ID was used to establish the relationships between the "Paper" entity and other types of entities and the relationships between different types of entities. The correspondence between the automatically extracted entities and the paper fields, as shown in Fig. 3.

Table 1. Entity relationships definition of the knowledge graph

Entity A	Entity relationship type	Entity B
Paper	paper_author	Person
Paper	paper_org	Organization
Paper	paper_fundproject	Fundproject
Paper	paper_journal	Journal
Paper	paper_conference	Conference
Paper	paper_keyword	Subject
Paper	paper_year	Year
Person	person_cooperation	Person
Person	person_org	Organization
Organization	org_cooperation	Organization
Organization	parent_org	Organization
Subject	keyword_year	Subject

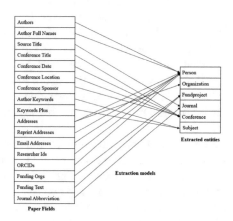

Fig. 3. Correspondence between automatically extracted entities and paper fields

3.2.2 Knowledge Fusion

The key technologies of knowledge fusion include entity disambiguation, entity link and knowledge merging. Knowledge fusion mainly needs to solve the problem of entity alignment. It will integrate the multi-source descriptions of the same entity to get a complete description of the entity, which can be realized by using the technologies of similarity calculation and clustering. There were some problems in the automatically extracted entities, such as duplication, irregular naming, incomplete attributes and so on. The automatically extracted entities need to be disambiguated and linked to the correct entity objects in the standard knowledge base through entity links. The disambiguation of scientific research institutions and researchers are relatively complicated.

The disambiguation of scientific research institutions mainly needs to solve the problem of multiple names, the names of the same institution are different in different papers, and the names extracted automatically have some problems such as abbreviation and inconsistency in writing order. Based on the algorithm of edit distance similarity, the similarity of organization name was measured by combining the calculation of string similarity and word set similarity. The algorithm of edit distance similarity, as below.

$$\text{sim}(s_x, s_y) = 1 - \frac{\text{minED}(s_x, s_y)}{\max(l_x, l_y)} \tag{1}$$

$\text{sim}(s_x, s_y)$ represents the similarity of string x and string y, $\text{minED}(s_x, s_y)$ represents the shortest edit distance of string x and string y, l_x and l_y represent the length of string x and string y, $\max(l_x, l_y)$ takes the larger of both.

The name of the organization to be disambiguated was calculated with the name of the organization in the standard knowledge base. When the string similarity or the word set similarity reached a certain threshold, the organization with the greatest similarity in the standard knowledge base and organization to be disambiguated were considered to be the same institution.

The disambiguation of Scientific Researcher entity can be divided into two situations. One is the phenomenon of duplicated name, different authors have the same name, and need to disambiguate so that the authors of the same name but different people belong to different entities [10]. The other is the phenomenon of multiple names, the names of the same author in different papers are different, and need to be disambiguate so that the same author with different names belongs to the same person entity. The author's characteristics include email, ORCID, ResearcherID, organization, cooperative relationship, etc. In order to increase the accuracy of disambiguation results, it is necessary to consider the influence factors of the author's characteristics to carry out disambiguation [11].

Email, ORCID and ResearcherID have the ability of accurate identification, which can quickly identify whether the authors to be disambiguated are the same person.

Authors of the same name may exist in the same institution, but it is not common to have different persons of the same name in the same secondary institution, the first-level and second-level institutional information contained in the author's address information can also assist in identifying whether the authors with same name to be disambiguated are the same person.

The author's cooperative relationship is also one of the characteristic elements with the degree of identification, which can be used to disambiguate the phenomenon of authors' renaming. The set of collaborators of the researcher to be disambiguated and the set of collaborators of the entity of the same name in the standard knowledge base were calculated by using the similarity algorithm of Jaccard coefficient [12], when the similarity of cooperative relationship reached a certain threshold, they were considered to be the same person. The Jaccard coefficient similarity algorithm, as below.

$$J(A, B) = \frac{|A \cap B|}{|A \cup B|} = \frac{|A \cap B|}{|A| + |B| - |A \cap B|} \tag{2}$$

J (A, B) was defined as the ratio of the size of the intersection of Set A and set B to the size of the union of Set A and set B.

3.2.3 Knowledge Storage

Neo4j graph database [13] was used to store entities, entity attributes and entity relationships in the scientific research knowledge graph in the field of hydrogen energy technology. The entities and entity relationships data after knowledge extraction and knowledge fusion were transformed into CSV format and imported into the Neo4j graph database using Cypher LOAD CSV statement. Some of the sample code, as below.

LOAD CSV WITH HEADERS FROM 'file:///paper_organization.csv' AS line.
MATCH (from: Paper{paperid:line.peperid}), (to: Organization{orgid:line.orgid})
MERGE (from)-[r:paper_org]-> (to)

A total of 345300 entities and 2167484 entity relationships were constructed in the scientific research knowledge graph in the field of hydrogen energy technology.

4 Knowledge Graph Analysis

4.1 Research Status Analysis

The distribution of annual publications in the field of hydrogen energy technology, as shown in Fig. 4. As can be seen from the figure, the annual quantity of publications in the field of hydrogen energy technology was on the rise in the past 10 years. The annual quantity of papers published in the field of "Hydrogen production by water electrolysis" was increasing rapidly, from 197 in 2012 to 3169 in the 2021. In the other 5 technology fields, the annual quantity of publications changed little, basically showing a stable trend. Among them, the "PEMFC" and "SOFC" technology fields had a relatively large number of publications, with more than 1100 each year.

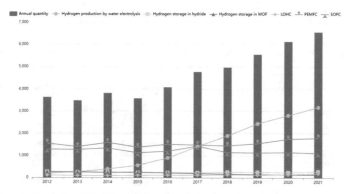

Fig. 4. Annual publication trends in the field of hydrogen energy technology

Tracing the hot topics in the technical field from the analysis of the subject words of the paper, searching the 5 most popular subject words in each year as the research hot spots, showing the knowledge graph between the hot topic words and the year, as shown in Fig. 5. As can be seen from the figure, "performance" was a hot research topic in the field of hydrogen energy technology every year for the past 10 years. Between

2012–2016 and 2017–2021, research hotspots changed significantly, the research focus changed from "hydrogen storage", "solid oxide fuel cell", "PEMFC" and "SOFC" to "nanoparticles", "oxygen evolution reaction", "electrocatalysts" and "catalysts". In the 2021 "nanosheets" as a new research focus emerged.

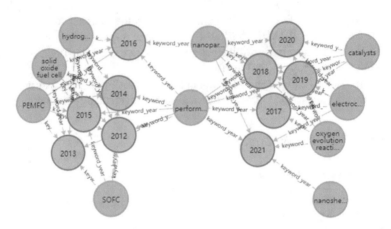

Fig. 5. Distribution of hot topics in the field of hydrogen energy technology in the past 10 years

There are more than 11000 first-level institutions in the knowledge gragh, which are mainly distributed in China, America, India, Germany, France, Korea, Japan, Italy and so on. Searched for the top 10 institutions in terms of number of papers, as shown in Table 2. With the exception of German Forschungszentrum Jülich, Nanyang Technological University and Technical University of Denmark, the rest are Chinese universities and research institutes.

Table 2. Distribution of the top 10 institutions in terms of number of publications

No	Organization name	Country	Paper number
1	Chinese Acad Sci	Peoples R China	2765
2	Tsinghua Univ	Peoples R China	681
3	Tianjin Univ	Peoples R China	583
4	Huazhong Univ Sci & Technol	Peoples R China	547
5	Forschungszentrum Julich	Germany	536
6	Harbin Inst Technol	Peoples R China	529
7	South China Univ Technol	Peoples R China	507
8	Jilin Univ	Peoples R China	487
9	Nanyang Technol Univ	Singapore	467
10	Tech Univ Denmark	Denmark	441

4.2 Knowledge Association Analysis

The "Paper" entity is the core entity of the knowledge graph in the field of hydrogen energy technology. A visual representation of the knowledge graph of a paper entity, as shown in Fig. 6. The entities associated with the paper entity and their relationships can be found.

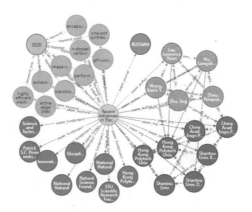

Fig. 6. Knowledge gragh of the paper entity

Organization cooperation and author cooperation are the main analysis objects of scientific research knowledge graph. Besides direct cooperative relationship, indirect cooperative relationship can be queried by depth query through knowledge graph.

Queried the cooperative relationship with a scientific research organization whose cooperative path length is less than 3, as shown in Fig. 7. The knowledge graph can be used to explore the potential cooperation between scientific research organizations. For example, "Chinese Acad Sci, Dalian Inst Chem Phys" and "Dalian Univ Technol, Sch Ocean Sci & Technol" do not have direct cooperation, but may have potential

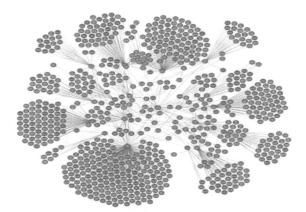

Fig. 7. Knowledge gragh of organization cooperation

cooperation, this is because they have two common cooperation institutions, "Dalian Univ Technol, Sch Chem Engn" and "Kyushu Inst Technol, Grad Sch Life Sci & Syst Engn".

Queried the cooperative relationship between two scientific research person entities whose cooperative path length is less than 4, as shown in Fig. 8. It can be found that although there is no direct cooperative relationship between two researchers, there are many paths of indirect cooperation, and the shortest cooperation path between two researchers is "Shao, Zhigang-Luo, Jiangshui-Pan, Mu".

Fig. 8. Knowledge gragh of scientific researchers' cooperative paths

Multi-dimensional analysis of complex relations can also be realized by the knowledge graph. For example, the annual distribution of papers published by researchers and the change of signatories can be found by the association analysis of person-organization-paper-year, as shown in Fig. 9.

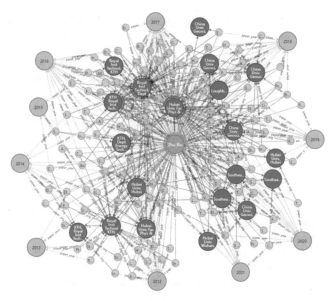

Fig. 9. Associated knowledge graph of person-organization-paper-year

5   Conclusions

In this paper, the method of constructing the scientific research knowledge graph in the field of hydrogen energy technology was explored. The construction of domain knowledge graph was realized by knowledge extraction, knowledge fusion and knowledge storage. Through the scientific research knowledge graph in the field of hydrogen energy technology, the relationships among various scientific research entities were demonstrated, and the visual analysis of complex knowledge system was realized, it provided some reference for researchers to understand the research status of hydrogen energy technology. Further research will be carried out around the construction of domain knowledge graph.

References

1. Zhuo, Y., Cao, Y., Nie, J.F., et al.: Analysis of key technologies of the hydrogen energy industry based on big data research. In: 3rd International Conference on Computer Information and Big Data Applications, CIBDA 2022, Wuhan, China (2022)
2. European Commission: EU Hydrogen Strategy (2020). https://ec.europa.eu/commission/pre sscorner/api/files/attachment/865942/EU_Hydrogen_Strategy.pdf. Accessed 18 Apr 2022
3. Department of Energy: Department of Energy Hydrogen Program Plan (2020). https://www.energy.gov/articles/energy-department-releases-its-hydrogen-program-plan. Accessed 28 Apr 2022
4. National Development and Reform Commission: Medium-and Long-term Plan for the Development of Hydrogen Energy Industry (2021–2035) (2022). https://www.energy.gov/articles/energy-department-releases-its-hydrogen-program-plan. Accessed 28 June 2022

5. Liang, H., Peng, X.J., Zhao, N.N., et al.: An approach of top-DOWN electric generation knowledge graph construction. In: 2nd International Conference on Energy, Power, Environment and Computer Application, ICEPECA 2020, Beijing, China (2021)
6. Hu, C., Xie, S.W., Xie, Y.F., et al.: Development of domain knowledge graph: a case study on flotation process. In: 6th International Conference on Robotics and Automation Engineering, ICRAE 2021, Guangzhou, China (2021)
7. Wang, M.Y., Hu, X.H., Xie, P., et al.: Automatic construction of a domain-specific knowledge graph for Chinese patent based on information extraction. In: 2021 International Conference on Management Science and Software Engineering, ICMSSE 2021, Chengdu, China (2021)
8. Yilahun, H., Hamdulla, A.: Review on the entity extraction methods for low-resource languages. In: 14th International Conference on Measuring Technology and Mechatronics Automation, ICMTMA 2022, Changsha, China (2022)
9. Mesbah, S., Bozzon, A., Lofi, C., et al.: Smartpub: a platform for long-tail entity extraction from scientific publications. In: Proceedings of Companion Proceedings of the Web Conference, Lyon, France (2018)
10. Zhang, S.Y., Xinhua, E., Pan, T.: A multi-level author name disambiguation algorithm. IEEE Access **7**, 104250–104257 (2019)
11. Zhu, G.G., Iglesias, C.A.: Computing semantic similarity of concepts in knowledge graphs. IEEE Trans. Knowl. Data Eng. **29**(1), 72–85 (2017)
12. Qin, X., et al.: Density peaks clustering based on Jaccard similarity and label propagation. Cognit. Comput. **13**(6), 1609–1626 (2021). https://doi.org/10.1007/s12559-021-09906-w
13. Neo4j: Neo4j Graph Database (2022). https://neo4j.com/product/neo4j-graph-database/. Accessed 8 Mar 2022

Problems and Strategies of Foreign Language Education Development from the Perspective of Linguistic Intelligence

Lin Han[✉]

Tianjin Vocational Institute, No. 2 Luohe Road, Beichen District, Tianjin, China
hanlin2013@foxmail.com

Abstract. This paper analyzes the problems in the development of foreign language education under the background of language intelligence, and proposes solutions to these problems. This paper holds that under the background of the great development of language intelligence, foreign language education should construct the smart foreign language education system by improving the construction of the teaching staff, constructing the teaching research and teaching management platform, and realizing the benign development of foreign language education.

Keywords: foreign language education · language intelligence · problems · strategies

1 Introduction

Language intelligence refers to the intelligence of language information. It is the science of analyzing and processing human language by using computer information technology to imitate the intelligence of human language. Language intelligence aims to use computer information technology to enable machines to understand, analyze and process human language and achieve human-computer language interaction. In recent years, language intelligence has entered a period of rapid development, which on the one hand brings historical opportunities for the development of foreign language education, especially provides new ideas and new technical means for the development of foreign language education. On the other hand, it also brings great challenges to the development of foreign language education and produces a series of problems. On the basis of sorting out the problems in the development of foreign language education from the perspective of language intelligence, this paper tries to put forward strategies for its future development.

2   Problems in the Development of Foreign Language Education from the Perspective of Linguistic Intelligence

2.1   The Development of Linguistic Intelligence Has Reduced the Demand for Foreign Language Professionals

In recent years, with the rapid development of artificial intelligence, the application level of language intelligent products such as speech recognition, speech synthesis and machine translation has been greatly improved. With the use of speech recognition and speech synthesis technology, the computer can read the foreign language text with the standard speech intonation, and provide the foreign language learners with the standard speech. With the help of machine translation system, we can not only cross the language barrier to communicate with people from other countries, but also translate foreign language articles into Chinese and understand the culture and local conditions of other countries. Machine translation in the treatment of the text, due to the increasing of corpus base, compared with the artificial can move more quickly and accurately professional vocabulary, which has significant advantage thanks to the wide use of computer processing and Internet technology, compared to human translation, machine translation, the use of low threshold, can provide translation services to customers at any time and place, the efficiency is also higher. In this sense, the development of language intelligence will inevitably reduce the social demand for foreign language professionals, especially for middle and low-end foreign language professionals.

2.2   Teachers Engaged in Foreign Language Education Generally Lack Language Intelligence Literacy

In the cultivation of foreign language professionals in China, too much emphasis has been placed on the instrumental and humanistic nature of foreign language education, and few science or science courses have been offered. As a result, foreign language graduates are inborn with insufficient scientific literacy. The survey on the background of science and technology subjects of foreign language teachers shows that foreign language teachers know little about science or technology characteristics of foreign language courses (such as mathematics, statistics, computational linguistics and machine translation, etc.) and rarely apply them in their work.

Due to the lack of language intelligence literacy, foreign language professional teachers have not done much in the field of language intelligence that they could have done (Table 1).

Table 1. Statistics on science and technology subject background of foreign language teachers

	Barely know, almost not used in the work	Appropriate use of common understanding, can be in work	Do preliminary master, can be used to study	Master proficiently, can use it to do research and guide teaching
Mathematics (Ratio:%)	62	23	15	0
Statistics (Ratio:%)	62	18	15	5
Computational Linguistics (Ratio:%)	70	22	8	0
Machine Translation (Ratio:%)	35	40	15	10

Source: from the author

2.3 Teaching Methods Fail to Achieve Organic Integration with Language Intelligence Technology

A questionnaire survey of foreign language teachers shows that teachers use intelligent technology in foreign language teaching mainly for dealing with routine work (such as checking in, collecting students' assignments, etc.), reviewing homework or testing, etc. There are few cases that use human-computer interaction function to provide personalized instruction for teaching. Therefore, the teaching of foreign language teaching content depends more on the qualitative description of teachers, and lacks the support of rich examples and data, so it is not intuitive and objective enough. In addition, we rarely use corpus technology and language intelligence technology to develop intelligent foreign language teaching platform, to construct data-driven, learner-centered discovery and exploration of foreign language teaching model. Due to the lack of intelligent language technology support, network courses and network textbooks can not achieve human-computer interaction, not to speak of the realization of teaching data and personalized (Table 2).

Table 2. Statistics on teachers' use of intelligent technology in foreign language teaching

	Barely use	Handle routine work (such as sign-in, collection of student assignments, etc.)	Evaluation function (homework correction, testing, etc.)	Human-computer interaction (personalized instruction for teaching)
Use of intelligent technology in foreign language teaching (Ratio:%)	10	60	25	5

Source: from the author

3 Strategies for the Development of Foreign Language Education from the Perspective of Linguistic Intelligence

At present, language intelligence has entered a period of rapid development, which not only provides new ideas and new technical means for the development of foreign language education, but also brings challenges and even impacts to the development of foreign language education. We can promote the organic integration between foreign language education and language intelligence development by building a compound teaching staff to meet the needs of language intelligence development and building a smart foreign language education system, so as to adapt to the development and solve a series of problems.

3.1 To Build a Compound Teaching Staff to Meet the Needs of Language Intelligence Development

As the main body of foreign language education, it is very important to build a compound teaching staff to meet the needs of language intelligence development.

3.1.1 To Integrate Resources from Multiple Related Disciplines

The interdisciplinary discipline of language intelligence needs to integrate the resources of multiple related disciplines, cultivate high-end compound talents in the new era who are "familiar with language, understand technology and new ideas", and serve the national development strategy. To be specific, it is necessary not only to have a solid language foundation, but also to be familiar with corpus construction and application, natural language processing, computational linguistics and other fields of knowledge. To meet the above requirements, on the one hand, rely on the introduction of compound foreign language professionals who master the relevant professional knowledge and skills of language intelligence; On the other hand, existing teachers can master professional knowledge and skills related to language intelligence and improve their teaching ability and level through on-the-job study of relevant professional degrees and domestic and foreign research. On this basis, the compound teaching staff oriented to language intelligence should also be composed of teachers with the background of experimental phonetics, computational linguistics, information science, statistics, mathematics and Chinese.

In fact, only with such a compound teaching staff can we realize the organic integration of foreign language education and language intelligence, and cultivate foreign language professionals who meet the needs of language intelligence development.

3.1.2 To Innovate the Model of School-Enterprise Cooperation

On the one hand, it should coordinate with the development plan of the school, be problem-oriented, break the disciplinary barriers, give full play to the advantages of multi-disciplinary and multi-language talents, and build teachers and research teams with complementary knowledge structure in language and computer disciplines. On the other hand, it attaches importance to the combination of industry, education and research, and strengthens the cooperation with enterprises related to artificial intelligence and language services by establishing school-enterprise joint laboratories. Strengthening school-enterprise cooperation is conducive to the cooperative development of multi-lingual language resources, intelligent teaching platforms, human-computer interaction systems, machine translation models and other fields.

3.2 To Build a Smart Foreign Language Education System

Intelligent foreign language education refers to the organic integration between artificial intelligence technology including language intelligence technology and foreign language education. Intelligent foreign language education takes the application of information technology and artificial intelligence technology as the core, and its main characteristics are as follows: openness, namely, the co-construction and sharing of educational resources; Intelligent, that is, intelligent teaching as the main content, to achieve personalized teaching and personalized learning; Integration, that is, the integration of educational processes such as teaching, management and evaluation; Diversification refers to the diversity of evaluation.

3.2.1 Teaching and Research Platform

To develop and apply an intelligent platform for foreign language teaching and research, and realize the wisdom of foreign language teaching and research. The application of corpus technology and language intelligence technology can promote the datumization and visualization of foreign language classroom teaching. Based on data analysis and corpus observation, the teaching will be more intuitive and objective, so as to effectively improve the teaching effect and quality.

Based on intelligent corpus of foreign language teaching and scientific research platform, constructing data driven, learner centered to discover and explore translation teaching mode, to promote personalized and adaptive learning, realize the automation of the whole teaching process, digital and visual, effectively stimulate students' interest in learning, fully activate students' learning enthusiasm and initiative, In this way, a data-driven, learner-centered, discovery and exploration foreign language teaching model can be constructed to promote scientific and intelligent foreign language research.

3.2.2 Teaching Management Platform

To realize the intelligent management and service of foreign language education and the evaluation of foreign language education, an intelligent foreign language teaching management and service platform is developed by using the relevant technology of language intelligence.

The platform includes language data such as students' papers and assignments, multi-modal data of students' classroom learning behaviors, various kinds of corpus and question banks related to course learning, and teachers' classroom utterances, embedded text data mining technology, speech recognition and speech synthesis technology, intelligent question answering and intelligent correction technology. Teachers can use the platform to comprehensively assess the language knowledge and ability of students, and based on this, determine the teaching content and teaching focus, formulate precise and personalized teaching plans, so as to carry out personalized teaching.

In a word, we should make full use of the technological advantages of artificial intelligence, start from the wisdom of foreign language teaching and research and foreign language teaching management, and strive to build a smart foreign language education system.

4　Conclusion

Based on the analysis of the problems in the development of foreign language education from the perspective of language intelligence, this paper puts forward Some strategies for its future development. We should actively respond to the challenges and seize the opportunities brought by the development of language intelligence technology. Thus, we should cultivate compound talents who not only have a solid language foundation, but also have a good knowledge and skills of language intelligence, and build a smart foreign language education system to meet the needs of development.

References

1. Hu, K., Wang, X.: The development of foreign language education in the context of language intelligence: problems and approaches. Foreign Lang. China (2021)
2. Hu, K., Tian, X.: The challenges of MTI education and their solutions in the context of language intelligence. Foreign Lang. world (2020)
3. Li, Z., Liang, G.: Interdisciplinary construction of language intelligence. Technol. Enhanc. Foreign Lang. (2022)

A Brief Analysis of AI-Empowered Foreign Language Education

Lin Mu[✉]

Tianjin Vocational Institute, No. 2 Luohe Road, Beichen District, Tianjin, China
moonlyn@163.com

Abstract. In recent years, artificial intelligence (AI), which advocates machine learning, has attracted the attention of many educational scholars with its unique deep learning technology. With the development of artificial intelligence technology, the research of linguistic intelligence has made remarkable achievements. These advances not only promote the development of intelligent foreign language education, but also pose challenges to traditional foreign language education in China. This paper will make a preliminary analysis of the current situation and development prospect of AI-empowered foreign language education in China.

Keywords: AI-empowered foreign language education · Current situation · Development prospect

1 Introduction

The research of linguistic intelligence is the research of artificial intelligence technology applied to the related field of natural language processing. It is an interdisciplinary subject of linguistics and artificial intelligence. The research of language intelligence aims to realize the intelligent language learning in the whole process by simulating human language ability with the help of modern technology, especially artificial intelligence technology. In recent years, remarkable progress has been made in the study of language intelligence such as machine translation, speech recognition and synthesis, intelligent correction, intelligent writing and intelligent question answering, which has effectively promoted the intellectualization of language teaching and language learning, expanded the new field of linguistic research, and played an increasingly important role in foreign language education. This paper will provide an overview of the current situation of foreign language education empowered by artificial intelligence and make prospects for its development.

2 The Current Applications of the Language Intelligence Research

In recent years, the results of language intelligence research have been applied in the whole process of the intelligent foreign language teaching. These applications can be divided into three main categories.

2.1 Intelligent Teaching System

As the best embodiment of artificial intelligence applied in language teaching system, intelligent teaching system aims to build a good student-oriented web-based teaching environment supported by all kinds of computer-aided technologies to enrich education resources, offer a full range of online and in-class language teaching activities and meet various demand of foreign language learning for students so that they can make reasonable use of fragmented time and enrich their knowledge reserve.

The intelligent teaching system can build highly personalized English learning programs for different students according to their individual development needs such as learning status, interest and ability, so that they can fully enjoy the convenience of artificial intelligence learning In addition, under the effect of data analysis and behavior simulation of artificial intelligence, the classroom teaching content can be reviewed anytime and anywhere, the classroom interaction between teachers and students can be recorded in detail, and automatic data can be generated for storage, so that teachers can accurately analyze the learning state of students and their own teaching effects as valuable scientific references to make effective teaching plans in the future. Finally, based on intelligent speech technology, a new style of language classroom can be formed with the model of "teaching training, evaluation feedback and strategy adjustment", which can develop and improve the comprehensive ability and technical literacy of teachers.

2.2 Intelligent Evaluation System

The study of linguistic intelligence is bringing about a revolutionary change in teaching evaluation of foreign language education. In addition to machine intervention evaluation systems such as automatic writing evaluation system and translation evaluation system to evaluate students' learning effect, and evaluation system to evaluate teachers' teaching effect based on questionnaire survey statistics, the intelligent evaluation can also be achieved by the big data based on students' learning process, teachers' teaching process, as well as education policy, textbook, syllabus and social evaluation. With the help of artificial intelligence and quantitative statistical analysis, objective and effective evaluation reports can be formed to provide a basis for subsequent decision-making at all aspects of education and teaching.

The intelligent oral ability assessment is an effective measure of learners' listening and speaking ability. Artificial intelligence can analyze the learner's pronunciation and the intonation. The artificial intelligence learning platform can give efficient, objective and immediate oral evaluation and problem diagnosis of each student's communication ability according to the actual performance of learners through the deep learning of machines, effectively reducing the influence of subjective factors in traditional manual scoring methods. At the same time, through the intelligent report automatically generated by the computer, teachers can timely learn about the learning situation and learning needs of students, so as to adjust the teaching content and strategy, and in turn, to increase targeted English instruction and intensive practice.

The intelligent writing evaluation system uses big data analysis technology, guided by theory, associative word mining technology and data science analysis technology to explore deep learning based on artificial neural networks. With language, content, text structure and technical specifications as evaluation constructs, it excavates associative lexicon from large corpora. The method of establishing Concept Nets realizes the judgment of the relevance and coherence of the composition, so as to realize the intelligent evaluation of the above four dimensions.

2.3 Intelligent Corpus System

Language intelligence research based on corpus big data is widely applied and practical, which can provide a set of effective methods and tools for foreign language teaching and research.

The study of language intelligence empowers intelligent teaching materials to be compiled based on language database. Relying on the intelligent editing engine technology, the main text material is selected and the textbook is arranged according to the syllabus of each learning stage, so as to achieve more efficient and high-quality textbook editing and production. Relying on the intelligent editing engine technology, the main text material is selected and the textbook is arranged according to the syllabus of each learning stage, so as to achieve more efficient and high-quality textbook editing and production. With the help of some relevant tools, we can measure the difficulty of the content of the textbook, analyze its language characteristics and grasp its regularity, so as to serve the teaching needs of different levels and types more scientifically.

In terms of reading teaching, corpus big data can contain reading corpora of millions, tens of millions or even hundreds of millions of words, from which people can get sufficient examples, verification or explanation of some reading teaching theories, and the generalization of the authenticity of the theories can be verified through empirical research.

The intelligent writing evaluation system realizes the judgment of the relevance and coherence of the composition by mining the association lexicon from the large corpus. In addition to the use of big data, it also generates massive and dynamic big data. After data processing, it forms learner corpus, learner typical error database, teaching case database, etc., providing rich and detailed data support for teaching and scientific research.

3 The Development Prospect of AI-Empowered Foreign Language Education in China

The development of language intelligence research will certainly bring about earth-shaking changes in the field of foreign language education. How to make use of the results of language intelligence research to truly empower foreign language education is a challenge that China's foreign language education community must face in the future. The field of foreign language education should first face this challenge, and combine the development of language intelligence with the impact of digital teaching empowering foreign language education, to open a new situation and seek new breakthroughs for talent cultivation and discipline construction. The goal of foreign language education in pursuit of AI is to better serve teaching. The joint effect of concept identification, expert guidance and school practice is needed to realize the goal of educational reform from inside to outside.

3.1 To Construct an Intelligent Teaching System of Foreign Language Education Based on the Research of Language Intelligence

Foreign language education needs to quickly adapt to the general trend of intelligent education, explore the intelligent foreign language education and teaching system based on language intelligence research, and construct educational big data and teaching big data.

Educational big data includes not only the macro educational policy data, but also the preparatory data such as teaching syllabus and textbooks, the actual teaching process data, and the post-hoc evaluation data composed of teaching evaluation and social evaluation. Teaching big data mainly refers to the related data generated from the two aspects of students and teachers in the actual teaching process. All these data, based on the study of language intelligence, constitute the intelligent system of foreign language education and teaching. Intelligent foreign language education and teaching system is to use computer and artificial intelligence technology to achieve accurate education and teaching based on language data science from macro and micro levels, and truly realize "teaching in accordance with students' aptitude".

3.2 To Truly Realize the Intelligentization of Foreign Language Education with Education Wisdom and Modern Technology

The empowerment of educational technology to the education industry is essentially an auxiliary to education work, which solves the problems of traditional education with teachers as the core, and then improves the efficiency of teaching and learning. In the future, foreign language education and teaching should break the tradition from the aspects of teaching concept, principle, approach, method, method, means and mode, and apply modern science and technology to improve the scientificity and accuracy of foreign language education and teaching.

However, machines do not have feelings and intelligence, so they have to be assigned selective work. Modern education pursues individualization and wisdom, but more importantly, teachers should bring their own brain to teaching work. The so-called smart education is the information reform of education system and the network support of resources. Smart education should be student-centered and maximize its growth potential. On the other hand, teachers should keep their ideas and experiences updated and make full use of smart technology – smart data empowerment education. The data value behind their behaviors needs teachers' attention. Teaching wisdom supported by intelligent technology is actually very simple: "integrates teaching content with wisdom, applies educational technology with wisdom, designs mixed teaching with wisdom, uses real-time data with wisdom." – That is, content-based, technology-oriented, personalized and refined learning.

3.3 To Promote the Transformation of Research Results and Realize the Effective Docking of All Aspects of Foreign Language Education with Language Intelligent Technology

In the field of foreign language education and teaching, we should adapt to the needs of the new era, closely combine advanced, mature and applicable scientific and technological achievements with front-line teaching and fully carry out a wider range of industry-university-research cooperation projects to truly realize the promotion of scientific research and teaching, and cultivate innovative talents with international vision, foreign language communication skills and analysis and problem-solving skills for the national development in practice. At the same time, we will make full use of the application of emerging technologies to promote the construction of high-quality foreign language courses, share high-quality resources with other colleges and universities, and actively promote the innovative development of foreign language education in China.

4 Conclusion

Artificial intelligence technology empowers foreign language learning, breaking the limitation of learning time and space. Learners can make use of fragmented time, access to cutting-edge information and master practical skills. Artificial intelligence technology empowers foreign language teaching, promoting the effective and precise development of foreign language education and teaching, changing the teaching ecology, and bringing many opportunities and challenges for foreign language teaching. Educators should follow the trend and take the initiative to promote the effective integration of artificial intelligence technology and foreign language teaching. In the field of foreign language education in the future, those who gain language intelligence will win the world.

References

1. Huang, L.: Language intelligence and foreign language education under the background of big data era. Foreign Lang. China (2022)
2. Hu, K., Tian, X.: The challenges of MTI education and their solutions in the context of language intelligence. Foreign Lang. World (2020)
3. Jin, C.: Research on the application of artificial intelligence in college English teaching under information environment. J. Changchun Norm. Univ. (2022)
4. Zhou, Q.: A brief analysis of the feasibility of artificial intelligence in College English Teaching (2018)

Complex-Valued Neural Networks with Application to Wireless Communication: A Review

Steven Iverson[✉] and Qilian Liang

University of Texas at Arlington, Arlington, TX 76019, USA
`steven.iverson@mavs.uta.edu, liang@uta.edu`

Abstract. We briefly review complex-valued neural networks (CVNNs) and compare them to real-valued neural networks (RVNNs). CVNNs allow for a richer representation of many wavelike based physics and engineering fields by retaining the data structure and correlation between real and imaginary parts of the signal. For example, most RF modulations use in-phase and quadrature components in which analysis benefits from the use of CVNNs. We then present state of the art in wireless radio applications utilizing CVNNs and/or complex-valued data. Wireless applications reviewed include modulation recognition, signal identification, channel sensing information and specific emitter identification. Finally, we present motivation for future exploration of CVNNs.

Keywords: neural networks · real-valued neural networks · complex-valued neural networks · complex numbers · wireless communication · deep learning · cognitive radio

1 Introduction

Neural network (NN) models, also known as artificial neural networks (ANNs), form an algorithmic structure loosely parallel to that of the brain's biological anatomy and dynamics [1, 2]. They are arguably a powerful computational tool at the vanguard of research and application in machine learning. With their ubiquitous use, there exists a plethora of ANN application and research fields. Some examples include: computer vision [3], biomedicine [4], speech recognition [5], and cryptography [6], among many others.

Real-valued neural networks (RVNNs) are ANNs that process data, compute internal parameters and apply activation functions that are real-valued. However, there are various physical and engineering domains utilizing models with a wavelike structure that can be represented with complex-valued data. Since RVNNs cannot directly process complex-valued data, complex-valued neural networks (CVNNs) have been proposed [7, 8]. Inclusion of complex-valued data for inputs and outputs, complex-valued weights, complex-valued activation functions, and complex-valued convolutional layers builds on traditional RVNN structure and by extension allows for the development of CVNNs. Modern wireless communications, audio and speech signal processing, biomedical imaging, and radar are example fields that utilize this representational tool.

Two recent critical and thorough surveys of the CVNN literature sufficiently reference the history and the latest in advances and applications in this domain [9, 10]. However, herein we focus on CVNNs in the wireless communications field, but first we discuss the basic structure of CVNNs and how they differ from RVNNs. Also, we present a short discussion of the structure of CVNNs, including activation functions, and network types. Finally, we review the use of CVNNs in wireless radio applications and close with some possible future research topics.

2   Why Use CVNNs?

A CVNN is an ANN that uses complex-valued inputs to compute neuronal outputs via these inputs into a complex-valued activation function and computes weights parameters and biases via a complex-valued backpropagation feedback process. The outputs of CVNNs can be real or complex-valued.

RVNNs have been at the forefront of the machine learning field for years. But while real-valued neural networks RVNNs have a well-established history, demonstrated performance and are used in many applications; current RVNN theory excludes the expressive power of complex-valued data. In other words, RVNNs built for real-valued data cannot directly process complex-valued data. There are some recent uses of complex-valued data with modified RVNNs regarding each component of the complex number as independent real numbers. Nevertheless, this double-dimensional RVNN, is not equivalent to a fully implemented CVNN. Additionally, this modification causes a doubling of the number of trainable parameters and increases computational load. This behavior is understandable due to ease of use, user familiarity and less computational processing load with RVNNs.

It has been shown that these two are not equivalent since the complex-valued weight multiplication operation at the neuron synapses results in amplitude magnitude scaling and phase rotation which reduces the degrees of freedom in the CVNN versus the RVNN [11, 12]. This reduction in the degrees of freedom in the network results in a more generalized system which assists in a reduction of overfitting of the network to the complex-valued data [11].

More specifically, favor of the use of CVNNs lies in their ability to represent the real and imaginary components of the complex-valued data more accurately without discarding the relationships between the parts. When processing complex-valued data, it is critical to handle the phase and amplitude of the signal directly. For example, researchers have found the emphasis on the phase component of the signal is found to be related to the understanding of human speech. Fine scale temporal structure of speech is affected by the data encoded within the phase component [12]. Information found in the phase portion of an image signal can be used to partially infer the magnitude of the signal [13]. Moreover, most complex-valued signals have at least partial statistical correlation. Therefore, using the relationship between these components and not just the structure is proper [14].

3 CVNN Activation Functions and Structure

Activation functions influence the ability of the neural network to converge. They also effect the rate of convergence, while also normalizing the output to a restricted range (bounded), typically [0, 1] or [−1, 1]. This output normalization is sometimes referred to as the squishing feature of the function. Activation functions should: 1) be differentiable over the entire domain, 2) contain nonlinearities, 3) be computationally efficient and 4) prevent the exploding or vanishing gradient problem. Although these activation function properties are desired, they are not strictly required nor exhaustive. In RVNNs many researchers use a standard sigmoid function or more recently are using the rectified linear unit (ReLU) activation function as one of the most popular, particularly for deep NNs [15]. However, in CVNNs, one of the most difficult tasks is determining the type of complex-valued activation function to use at the neuron.

3.1 Complex Activation Functions

Activation functions for CVNNs are desired to be holomorphic (that is a function that is differentiable everywhere in the complex plane/hyperplane) to maintain compatibility with the backpropagation process. However, the desire of boundedness and holomorphicity into the complex activation function runs afoul of Liouville's Theorem.

Since achievement of both objectives is impossible, the researcher must carefully choose between boundedness or differentiability. A review of the research literature reveals that there is no agreement on choice or application of complex-valued activation functions, unlike that of RVNNs.

3.2 CVNN Structure

Developing a fully CVNN is a demanding task. Researchers [16] have used three approaches to tackle it. The first consists of splitting the complex-valued input data into real and imaginary parts and applying them to a RVNN as independent real-valued components. The split-CVNN/split-complex-valued multilayer perceptron model is still very much in use today because of its popularity and simplicity. The network solves the above activation function problem. However, it lacks the approximation capability of complex-valued functions. The next approach builds a real-valued activation function in which the inputs to the function are complex, but the outputs are real-valued. This approach suffers from poor phase approximation of the input signal. The third approach defines a fully CVNN, using the sigmoid activation function defined by

$$f(z) = sgm[Re(z)] + i * sgm[Im(z)]$$

where $sgm(.)$ represents the sigmoid activation function and z is a complex number. Performance and computational complexity results in a more efficient structure for this approach in some communication applications. However, the choice and identification of the real-valued function for each of the real and imaginary components is application dependent and is full of challenges.

4   Wireless Radio CVNN Applications

In this section, we introduce and discuss the wireless radio applications of CVNNs. We also review wireless radio applications that uniquely utilize complex-valued data within a traditional RVNN structure. A summary of these applications follows, highlighting their strengths and weaknesses.

4.1   Modulation Classification

Wireless RF spectrum is a time dependent limited resource. This spectrum has mostly been reserved through an auction process through the FCC. Legitimate owners of this spectrum do not need to share it. However, studies have shown that the spectrum is being inefficiently used. A proposed method to combat this inefficient use is cognitive radio, in which primary users (spectrum owners) can co-exist with secondary users (non-spectrum owners) without one interfering with the other during communication.

A necessary step to spectrum access is the requirement to identify signal modulation type to assist in choosing the targeted user's communication scheme so that they can be classified for further processing. The following discusses recent modulation grouping techniques.

Recently, Krzyston et al. [17] have taken existing convolutional RVNN frameworks and applied in-phase and quadrature RF data to them for modulation recognition. Their linear transformation algorithm of the data increases the strength of the representation in the network. It also allows for increased filter size to evaluate the I/Q components more completely and to learn the relationships between them. Their linear processing algorithm improved accuracy of modulation classification at each SNR by approximately 35%. Although they did not strictly use CVNNs, the process of linearly combining complex-valued data did simplify the network's processing.

The same authors [18] leveraged complex convolutions into updated state of the art convolutional neural network structures: Residual Networks, Dense Connected Networks, and combinations they call Dense ResNets. This use of the improved frameworks increased I/Q modulation recognition over convolutional NN in the literature for all SNRs above −6 dB. For SNRs above −2 dB the recognition increased at least 12%.

Peng et al. [19] have also used a traditional convolution RVNN framework for processing of raw modulated RF signal data for modulation classification. They convert complex-valued signal points into constellation diagrams and then apply these to the AlexNet framework. Given four modulation types; QPSK, 8PSK, 16QAM and 64QAM, at SNRs of 4 dB and 8 dB, the identification accuracy for the PSK modulations is near 100% and the amplitude modulations average nearly 72%. Compared with the traditional algorithm, the convolutional NN based approach attains comparable outcomes, while avoiding expert feature selection.

4.2   Signal Recognition

Zhang and Liu [20] use multi-modal deep learning along with CVNN architecture to combat interference in the form of jamming. They exploit stellar I/Q waveform images using multimodal fusion to obtain the interfering signal. Their architecture is based on

AlexNet networks and deep CVNNs. They also apply decision-based fusion to increase feature distinction and improve recognition performance. Quantitative results were not revealed.

Xu et al. [21] use two complex-valued network architectures for radio signal recognition. A CvVGG convolutional NN model is proposed that has smaller receptive convolutional fields (3 × 3) than AlexNet with large receptive fields (11 × 11) upon which it is based. Due to smaller sized filters a VGG model allows for larger weight layers. Additionally, they replace the 2D convolutional layer with a 1D layer. In the second architecture, they propose a CvRN developed from a Deep Residual Network. In the multilayer model, again they replace the 2D convolutional layer in the residual block with a 1D layer. Additionally, they also replace the real-valued convolutional and batch normalization layer with their complex-valued counterparts. At or below 0 dB SNR, the two proposed new networks perform similarly to real-valued NNs. However, at higher SNRs, the complex-valued convolutional NNs outperform their real-valued counterparts.

4.3   Channel State Information (CSI) - Sensing and Prediction

Ding and Hirose use a chirp z-transform (CZT) along with two different CVNN architectures to predict long term channel fading. Due to a desire to decrease complexity cost and extract a smoother frequency domain approximation of Doppler shift fading, in a previous work, they windowed the RF signal in the real time domain using a CZT and applied Lagrangian extrapolation of the frequency domain parameters [22]. Improvements in the algorithm are made by initializing the weights in a multilayer CZT-CVNN with the output of the CZT with extrapolation and then applying new input signals to the CZT-ML-CVNN to update the previously applied weights from NN output. In the second framework, a real time recurrent learning (RTRL) neural network with complex-valued inputs CZT-RTRL-CVNN is applied to predict frequency fading characteristics. Results show higher accuracy in prediction of the channel characteristic. The bit error rate of the CZT-ML-CVNN model reveals better prediction of the channel fading [23].

Ji and Li focus on a novel lightweight NN with complex-valued inputs with an encoder-decoder framework they call CLNet. In a deep NN and a massive MIMO RF environment the increases in computational complexity historically have caused bottlenecks, especially at the user interface. They propose a reduction in complexity by forging a new complex-valued 1 × 1 point-wise convolutional input that combines the real and imaginary parts into to preserve the physical CSI. A new signal clustering mechanism is also provided in the form of an informative decoder, which concentrates the more informative laden signal paths. It is noted that they decline to use CVNNs but do acknowledge they are one of the first to broach the use of complex-valued data in this domain. CLNet increases accuracy by 5.41% and decreases computational overhead by an average 24.1% [24].

Scardapane et al. apply uniquely designed nonparametric complex activation functions (kernels) that are flexibly suited to the complex-valued data [25]. This is related to the difficulty of using bounded and holomorphic complex-valued activation functions in CVNNs. In one example domain, they examine wireless channel identification by applying various CVNN architectures using a variety of experimental kernels. Influencing their outcomes was the ability of their model to adjust the complex-valued data to either be circular, highly non-circular or somewhere in between. Favorable outcomes were found for the kernel-based complex-valued activation functions, however overall results were dependent on the degree of circularity of the complex-valued data.

4.4 Specific Emitter Identification (Aka RF Fingerprinting)

In a seminal paper, Kohno et al. [26] introduced the identification of a transmitting physical device on a network by evaluating software packet headers and computing the clock skew due to the hardware characteristics of a specific device. More recently, "fingerprinting" at a distance, also known as, specific emitter identification (SEI) determines the exact transmitter, not by software identification features, but by the unique transient artifacts inherent in hardware devices like power amplifiers. This specific technology has many security and tracking applications.

Agadakos et al. [27] have developed a recent, state of the art, platform independent Deep CVNN that uses novel techniques to processes raw complex I/Q wireless signals. Their technique uses learnable characteristics from passively captured transmitted wireless signals over a wide range of frequencies. Also, their experimental evaluation is designed to focus on the task of identifying devices based on their transmitted RF signals independent of the specific deployment environment. Evaluation used complex-valued data from Wi-Fi and ADS-B applications. Although their test evaluated many architectures with many variable parameters, in general recurrent deep complex-valued and convolutional deep complex valued NNs performed above standard recurrent and convolutional NNs, albeit with increased processing time.

An efficient SEI method based on a combination of CVNNs, and network compression is proposed by Wang et al. [28]. The inclusion of CVNNs proposes to increase emitter identification performance to nearly 100% at high SNRs and increase convergence speed. Additionally, a proposed modification to the SlimCVCNN algorithm inserts a hyperparameter that balances network compression and emitter identification performance. Compression reduces network size to between approximately 10%–30% the size of a basic CVNN. With proposed improvements, computing complexity generally declines at various SNRs.

5 Future Research

5.1 Metaheuristic Learning Algorithms

The learning process in NNs is traditionally achieved via optimization (minimizing) of the loss function during backpropagation using stochastic gradient descent (SGD). This is a hurdle for a couple of reasons. First, there is the well-known problem of capturing a localized minimum/maximum during SGD, where a global optimum is desired. Second, the calculation of the complex-valued gradient for use in SGD is a complication as well. Researchers [29] have proposed non-traditional metaheuristic evolutionary algorithms which may provide optimization routines that are superior to traditional gradient based methods.

5.2 Complex-Valued Activation Function Design

One of the most challenging aspects of CVNN usage is the essential application of the complex-valued activation function. It is known that CVNNs are more difficult to train than RVNNs. One reason for this difficulty is the constraint that the activation functions in CVNNs cannot be holomorphic and bounded simultaneously. Although, a fully complex activation function is desired some have suggested [30] that activation functions could be independent with respect to the real and imaginary parts of the complex-valued function. Piecewise linear and kernel-based complex-valued activation functions are also an area of active research [24].

Acknowledgement. This work was supported by US National Science Foundation under Grant CCF-2219753.

References

1. McCulloch, W.S., Pitts, W.: A logical calculus of the ideas immanent in nervous activity. Bull. Math. Biophys. **5**(4), 115–133 (1943)
2. Rosenblatt, F.: The perceptron: a probabilistic model for information storage and organization in the brain. Psychol. Rev. **65**(6), 386–408 (1958)
3. Krizhevsky, A., Sutskever, I., Hinton, G.: ImageNet classification with deep convolutional neural networks. Commun. ACM **60**(6), 84–90 (2017)
4. Mamoshina, P., Vieira, A., Putin, E., Zhavoronkov, A.: Applications of deep learning in biomedicine. Mol. Pharm. **13**(5), 1445–1454 (2016)
5. Abdel-Hamid, O., Mohamed, A., Jiang, H., et al.: Convolutional neural networks for speech recognition. IEEE Trans. Audio Speech Lang. Process. **22**(10), 1533–1545 (2014)
6. Kinzel, W., Kanter, I.: Neural cryptography. In: International Conference on Neural Information Processing, vol. 9, no. 3, pp. 1351–1354 (2002)
7. Hirose, A.: Proposal of fully complex-valued networks. In: International Joint Conference on Neural Networks, vol. 4, pp. 152–157 (1992)
8. Hirose, A.: Complex-valued neural networks: the merits and their origins. In: International Joint Conference on Neural Networks, pp. 1237–1244 (2009)
9. Bassey, J., Li, X., Qian, L.: A survey of complex-valued neural networks. arXiv: 2101.12249v1
10. Lee, C., Hasegawa, H., Gao, S.: Complex-valued neural networks: a comprehensive survey. IEEE/CAA J. Automatica Sinica **9**(8), 1406–1426 (2022)
11. Hirose, A., Yoshida, S.: Generalization characteristics of complex-valued feedforward neural networks in relation to signal coherence. IEEE Trans. Neural Netw. Learn. Syst. **23**(4), 541–551 (2012)
12. Sarroff, A.: Complex neural networks for audio. Ph.D. thesis, Dartmouth College (2018)
13. Zhang, Z., Wang, H., Xu, F., Jin, Y.: Complex-valued convolutional neural network and its application in polarimetric SAR image classification. IEEE Trans. Geosci. Remote Sens. **55**(12), 7177–7188 (2017)
14. Hirose, A.: Complex-Valued Neural Networks, 2nd edn. Springer, Heidelberg (2012). https://doi.org/10.1007/978-3-642-27632-3
15. Glorot, X., Bordes, A., Bengio, Y.: Deep sparse rectifier neural networks. Proc. Mach. Learn. Res. **15**, 315–323 (2011)
16. Savitha, R., Suresh, S., Sundararajan, N., Saratchandran, P.: A new learning algorithm with logarithmic performance index for complex-valued neural networks. Neurocomputing **72**, 3771–3781 (2009)

17. Krzyston, J., Bhattacharjea, R., Stark, A.: Complex-valued convolutions for modulation recognition using deep learning. In: IEEE International Conference on Communications Workshop, pp. 1–6 (2020)
18. Krzyston, J., Bhattacharjea, R., Stark, A.: High-capacity complex convolutional neural networks for I/Q modulation classification. arXiv: 2010:10717 (2020)
19. Peng, S., Jiang, H., Wang, H., Alwageed, H., Yao, Y.-D.: Modulation classification using convolutional neural network based deep learning model. In: 26th Wireless and Optical Communications Conference, pp. 1–5 (2017)
20. Zhang, X., Liu, X.: Interference signal recognition based on multi-modal deep learning. In: International Conference on Dependable Systems and Their Applications, pp. 311–312 (2020)
21. Xu, J., Wu, C., Ying, S., Li, H.: The performance analysis of complex-valued neural network in radio signal recognition. IEEE Access, 48708–48718 (2022)
22. Tan, S., Hirose, A.: Low-calculation-cost fading channel prediction using chirp z-transform. Electron. Lett. **45**(8), 418–420 (2009)
23. Ding, T., Hirose, A.: Fading channel prediction based on combination of complex-valued neural networks and chirp z-transform. IEEE Trans. Neural Netw. Learn. Syst. **25**(9), 1685–1695 (2014)
24. Ji, S., Li, M.: CLNet: complex input lightweight neural network designed for massive MIMO CSI feedback. IEEE Wirel. Commun. Lett. **10**(10), 2318–2322 (2021)
25. Scardapane, S., Vaerenberg, S.V., Hussain, A., Uncini, A.: Complex-valued neural networks with nonparametric activation functions. IEEE Trans. Emerg. Top. Comput. Intell. **4**(2), 140–150 (2020)
26. Kohno, T., Broido, A., Claffy, K.C.: Remote physical device fingerprinting. IEEE Trans. Dependable Secure Comput. **2**(2), 93–108 (2005)
27. Agadakos, I., Agadakos, N., Polakis, J., Amer, M.R.: Chameleons' oblivion: complex-valued deep neural networks for protocol-agnostic RF device fingerprinting. In: IEEE European Symposium on Security and Privacy, pp. 322–338 (2020)
28. Wang, Y., Gui, G., Gacanin, H., Ohtsuki, T., Dobre, O.A., Poor, H.V.: An efficient specific emitter identification method based on complex-valued neural networks and network compression. IEEE J. Sel. Areas Commun. **39**(8), 2305–2317 (2021)
29. Wang, P., Zhou, Y., Luo, Q., Han, C., Niu, Y., Lei, M.: Complex-valued encoding metaheuristic optimization algorithm: a comprehensive survey. Neurocomputing **407**, 313–342 (2020)
30. Trabelsi, C., et al.: Deep complex networks. arXiv: abs/1705.09792 (2018)

Author Index

Printed in the United States
by Baker & Taylor Publisher Services